ORKNEY FARM-NAMES

by

Hugh Marwick

Kirkwall
W. R. Mackintosh
1952

CONTENTS

	Page
PREFACE ...	iv
PHONETIC SYMBOLS USED ...	vi
PART I—Island and Parish Farm-Names ...	1
PART II—A. Skats and Rents ...	191
B. Nature and Origin of Orkney Skats ...	205
C. The Orkney Tunship ...	216
PART III—Farm-Name Chronology ...	227
INDICES—1. Alphabetical List of Names ...	253
2. Personal Farm-Names ...	262
3. Index to Parts II and III ...	264
SOURCES CONSULTED ...	265
ADDENDA ET ERRATA ...	267

PREFACE

THE present work was first projected many years ago, when through the kind co-operation of my friend Mr. Robert Rendall a start was made by collecting and listing all the farm-names to be found in our early county rentals and other early records. Owing however to the pressure of professional duties, and *inter alia* the outbreak of war, it was not until after my retirement that I was at liberty to study and work up the material.

A complete survey of Orkney farm-names has long been a desideratum of place-name students in Scandinavia, and the names herein recorded should go far to meet that want. It is not a complete catalogue of all present-day names; the relatively few (more or less modern) names of Scots origin I have ignored except when they chance to be of some historical interest, and some names which have been applied to small crofts in recent centuries have likewise been omitted, even though such names may sometimes be of Norse origin. On the other hand, every name (so far as I am aware) which occurs in the early rentals is included, and these names appear under their respective parishes in alphabetic lists by themselves. And after each list, under the heading of "Other Farm-Names," I have added names of presumably Norse (or other early) origin which do not happen to appear in early records; thus, taken together, these lists may be regarded as being very fully representative of the early farm nomenclature of Orkney. Under each name also I have appended such details as are available for shedding light on the history of the farm or the origin of the name. And, whatever value the book may have otherwise, it is to these lists of names and the accompanying notes that I would attach most importance, since they are, in general, objective statements of fact, unbiased by personal opinion of the author, and they provide a factual basis on which any interpretation, present or future, can be based.

In regard to derivation the casual reader will no doubt be disappointed at the number of names for which no origin at all is suggested, or of which the origin is stated to be uncertain or obscure. The place-name specialist on the other hand will almost certainly consider that far too many suggested derivations appear. I am fully conscious of the risks I have taken in dealing with names such as these—i.e. names for which original or really early forms are not on record—and am quite alive to the criticism that is pretty sure to follow. But competent and pertinent criticism can but serve to correct error and advance real knowledge, which is the only thing that matters.

Part II—The Farm Background—may be found somewhat irrelevant, but it contains a deal of miscellaneous information not hitherto brought together, and it draws attention to various facts in connection with early farming, some of which are not generally familiar, and some awaiting further clarification. And it all has a bearing on Part III, in which I have attempted to interpret the chronology of the main types of names with a view to casting light on the early Norse settlement. This is admittedly a pioneer venture

PREFACE

only, and here once again competent and constructive criticism is not only necessary, but to be whole-heartedly welcomed.

For generous help received I have first and foremost to record my indebtedness to my above-mentioned friend Mr. Rendall. When the work was first projected he undertook the laborious task of collecting, collating and listing parish by parish all farm-names to be found in early records, a task he performed with meticulous thoroughness and care. At a more recent stage he was kind enough to read the proof sheets of Parts II and III. Such invaluable help I can never hope to repay. To two friends in Scandinavia I am similarly indebted. For over a quarter of a century Professor Magnus Olsen of Oslo has been unfailingly kind in answering my numerous queries on place-names and other problems, and his masterly work *Ættegaard og Helligdom* was to me, as no doubt to others, a real revelation of what place-names adequately interpreted have to teach us. I have likewise to thank Professor Matras of Copenhagen University for quite invaluable help. He was kind enough to read over the proof sheets of Part I, and by so doing was able through his expert linguistic knowledge and familiarity with Færoese place-names to make suggestions and save me from slips of various kinds. He did not, however, see the final proofs, and must in no case be held responsible for such mistakes as may appear, for all of which I alone must assume full blame.

My sincere thanks are also due to my friends the local printers, and more particularly to Mr. P. Thomson and Mr. J. Marwick, for the pains they took in producing a work involving the printing of so many names of which every letter demanded separate attention, and of so many unfamiliar Norse words with unfamiliar letters—a task that might well have taxed patience and temper beyond endurance.

And last, but not least, I have to record my deep and grateful thanks to the Carnegie Trust for the Universities of Scotland for their very generous grant which helped to make publication possible.

NOTE.—Neither in records nor in present day usage is there any standardised spelling of Orkney farm names. A name may thus appear spelt Air or Ayre; Bay or Bae or Bea; Houseby or Housbae, etc., etc., and I have not attempted herein to attain uniformity.

PHONETIC SYMBOLS USED

These have been restricted to a minimum as follows:—

VOWELS:

a is used for both back and front a; e.g. mar or man.

ɛ	heard in	met	thus	mɛt
e	,,	,, mate	,,	met
i	,,	,, leap	,,	lip
ɪ	,,	,, mid	,,	mɪd
ə	,,	,, over	,,	o·vər
ʌ	,,	,, cut	,,	kʌt
ɔ	,,	,, lot	,,	lɔt
o	,,	,, note	,,	not
u	,,	,, rood	,,	rud
ø	,,	,, { deux (Fr.) { do (Ork.)	,,	dø

CONSONANTS:

ʃ	,,	,, shun	,,	ʃʌn
ʒ	,,	,, measure	,,	mɛʒ·ər
χ	,,	,, loch (Scots)	,,	lɔχ
þ	,,	,, thin	,,	þɪn
ð	,,	,, then	,,	ðɛn
j	,,	,, you	,,	ju
ŋ	,,	,, song	,,	sɔŋ

A long vowel is indicated by a colon following—thus ru:d (rude); a stressed syllable by a dot following—thus ou·skɛri (Auskerry).

Part I
Parish and Island Farm-Names

NORTH RONALDSAY

Rinansey and **Rinarsey** (O. Saga).

This has been usually regarded as a hybrid name—signifying 'Ringan's isle,' Ringan being a variant of the name of St. Ninian. Some support for that conclusion may be found in the fact that in the excavation of the Broch of Burrian in the south-east corner of the island certain relics of undoubted Celtic origin were recovered, viz. a small iron Celtic-church bell, a stone slab inscribed with a Celtic cross and ogam runes, and a bone bearing two of the well-known Pictish symbols—the mirror-case symbol and the crescent with V-shaped rod. While the Ninianic origin of the name is by no means certain, any better origin is difficult to suggest.

By the 14th century the island name was becoming confused with Rognvaldsey, the name of the most southerly of the Orkneys, and the prefixes North- and South- gradually came into use to discriminate them.

North Ronaldsay does not figure in either the 1492 or the 1500 Rental, but in the 1595 Rental it does appear, being skatted as 72¾ pennylands (Sailnes and Sand=24d. lands, Linklet=18d. lands, Nes, Busta and Sand together as 30¾d. lands). Originally it was skatted no doubt as four urislands (72d. lands).

OLD RENTAL NAMES
(INCLUDING NAMES FROM PRIVATE ESTATE RENTALS OF 1733 AND 1785)

Ancum (æŋ·kʌm: N. Ry. 'a' is practically the same as Southern English 'a' in e.g. fat): Ankum, and The Valtenes of Ankum (1653 Val.); Ancum 1733; Anchum and Vaultnes of Anchum 1785 R.

The present farmhouse stands on an old site on the Brae of Ancum, while the land around is low-lying meadow ground which on the south side of the adjacent Loch of Ancum is called Enyan (ɛn·jən), i.e. O.N. *engin*, 'the meadow.' The first element of Ancum itself is probably also O.N. *eng*, meadow, but the termination is puzzling: Cf. however Holm *infra*.

Valtenes is almost certainly the same term as the Shetland Veltins which Jakobsen derived from O.N. **velturnar*, def. pl. of **velta*, a word which in Norway and Færoe is used of ploughed or upturned soil. Cf. Waltness (Shapansay).

Busta: 8d. land of Bowsta (1653 Val.); Kirbist in the 8d. land of Buista, 1733. Buistie-tun (bøst·i·tun) is still in use today for this old tunship. O.N. *bólstaðr* or *bústaðr*, farm-settlement.

This is a curious survival as in Orkney *bólstaðr* is very rarely found without a prefix attached—the only other exception known to me being Bousta in Flotta, where the name is also applied to a small district (probably a former tunship). Here, however, in North Ronaldsay, in this very tunship is a farm Kirbist, a name which certainly represents an O.N. *kirkju-bólstaðr*, 'kirk-bister,' and such a duplication is difficult to interpret. Perhaps the simplest explanation would be that Kirkjubólstaðr was originally the name of the whole tunship, but later (after division of the tunship) the name became restricted to one farm, while the name in a truncated form—Bólstaðr (Bústaðr)>Buister (bøst·ər) >Buista (bøstə) remained in use for the tunship as a whole.

Conglibist (kɔŋ·libəst): id. 1733. In Sennes-tun near part of beach known as Conglibanks where the shore is piled up with huge boulders of rock which in parts of Orkney (N. Ry., Rousay, etc.) are termed kongles or kungles (koŋls, kʌŋls). That word has acquired a specialised sense in Orkney; O.N. *kǫngull* meant a cluster (of berries, etc.), but the modern Norw. derivative *kongla (kangle, kaangul)* means a conifer cone. O.N. *kǫngla-bólstaðr*, ' k——farm.'

Disher: id. 1733. Only the site of house now remembered as having been on the Brae of Disher in Busta-tun. O.N. *dysjar*, pl. of *dys*, a mound.

Gerbo: Gerback 1733. Farm in Ness-tun. O.N. *garð-bǫlkr*, a dividing wall. As this farm is adjacent to the Muckle Gairsty the ' garth ' in question must have been that ridge; see further *sub* Treb.

Gravity (grev·ǝti): id. 1733. Now two farms in Linklet-tun, North and South G——. First element pretty certainly O.N. *grǫf (grafar)*, a hollow or depression : influenced in sound perhaps by the Eng. cognate *grave*. Termination obscure; in Orkney -ty or -dy (ti, te, di, de) is a very common ending and can only rarely be referred to any certain source. Cf. Graves (Holm).

Harga: 1653 Val. Now vanished, but in old records appears also in form Hargar, and usually in combination with Howar—" Howar and Hargar." In Busta-tun.

It would seem clear that Hargar represents an unmutated form of O.N. *hǫrgar*, pl. of *hǫrgr*, a term which was used in two senses : (1) a cairn of stones or stony hillock; (2) a (heathen) temple-site. Fritzner notes how in the Old Gulating Law the phrase *haugar og hǫrgar* was used in the same sense as *haugar og hreysar* (heaps of stones). It is thus of interest to note the same combination here in Orkney. This is the only example known to me of the term *hǫrgr* in Orkney place names, and (even apart from the plural form) it is extremely doubtful whether it can indicate the existence of a heathen Norse temple here.

Holland : id. 1595. Now the chief farm on the island, and the proprietor's residence. A very common Orkney name—O.N. *hó(há)-land*, high land; it stands near the highest spot on the island.

Holm (hom): id. 1733, 1785. The farmhouse is built on higher ground (between an old dam and the Loch of Hookin) which has probably at one time been an actual islet or at least surrounded by marsh. O.N. *hólmr*, islet.

Howar (hou·ǝr): id. 1653 Val., 1733. In Busta tunship. O.N. *haugar*, pl. of *haugr*, a mound. See also Hargar.

Howatoft (hou·tu): Howtowa 1733; Howatoft—1798 R. In Busta tunship. The house stands on a brae or rising ground, and the first element may thus represent O.N. *haugr*, a mound. The variant spellings make it uncertain whether the second is O.N. *topt*, a house-site, or *púfa*, a small mound, though ' mound-mound ' would be an absurd name. It is not impossible that we may have here an O.N. *hof-topt* (heathen) temple-site. It may be noted that Hargar was in this same tunship.

Kirbist: see Busta.

Linklet: id. 1595. An urisland (18d. land) tunship, but there is no record of any separate farm of this name. O.N. *lyng-klett[a]r*, ' heath-rock[s].

Linney (lɪn·i): Lynie 1646 (RMS); Laynie 1646 (quoting deed of 1594); Nether Linnay, 1653 Val.; Linay 1733, 1785. Two farms today—Upper and Nether L———, in north-west of island. There is no such pronounced slope here as to justify derivation from an O.N. definite dat.—*hlíðinni*; origin therefore uncertain.

Ness or **Ness-tun**: Nes 1595. The original head-house of this tunship is now represented by two farms—North and South Ness. It was an old 16d. land tunship taking its name from the adjacent headland of Brides Ness. O.N. *nes*, a headland.

Sand: id. 1595; Sander 1653 Val. and 1785 R. There were two tunships so called—one in the north, the other in the south of the island: distinguished formerly as Sander-benorth and Sander-besouth. The latter included Hollandstun; the former name survives as a field-name—Sander. O.N. *sandr*, sand; the form Sander derives from the pl. *sandar* which was applied to a sandy beach or sandy ground.

Scottigar (skʌt·igər): Scottiger 1733. Situated close to beach in north of island. Tradition says it was built by a Scotsman—an obvious bit of folk-etymology. Origin uncertain.

Sennes (sɛn·ez): Sailness 1595, 1733. The most northerly tunship in island. Sailness and Sand(benorth) together comprised 24d. lands.

The present houses of Sennes together with those of Sholtisquoy and Westhouse (which in 1733 was termed West Sailness) form what is called the Bigging o' Sennes. Offshore near here is an outlying skerry called Seal Skerry, which from of old has been a favourite haunt of seals. Probably the whole neighbouring peninsula was originally called *Sel*[*a*]*nes*, 'the ness of seals,' from O.N. *selr*, a seal.

Sholtisquoy (ʃalt·əsko, ʃalt·sko): Salkisquoy 1733. Final element O.N. *kví*, an enclosure for animals, but first element obscure; it may possibly indicate a man Hjalti, though if so the Eng. genitive 's' has replaced the O.N. 'a.'

Sugarhouse (ʃʌg·ər-hus): Suggarhous 1653 Val.; Sugarhouse 1785. As this farm is situated at the south side of Linklet-tun the name probably represents O.N. *suðr-garð-hús*, 'south garth house,' i.e. near the Muckle Gairsty which runs past near by. See further *sub* **Treb**.

Treb: id. 1653 Val.; Treb in Holland, 1733. This farm, one of the most fertile in North Ronaldsay, is adjacent to the Muckle Gairsty. A specially instructive name.

All over Orkney one may come across the remains of prehistoric earthern dykes or ridges which are usually known by the name of 'gairsties.' In the course of centuries many of these have been levelled down or carted away to lay on poor shallow soil, but many portions still survive, and their origin and purpose have been and still are quite obscure. They vary greatly in size, some being as much as 5 or 6 yards in breadth at the base and 3 or 4 feet in height, but the majority are now much smaller. Only rarely can they be shown to have any direct connection with tunship boundaries.

The name *gairsty* represents O.N. *garð-stœði*, 'dyke-steethe,' or dyke-foundation, and Garson (*garðs-endi*) is a common place-name for a house or place where a gairsty ends (usually at a shore).

In North Ronaldsay there are two of those old gairsties running across the island and dividing it into three rather unequal portions. The more southerly and more massive is called The Muckle Gairsty, the other by the unexplained name—Matches Dyke. In explanation of these structures an old island tradition tells how the island once belonged to a man who had three sons. He divided the island by means of these dykes, giving a part to each son, one of whom resided at Holland, the second at Finyerhouse, and the third at Breck. The northern third is still known as North Yard (*norð-garðr*) and the southern as South Yard. The middle portion embraces Linklet-tun and part of Ness-tun as well. It may be added that Papa Westray also has a North Yard and a South Yard—the dividing line there being an old gairsty also.

In Sanday there are many remains of those old dykes, but there the name applied to each is The Treb, and they are so common that 'treb-dyke' has become a generic term for them. In old Sanday records references may be found to a house called **Treb** which appears on Blæuw's map of Orkney (c. 1650) at or near a spot where one of those treb-dykes may still be seen today. In Sanday also there are three different houses of the name of Thrave (pron. trev), and one of these, West T—— in Burness, is built on the actual line of a treb-dyke which is still called The Treb today.

That Trave was an alternative form of Treb would thus seem clear, but confirmation can fortunately be adduced from a Sheriff Court record preserved in Kirkwall in regard to a dispute about property in Stronsay, the island immediately south of Sanday. In a 1738 perambulation of the farm of Haghquoy it was stated that "Mr. John Scollay (parson of Stronsay) had in harvest or winter last built a house . . . upon the plate [plot] of ground lying near to the *Treve* which is a part of said balk of North Strenzie." And it was further declared that the quoy of land in dispute had its dykes built "upon old *Gaistys* [the Stronsay form] or the Steith of an old Dyke." Today, however, the term Treve seems quite forgotten in Stronsay, and I have not come across any record of *trave* or *treb* in any other part of Orkney beyond those isles just mentioned and Westray.

That these *treb* and *trave* names are relics from Pictish times in Orkney may be regarded as certain, as no doubt are the dyke-structures to which they refer. In his *Celtic Place-Names of Scotland* the late Professor W. J. Watson cited a large number of cognate place-names, all derived from an Early Celtic *trebo*, a dwelling, perhaps village. He pointed out that in Welsh *tref* is a homestead or hamlet (technically a division of land of a certain extent), while the Irish cognate *treb, treabh* means a place of abode, home, region, family. In Scottish Gaelic *treabh* is used as a verb only—"to plough," while *treabhar* means "houses" collectively, especially farm-buildings. Of Scottish place-names referred to by him as containing the same root we may mention the following: Threave (the Douglas stronghold in Galloway)—older forms of which are Trefe, Treffe, Tref; Tranent (Treuernent 1150); Ochiltree; Fintry; Rattray, etc.

At this late date it is impossible for us to interpret the precise original significance of those treb-dykes here in Orkney. It is possible that they may have been division or enclosure walls in connection with Pictish settlements, though, if that be so, they have been constructed on a far more massive scale than the hill-dykes or tunship-dykes erected by Norse inhabitants later. As yet no similar structures seem to have been put on record from other Pictish areas in Scotland though it would be strange if they are not to be found there also. And it is to be hoped that these few farm-names in North Ronaldsay and Sanday to which reference has just been made will long be preserved as memorials of a distant past.

Though somewhat irrelevant, reference might be made here also to another North Ronaldsay place-name connected with one of those treb-dykes or gairsties. At the western end of the Muckle Gairsty is a 'geo' or cleft in the cliffs known as Gersna Geo, i.e. O.N. *garðs-enda-gjá*, 'dyke-end geo.' At the western end of the other—Matches Dyke—is another geo called Himmera Geo. The pronunciation, as well as the nature of the shore there, rules out the possibility of any connection with the common Orkney term 'hammer' from O.N. *hamarr* (projecting rocks on a hillside or cliff-line), but the name may very well represent the Celtic (Gaelic) *iomair*, a ridge of land, or 'rig' in a field: Early Irish *immaire*, *imbaire*. In Shetland Jakobsen considered he found that term in Imrie, Immeri—the name of a ridge jutting out into a loch in Aithsting. (*Shetland Place-Names*, p. 191.) Its occurrence here in North Ronaldsay at the end of a treb-dyke need thus cause little surprise.

Finally, it may be noted that one small portion of a Treb at Hillhead in Sanday is traditionally known as "The Trows' Buil," i.e. the abode of trolls. That association with the supernatural is an indication of the mystery and dread of the unknown with which these structures were associated in the minds of the succeeding race.

OTHER FARM-NAMES

Antabreck: farm in northern part of island. Second element O.N. *brekka*, a slope, but the first is obscure as it is also in Purtabreck, a neighbouring farm.

Breck: in north part of island. O.N. *brekka*, a slope.

Finyarhouse (fɪn·jər-hus): an old house—now vanished. This name can hardly be other than O.N. *vinjar-hús*, the first element being the genitive of *vin*, pasture. For change of 'v' to 'f' cf. Fitty Hill (Westray) from O.N. *viti*, beacon, and the name following.

Fisligar (fɪz·ligər): this probably represents O.N. *vesli-garðr*, from *vesall*, poor, worthless, and *garðr*, farm.

Garso: farm in Sennes-tun. Name refers to a neighbouring mound situated close to beach and adjacent to the sheep-dyke (garth) that encircles the island. O.N. *garðs-haugr*, 'garth-mound.'

Hookin (hu·kin, hu·ken): Hooking 1796. This farm-house is built on the bank of a small stream flowing down from The Loch of Hookin to the sea. This is a somewhat puzzling name which occurs again in Sanday and in Papa Westray, in each case applied as here to a house on the bank of a stream where in olden days there appears to have been a water-mill. The Papa Westray name appears on Blæuw's map (c. 1650) as Ouquin, and in old estate papers as Uquin and Howkin. The name may probably represent O.N. *á[r]-kvern*, 'burn-mill,' from *á*, a stream or river and *kvern*, which was the Norse word for a grinding mill, though remembered in Orkney only of the hand-grinding implements of which so many of the stones survive still today.

Whunderless: an ancient house-site in Busta-tun. The first element can hardly be other than O.N. *hvannar*, gen. of *hvǫnn*, angelica, a plant highly prized by Norsemen for its food or medicinal value. The termination is probably the same as appears in a field-name in this island—The Lint Lues (løz), where flax has apparently been grown at one time. That term would seem to be O.N. *ló*, a low-lying flat; No *lo*, a meadow-flat—cognate with Eng. *lea*, O.E. *leah*.

SANDAY

Sandey, O. Saga. 'Sand-isle.'

For the most part this island is low-lying, rising little above sea-level, and, as the name implies, is composed largely of sand. The soil is thus light and easily cultivable, a fact sufficiently attested by the abnormally high valuation for skat purposes imposed on the island in early Norse times. In Sanday alone there were approximately 36 urislands (ouncelands), practically a fifth of the total number in Orkney. Taking mere area into account that scale of taxation was about three times what one might have expected. The obvious explanation is that the island was already extensively cultivated when Norsemen first arrived, and a few of its surviving place-names can still be regarded as memorials of that pre-Norse age.

In the earliest of the Orkney Rentals—that of 1492—Sanday unfortunately is not included, an omission doubly regrettable as the farm-names appearing in the 1500 and 1595 Rentals are in some cases manifestly misleading, while others are of a type unknown elsewhere in Orkney, a fact which makes the absence of really early reliable forms especially disappointing.

The island was divided into three parishes—Cross, Our Lady and Burness or St. Columba.

RENTAL NAMES

Airy: Arie, U.B. 1601; Airay, Sas. 1618. Freq. appears in sasines and other records with diverse spellings—Airie, Eary, etc. It usually appears as a tumail going along with Grindally which is still remembered as situated between Warsetter and Stove. Elsewhere in Orkney (e.g. Stronsay and Westray) the same name Airy occurs three or four times, and it almost certainly represents the Gaelic *airigh*, a shieling (E. Irish *airge*).

In Sanday the name is now forgotten, but (near its old site) a little to the east of Grindally and on the present farm of Warsetter is a ridge of high ground known as Airafea (ɛrˑəfi) which must contain the same element with the suffix—fea, (O.N. *fjall*, hill). Airy and Airafea both occur again as farm-names in Stronsay.

Arstais: 1500; Arstas 1595. Freq. on record in 17th cent. and later sasines—sometimes in form Erstas. It formed a whole urisland or 18d. land by itself, but is remembered now only as a field name Erstas (ɛrsˑtəs) on the farm of Newark in which it is incorporated.

Origin of name quite obscure, but there were three other old tunship farms in Sanday having names with which this must be compared : Langtas, an 18d. land now also incorporated in Newark, Houstais, an 18d. land in the district of Evirbist, and another Langtas—a 36d. land—also in Everbist. The prefix Ars- in Arstais probably indicates a personal name.

Brabustir: 1500; Brabustar 1595. A 9d. land in Burness, no longer a farm name though the area still goes under the name of Brybist (braiˑbəst) or Brybistun.

O.N. *breiði-bólstaðr*, ' broad-bister ' or ' broad-farm.'

Breck: Brek 1500; Breck 1595. A vanished farm-name in Lady parish. O.N. *brekka*, a slope.

Bressigar: Brusgarth 1500; Bressigair 1595. A 9d. land or half-urisland in Lady parish. A *garth* name, the prefix denoting a personal name—most probably *Brúsi—Brúsa-garðr*—though the 1595 spelling and present pronunciation suggest the possibility of a metathesis of *Bersi*.

Brough (brɔχ): Brugh and Brught 1500; Brught 1595. Now known as West Brough in distinction from East Brough, a former name for Newark.
O.N. *borg*, a fortress; in the stackyard of this farm are the remains of a prehistoric broch. The neighbouring district is still called Broughston (Brough's-tun). The metathesis may be due to Scots influence.

Burness: Burness 1595. A parish—not a farm name. Applied to a large peninsula so named probably from the broch at West Brough which is situated at the inner end of the peninsula. O.N. *borg[ar]nes*, 'fort-ness.'

Clamer (kle·mər): id. 1595. A ¼d. land in Burness. Still remembered as a place-name, but no longer a farm.

Cleat (klet): Clat 1500; Cleat 1595. An old urisland, and a large farm still today.
O.N. *klettr*, a rock, rocky eminence, hill-nab, etc.
A common farm name in Orkney, occurring (uncompounded as here) in Stronsay, Westray, St. Ola and S. Ronaldsay, and in several compounds elsewhere—e.g. Linklet or Linklettar, Whitecleat, Hunclet, Manclet, etc., and the precise reason for the application of such names is obscure. In Orkney *klett* is a well-known generic term for a large boulder of rock, and is pronounced klet, whereas in farm names it is usually klet. That is probably due to the influence of the written name *Cleat*. On the analogy of the local pronunciation of such words as *beat, meat, peat, cheat*, and the nautical word *cleat* itself the *ea* would tend to be pronounced *e*.
Apart from pronunciation, however, there is rarely, if ever, to be seen in the vicinity of a *-cleat* farm any boulder or rocky protuberance that might have given rise to the name. And though two or three of those farms stand on rising ground, others—as Cleat here in Sanday—are on quite low-lying flattish ground, as are Benzieclett and Hammercleat in Sandwick.

Colligar (kɔl·igər): Cuthilgarth 1500; Colzegarth 1595. A half-urisland unit, and a farm name still today. Part of the lands form a peninsula projecting into Otterswick, and known as Colliness.
Another *garth* name, but the prefix is obscure; the 1500 spelling might suggest a Celtic personal name *Kathal* which was borrowed by the Norse. Lind cites two Icelandic farm-names which contain that word—*Kadalstaðir*. The 1595 spelling, however, and the pronunciation of today render such an explanation hardly possible.

Elsness (ɛls·nes[z]): Hellisnes 1500; Ellisnes 1595. A large peninsular farm which was skatted as 2 urislands.
Origin of first element obscure. The 1500 spelling at once suggests the genitive of O.N. *hellir*, a cave, but such an explanation seems ruled out by the fact that the shores of this ness are flat and shelving with no cliffs of any consequence in which a cave could have been formed.
There is just a possibility that the original name may have been O.N. *Elds-nes* or *Elz-nes*, 'fire-ness.' In the *New Statistical Account*, in a description of this peninsula, we read of the remains of a wall running across and enclosing about a third of the ness. "The enclosed space is literally covered with tumuli and heaps of ruins. . . . A number of the smaller heaps . . . are formed of what the country people call cramp, and are said by them to have been used as places of sacrifice.

For whatever purpose they were used it is plain they must have been the sites of strong and long-continued fires. . . ."

Everbist: Langta in Evirbuster 1500; Langtes in Overbuster U.B. 1601. O.N. *Efri-bólstaðr,* ' Upper-bister.'

Though bearing a typical farm-name this was not skatted as an individual farm or tunship, but embraced an extensive area of the island in which there were at least two skatted units—Langtas, a two-urisland, and Houstais a one-urisland tunship. Evirbist has therefore to be classed as one of the larger areas which are probably to be regarded as original settlements or ' land-takes ' by early settlers from Norway and which even at the early period when skat was imposed had in some cases been already split up into separate component units. See Part II.

The name Everbist is still in general use in the island for the area in question.

Fea: Houbustar and Fea, 12d. terre, 1595. Not mentioned in 1500 R. where Hobsta (Hobbister) is entered by itself as a whole urisland. Fea is no longer in use as a farm name. O.N. *fjall*, hill.

Frow and **Wattin**, and **Frowattin** 1500; **Frowantown** 1595. See **Stove**.

Gardemeles: there were two farms of this name in Sanday—
1. An urisland at Stove : Gardemeles, Gardymelis and Gardemellis 1500; Gardumellus 1595.
2. An urisland at Lady parish kirk : Gardemeles 1500; Gardemellis alias Marykirk 1595.

In each case the name is now long obsolete, but it is especially interesting as of a type unique in Orkney. From each of these farms skats were payable to the bishop, and it is possible that the name may be due to churchmen.

Each of those farms also was on sandy soil adjacent to a vast sandy beach, acres of which ebb dry at low tide; there can thus be little doubt that the termination of the name points to an anglicising of O.N. *melr* (or its plural), sand(s). Nor can the first element be other than O.N. *garðr*, a farm, and one might therefore suggest that the original compound may have been *garðr á melum*, ' farm at (the) sands.'

Interesting parallels can be cited from England and Wales. In Charles's *Non-Celtic Place-names in Wales* (London 1938) a Meles Grange is cited from Glamorgan, of which earlier forms were *Grangia de Melis* (1186), and *The Meles* (1633), and the term is referred to O.N. *melr*. Cf. also Argarmeles, North Meols and Ravensmeols—forms which Ekwall records in his *Place-Names of Lancashire*, and which he likewise refers to the same origin.

Garth : 1500 and 1595. An old urisland; name now applied as a field-name on Warsetter (gert). O.N. *garðr*, farm.

Goir (go'ər) : 1595. Two ' tumails ' of this name are entered—sites not stated.

The two foregoing names are examples of a large group of farm or field names in Orkney, all deriving from O.N. *garðr* or its cognates, but the history of which is exceedingly difficult to trace. As the second (unaccented) element in farm-names (where for the most part there is a personal name as prefix, e.g. Hermisgarth) the usual pronunciation is simply—gər. When used by itself (uncompounded) as Garth the native pronunciation is usually gert. In Evie, however, the local pronunciation of an old tunship name which appears in the 1500 R. as Garth is gjort or more often djɔrt: in Orkney initial gj- often passes into dj- as e.g. O.N. *gjá* (a ravine or indentation in the coast line) which is pronounced indifferently gjo or djo.

In Rousay are several examples of an expanse of rough pasture, each of which bears the name of The Goard (gjoərd, djoərd) of such and such a farm, and in my *Place-Names of Rousay* I referred these to No: *gjorde*, meadow-field, piece of enclosed pasture. A house built adjacent to one of these is called Goarhouse (gjoər-hus, djoər-hus).

Lastly there are several examples throughout Orkney of the name Corn, both as a farm-, and as a field-name. In Rousay one old farm bears the name Stennisgorn, older forms of which are: Stannesgair 1565; Stainsger 1570; Stennisgar 1578; Stennisgair 1661; Stennisgarth and Stennesgorne 1624; Stennisgoird 1629; Stenhousegorne 1664. There can be little doubt that all these variants must be derived from an O.N. *steins-garðr*, [standing]-stone farm. Though no standing-stone is now apparent at this farm two other Orkney farms—each having an old standing-stone adjacent—are similarly named. One in Birsay is now spelt Stanger (pron. sten·ʒər) but earlier spellings were Stansgar and Stainsgair (1595 R.); the other is Stensigar (stɛns-i-gər) in S. Ronaldsay—Stansgair, Stensagair (1595 R.).

In my *P.N. of Rousay* I suggested an explanation for the change to *-gorn* which I have since found to be quite misconceived, and I would now suggest that *-gorn* represents the old dative plural *gǫrðum*. In Færoese an ending *-um* becomes *-un*, and a parallel change may be seen in the many Orkney *-staðir* names where that element now appears as *-ston* (apparently from the dat. *stǫðum*); e.g. Corston.

That still leaves Goir or Goar (go·ər) unexplained, and I can only suggest it may be the same name as Goard *supra*.

Grindally (grɪnd·əli): Gryndleith 1500; Grindlayth 1595; Grindlie U.B. 1601. Name of former urisland later incorporated in lands of Stove, and now employed merely as a field name. Grindally originally formed part of "the xxxvi d. terre of Southwall" as appears from 17th cent. Sasines.

O.N. *grindar-hlið*. In Ork. dialect, as in O.N., *grind* (pronounced with 'i' sound as in *grin*) is a name for a gate (e.g. "Close the grind!"), and the O.N. *grind-hlið* or *grindar-hlið* was applied to a gateway or opening that was fitted with a gate to close it. The compound appears frequently in Ork. pl.-names, usually as a house- or farm-name—Grindally, and was so applied by reason of proximity to one of the 'slaps' or gateways through the old hill-or tunship-dykes through which animals were driven to and from the out-pasture or common.

Gruttill: 1595. A 'tumail' in Sutherbie. Now remembered as a field (grɪtl) on Backaskaill. The first element is O.N. *grjót*, stones; the suffix is uncertain but probably *hóll*, hill, or its deriv. No. *hol*, a low mound. The site here is on a steep hillside: hence probably = 'Stony-hillock.'

Hacksness: Halkisnes 1500; Halksness 1595. Applied to the ness that forms the east side of Bay of Stove. A former urisland and bordland (i.e. private earldom estate, and as such unskatted). Was later incorporated with lands of Stove.

A puzzling name which occurs again in Shapansay and is found in old sasines as the name of what is now the Head of Work near Kirkwall. *Aks-* is a frequent element in Norw. place-names, and in N.G. XIII, p. 82, are to be found several possible interpretations, the only one of which that could possibly hold good in Orkney being O.N. *akrs-*, genit. of *akr*, arable field. Two facts however would seem to rule out that possibility: 1. *akr* (or *ekra*) appears in two well-known Orkney *ness* names—Aikerness in Evie and Westray—where the 's' is absent. 2. Though parts of the Sanday and of the Shapanasay nesses may have been, the Head of Work has certainly never been, cultivated.

Ax-, however, is found in Orkney as the first element of a name Axni-geo which occurs often, and which is almost certainly to be explained as O.N. *øxna-gjá*, oxen-geo, i.e. a geo where cattle probably went for shelter or perhaps to be shipped. Jakobsen also records the name *Eksnabø* from Shetland and derives from O.N. *øxna-bær* in the sense of ox-pasture.

While *øxna-nes* would seem the most probable explanation of the name Hacksness the presence of initial 'H' in all three examples of the name, and its absence in all the many Axni-geos, make such an interpretation at best doubtful.

Hermisgarth: Harmannisgarth 1500; Hardmundigarth 1595. Went along with Stangasetter to make up an urisland. Still a considerable farm today.

O.N. *Hermundar-garðr*, farm of a man Hermundr. It may be noted that Lind records the appearance of a possessive '-z' instead of the old '-ar' as early as 1397 in Norway. It is curious how the later Rental preserves the older form of the name.

Hellihowe: Hellehow 1500; Hellihow 1595. A former urisland, and a moderate-sized farm today.

Evidently O.N. *helgi-haugr*, 'holy mound.' Association with the supernatural survived long in connection with a mound here. Tradition records how it used to be the abode of a *hogboon* or howe-dweller (O.N. *haug-búandi*) who was a cause of such annoyance to the people at the farm that they resolved to leave and live elsewhere.

Hobbister: Halfynscoffis al[ia]s Hobsta 1500; Houbustar 1595. An old urisland; name now applied to a small croft.

In Peterkin's *Rentals* the middle 's' of the MS. Halfynscoffis has been omitted. The exact reason for the appearance of such a curious term is obscure, but it probably has reference to the statement in the Rental that the skats of 4½d. land in this tunship " are free in the handis of Symoun Ramsayis airis becaus of the landmale that he coffit (bought) and gaif to erle William in the bull of Karstan (Cairston in Stromness)."

In the name Hobbister the second element is O.N. *bólstaðr*, a farm, and the prefix has reference to the adjacent bay or 'hope' which is now called the Bay of Brough but which at one time must have had a hope-name. Hence O.N. *Hóp[s]-bólstaðr*, 'hope'-farm. Near the head of this bay is an expanse of low-lying ground known as the Meadows of Whup (hwʌp), a term which shows a local consonantal change of 'h' > 'hw.' The same development is found in the name Whupland which is applied to the land along the shore of a bay in the south end of this island, which, though now called Braeswick, must at one time have had a 'hope' name like the Bay of Brough.

In old Norwegian the term for such a bay was masculine—*hópr*, and Rygh states in his *Indl.* that there is no trace in Norway of the neuter Icelandic form *hóp*; but oddly enough both forms must have been current in Orkney as both Hubbin and Hubbit occur as place-names.

Holland: 1500; when it was entered as a tunmail under Gryndleith. There is a small farm of this name in the same neighbourhood still today. It is a common name in Orkney and usually regarded as O.N. *há-land*, 'high-land,' but the present Holland is definitely not on high land. A few hundred yards to the south-west, however, is an elevation known as

Hool (see *infra*), and it is probably that name which appears also in this Holland : hence O.N. *hól-land,* ' hillock-land.'

Housegarth : 1500; Howasgarth 1618 (Viking Club Sas.), 1½ urislands. In the 1595 R. from some inexplicable reason this tunship is represented only by the 9d. land of How.

In 17th and 18th cent. sasines frequent reference is found to the lands of " Housgarth *alias* Bea "; in Heart's Sasines we read of the lands of " Nearhouse otherwise called Housgar," and again of " Overhow and Housegarth *alias* Bea "; in the Viking Club's *Ork. Sasines* (p. 115) we read of " Nearhous *alias* Netherhow," and again of " Nearhous *alias* Nethirhous," and finally of a witness James Burgar " in Bea."

From these entries it is obvious that this old tunship had been broken up into several farms, but now again it has for long been reunited into the one large farm of How. In the stackyard of this farm is a large mound—almost certainly an old broch mound—from which the farm has its name : O.N. *haugr,* a mound. In the Rental form Housgarth the first element pretty certainly refers to this mound and not to any house. But as is obvious from the above references to " Nearhous *alias* Nethirhous," etc., confusion had arisen. The tunship had split up, and there was an Overhouse or upper-house, and a Nether-house or lower-house which became contracted to Nearhouse; and there was at least one other house called Bea.

These houses have all disappeared but the old name Bea (pronounced be :) survives. The farm of How occupies a peninsula called Bea Ness, the sandy beach running west from this is called the Sand of Bea, a loch behind this beach is Bea Loch, and on the bank of a stream between the loch and the sea is an old watermill called the Mill of Bea. O.N. *bœr,* a farm settlement.

The old Rental tunship adjacent to this was Southerbie (though it is rather to the west than south of Housgarth *alias* Bea). Here we have another Bea, and though it may possibly have been a separate unit from the beginning, there is strong probability that it may have been part of one large original Bea, which in that case would have embraced all the lands lying around the present bight of the sea known as the Bay of Backaskaill.

Here then we have a suggestion of another large original settlement similar to what was noted in Evirbist. A third and still more remarkable example will be referred to *sub* **Stove.**

Hool : Houlle 1500. A 9d. land which in the 1595 R. is entered under the name of Toftis. The name Hool is today applied to an elevation or ridge of land quite near the old field of Stove called Tofts. O.N. *hóll,* a rounded hillock or hill.

Hoosay (hus·e, hus·i) : Houshay 1500; Houssay 1595; Howssa U.B. 1601. A 9d. land or half-urisland of old; name applied now to a large cultivated natural mound or hillock in the crofters' land below Warsetter.

This could be simply O.N. *húsi,* dat. sing. of *hús,* a house, but from the ending -*hay* in the earliest recorded form it is more probably an original O.N. *hús-haugr* ' house-mound.' For the phonetic change cf. Horraldshay (Firth).

Hushaug is recorded as a Norwegian farm-name in G.N. XI, 79.

Houscrow : 1500; Howstrow 1595. A half-urisland of which both the name and situation are forgotten. From its place in the Rentals, however, it was probably immediately north of Grindally, and if so it would appear to have coincided roughly with three or four small farms of today called Howar (North H., West H., etc.), a name derived from hillocky ground nearby—O.N. *haugar,* mounds.

That fact makes the derivation of Houscrow doubly uncertain. The 't' of the 1595 form is a manifest scribal misreading of 'c' for 't' in an older MS., as in later sasines the name is regularly spelt with a 'c' or 'k.' At first blush the name would seem to be the same as is on record from Norway (G.N. VII, 364)—Huskroen—where the second element is explained as *Kro* (a nook or corner, etc.). No explanation is offered as to the sense of such a curious combination for a farm name.

Here in Sanday, however, we have noted above in Housgarth how an initial *hous-* may indicate an original O.N. *haugs-* (genit. case) = mound, and from the situation suggested above the origin here may well be the same: hence probably O.N. *haugs-kró.* In Orkney, however, *kró* usually indicates an enclosure or pen, and what actually gave rise to the name must be left obscure.

Houstais: 1500; Howstes 1595. An urisland in Everbist q.v. The name has been quite forgotten as a tunship name, but seems to persist in the local place-name Houstiebrig. or Houstie-breck, the northern end of a ridge in Evirbist of which the southern end is called Hool.

A puzzling name: cf. Arstais.

How: id. 1595. See Housgarth.

How-Benorth: 1500 and 1595. This farm in the north of the island has had the suffix—"Benorth" added in the rentals to distinguish it from the other How. In both rentals it is regarded as a 16¼d. land, an odd skatting which suggests that it must originally have been skatted as a whole urisland. O.N. *haugr,* a mound.

Lamaness: Lambnes 1586 (R.E.O.); Lamminess 1595; Lambnes U.B. 1601. Garth and Lamaness together formed an urisland. O.N. *lamba-nes,* 'lamb-ness.'

Langtas: there were two tunships of this name:
(1) Now incorporated in Newark; Langta 1500; Langtas 1595; Langtes U.B. 1601.
(2) In Evirbist; Langta in Evirbuster 1500; Langta 1595; Langtes in Overbuster U.B. 1601.

The former was an urisland, the latter a 2-urisland tunship. Origin of name doubly obscure because this tunship name appears to occur repeatedly as a field-name in Sanday. On the farm of Newark a field-name—The Linties—indicates the approximate situation of the old tunship there; in a 1797 list of Evirbist field-names one notes The Lays of Houstie and The Lays of Lawntie or Lanti; among 18th cent. field-names on Stove we find Lantea; near Hellihowe in Burness are fields still remembered as The East and the West Lanties; and an old track west of Scar is known as The Lantie-gate (road).

If it were not for the early spelling *-ta, -tas* one would be disposed to regard the name as from O.N. *langi-teigr,* 'long-tie'; though now obsolete that word *teigr* in the form *tie* was a common Ork. dialect term for a kind of field or strip of arable land (see my *Ork. Norn*). *Langeteigen,* indeed, is an actual farm-name in Norway. But one would scarcely expect such a name to be applied as a tunship name; moreover, the name Langtas cannot be disassociated from Arstas and Houstas, q.v.

Leavsgarth (livz·gər) : Lemsgarth (an obvious mistake) 1500; Levisgarth 1595. A half-urisland. O.N. *Leifs-garðr*, Leif's farm.

Leyland (li·lən[d]) : Lyrland al[ia]s Leyland 1500; Leyland 1595. An urisland tunship. O.N. *hlíðarland*, ' slope-land.'

Lopness : id. 1595. A bordland urisland.
  Probably the same name as is on record in *O. Saga* as the residence of a chieftain Amunde, who lived "i Hrossey i Sandvik a Hlaupandanese" i.e. at H—— in Sandwick [tunship in the parish of Deerness] on the Mainland of Orkney. There may well have been another farmstead of the same name in Sanday. The meaning of the term would seem to be—" [Out-] leaping ness."

Neebister (ni·bestər) : Nybuster 1500 (Peterkin has printed Wybuster through misreading of MS.); Nesbustar 1595; Nibuster U.B. 1601. A half-urisland of old; now incorporated with Warsetter, but the old farm-buildings are still standing. O.N. *Ný-bólstaðr*, ' new-bister.'

Noltland (nout·lən[d]) : id. 1500 and 1595. An old urisland in Burness parish : no longer a farm-name though it still survives as a name for a small district. O.N. *Nautland*, ' cattle-land.'

Sand : 1500 and 1595. An old urisland in Burness. Obsolete as farm-name, but Sander is a name still applied to an area of ground here. O.N. *Sandr*, sand, or its plur. *Sandar*; cf. Sand, N. Ry.

Skelbrae (skɛl·bre) : Scalbray 1500 (Peterkin has misread the MS. 'lb' as ' w,' and printed Scawray); Scabra 1595. Another Burness urisland, and a medium-sized farm still today. O.N. *Skjaldbreiðr*, lit. ' shield-broad.' This was a common term in Norse lands, and was applied to any rounded feature with a flattish or slightly swelling surface like that of a shield. In Orkney the name occurs in Rousay and Sandwick as well as here in Sanday.

Sellibister : Sellebustar 1500; Sellibustar and Salibuster 1595. An urisland in Lady parish; name still in use for a district, but not for any farm.
  Origin rather uncertain. *Selle-* is a frequent prefix in Norw. place-names and is usually referred to O.N. *selja*, a willow. Hence here probably O.N. *Selju-bólstaðr*, ' willow-bister.'

Skaill : Scale 1500; Skaill 1595. Entered as a tumail both times. The name is still that of a farm in Evirbist. O.N. *skáli*, a hall.

Skelbister : *scale butter* (sic) 1500; Skelbustar 1595. A 3d. land in Leyland tunship; a medium-sized farm still today. Fairly certainly O.N. *skála-bólstaðr*, farm having a *skáli* or hall thereon. Cf. Skelwick, Wy.

Suthirbie : 1500; Southerbe and Sutherbe 1595. An old urisland. The udal 1d. land of Treb (see Treb, N. Ry.) was in this tunship, but oddly enough the farm of Backaskaill, which must have been the head-house, never appears in the early Rentals. O.N. *suðr-bœr*, ' south-bae or -farm.' See also Housgarth.

Southwall : 1595, where all that is entered is 3¼d. land that paid skat to the Bishopric. In a sasine of 1664, however, we find a conveyance of "the lands of Grindilla, Boirland, Leata, Houskrow and Holland called in the old rental the **XXXVI** pennyland kingsland of Southwall . . . ." The inclusion of the word 'Boirland' here does not mean that there was a farm of that name (though it was probably so understood by the notary); it merely implied that Grindilla was bordland, i.e. private earldom property and as such free from skat. Actually, as we see from the 1500 R. only 6d. land of the 18d. land of Grindilla was bordland.

In other sasines of Southwall we find one of the farms therein specified as "Houskrow or Housnea," an addition which suggests that the former name had become more or less obsolete long before and been replaced by another—O.N. *Hús-nýja*, i.e. 'Newhouse.'

Thought the name is now extinct, the location of Southwall is sufficiently clear from the names of its constituents—Grindally and Holland, and though Houskrow is now lost and represented probably by Howar, Leata (sometimes recorded as Leatan) can still be identified in the name "The Lettan" which is applied to a flat area above Grindally on the links of Warsetter. The *-wall* in Southwall (as e.g. Kirkwall) is a faulty emendation of the true Ork. form *-waa*, deriving from O.N. *vágr*, a bay—in this case the present Braeswick, and *South-* has been added to distinguish this district of Waa from another in the north end of the isle—Northwaa—a name still in general use today.

The local name for Southwall seems to have been Voy, and on one of Mackenzie's charts (1750) it is entered under that name together with its boundaries which show that in addition to the farms indicated above it included all the ground along the south shore of Braeswick now occupied by a row of crofts. This latter area, as noted above under Hobbister, is known now as Whupland, i.e. O.N. *hóp-land*, land along a 'hope' or bay. Voy is of course the same name as Waa, but from the old dative case—O.N. *vági*, and it occurs again in Sandwich and St. Andrews. And it may be noted as a curious fact that this small bay from which the tunship had its name must from time to time have been referred to by each of the three usual O.N. terms for a bay—*vágr*, *hóp* and *vík*.

Before leaving this interesting name it should be noticed that here again we have a unity of a larger kind than the normal tunship. It contained 2 urislands and embraced the whole area on the south and east sides of the bay of Braeswick. Its constituent farms appear in the earliest Rental, but as a whole unit it is not mentioned. See further, however, *sub* Stove.

Stangasetter : 1500 and 1595. Name applied today to a small farm and spelt in the same way still. In old rentals it went along with Hermisgarth (which was some distance apart) to make up an urisland.

As for the second element in the name it is one of the commonest in Ork. farm-names, and as such may probably derive from O.N. *setr*, a dwelling, rather than from O.N. *sætr*, which generally is applied to a summer shieling up in the hills. The first element *Stang-* is also common in Norse place-names but its exact significance is quite uncertain. (See Magnus Olsen's remarks in G.N. X, 126). O.N. *stǫng* (genit. *stangar*) meant a pole, but one cannot be sure that it had any such meaning in place-names.

This Sanday farm is inland, and part of the low flat lands of Burness parish.

Stove : Stoif 1595. The name does not appear in the 1500 R. nor in the Earldom section of the 1595 R. In tthe 1500 R., however, we find Halkisnes, an 18d. land, Gardemeles, an 18d. land, Frow and Wattin, a 9d. land the last also spelt Frowattin. In the Earldom part of 1595 R. we find Halksness, Gardumellus and Frowantown = 2½ urislands = 45d. lands, of which "the bishop takis the third of the scats of 27d. terrae, viz. of 9d. terræ." Then in the Bishopric part of that Rental we find Stoif, a

9d. land paying skat to the bishop. Our first problem is to discover what part of the 2½ urislands was represented by the 9d. land of Stove.

Hacksness was all bordland and thus skat-free. Hence the 27d. lands from which the bishop drew a third of the skats must have been Gardemeles and Frowantown, and we learn indeed from the 1500 R. that the bishop's skats were drawn from 6d. land in Gardemeles and 3d. land in 'Frow and Wattin.' These then must have been the 9d. lands entered as Stove in 1595.

The facts here are unusually difficult to unravel for the names Gardemeles and Frow and Wattin are entirely unknown locally today, and only Stove survives. For some centuries down to about 1920, when it was bought by the Scottish Board of Agriculture and split up among small-holders, Stove was one of the largest farms in Orkney and embraced all the old Rental lands of Hacksness, Frow and Wattin, Gardemeles, Houlle and Grindally. The farm-houses of Stove stand at the head of a picturesque bay flanked on the east by the ness of Hacksness. The site of the old Gardemeles is quite unknown but it was probably near or at the present houses of Stove. The names Frow and Wattin and Frowantown are almost certainly due to scribal errors in compiling the 1500 R. from an earlier rental, and for some time I thought that the *Frow-* might be a misreading of Stove. But after mature consideration I have given up that idea. Between the present fields of Hacksness and the farmhouses of Stove are fields which go under the name of How, and there I now believe the old so-called Frow- farm is to be located. O.N. *haugr* with the def. article attached is frequently found in Orkney in the form Howan, and if we assume an original Rental name *Howan-tun* I believe we can have a satisfactory explanation of the Rental's *Frowantown*.

We should thus have the two farms Howan and Gardemeles close together, and perhaps intermixed, and I think it probable that Stove may have been an alternative name for the rather peculiar Gardemeles, and one that has certainly outlived it.

O.N. *stofa*, a room; a fairly common farm-name in Orkney, and applied probably, like *skáli*, to a farm with a house of special or superior type.

Before leaving this group of farms two facts must be noted which indicate fairly conclusively that here again we are face to face with another of those large original settlement units which have been referred to under Evirbist and Housgarth. Hacksness, being bordland, was skat-free, but in the 1500 R. we find that the bishop had right to a third of the skats of Gardemeles, Frowantown, Houlle, Grindally and Houskrow which together amounted to 3½ urislands. And at the end of that group the Rental adds:

" And heir endis all the scattis that the bischop takis in the parrochin and I traist that the bischop hes richt to thir scattis becaus I found in the old parchment rentall that notwithstanding the land is all the kingis to the male [i.e. landmail or rent] yit ilk iiid. terre is laid by to pay scat to the bischop."

That assignment of skat at a uniform rate from a group of lands that stretched in continuity for about two miles can hardly be explained except on the basis that they formed one large estate or settlement when the grant was made.

That conclusion is strongly buttressed by a curious document of 1707 which is a copy of the decision of a court held at Warsetter, Sanday, by Earl Robert Stewart on 15/6/1573. An assize of 24 men sat to consider a complaint lodged against a servant of Wm. Sinclair of Warsetter—one John Williamson by name—who " had cumed within the bounds and ground of Stoiff . . . and sett down at his owne hands march stones beside Grindillay, etc." Whereupon " six famous men" whose names are given, declared that the ground of Stoiff came to the south dyke of Howar, and past out through the slack of the links. . . ." The result was that Warsetter was warned against further trespass, but for us the vital point to note is the testimony as to the old bounds of Stove which evidently extended so far as to include all the lands referred to above from which a third of the skats went to the bishop.

Now as already noted under Southwall several of those lands—Houlle or Tofts, Grindally and Houskrow—were included in that Southwall tunship. Here then we catch a glimpse of an original unity of a kind that went clean beyond the bounds of a single tunship.

Simbister: Synbustaith 1500; Symbustar 1595. An urisland on west side of Burness; no longer a farm-name but it is still remembered as a tunship name. The same name occurs in Whalsay, Shetland. O.N. *sunn-bólstaðr*, 'south-bister' or farm.

Tafts: Toftis 1595. One of the five earldom bu-s in this island—each of which extended to an urisland. (The word was usually spelt 'bull' but was pronounced 'bu'; see further Bu—Part III.)

Tafts is still a farm—on the most northerly point of the island—viz. Tafts Ness. O.N. *toptir*, plur. of *topt*, a house-site. In Orkney it is a fairly common name, and is usually, if not always, found applied to a site where there are traces of prehistoric building. For another Tofts v. **Hool.**

Tresness: Trosnes 1595. An evident misspelling as Tresness is regularly the form found in old records, and is the name still today. This was another old earldom bu, a whole urisland, and is still a single undivided farm today.

Origin of first element of name uncertain. In N.G. VIII, 79, a Norw. farm Træsnes is recorded, and the first element is there compared with another farm-name Træ which is regarded as a contracted form of No. *trœe*, a cattle-fold, etc. But one can also compare the Caithness Freswick of which the O. Saga form was *prasvik*, the origin of which is also obscure. It may be indeed that Tresness is merely O.N. *trés-nes*, 'tree ness,' and so named from driftwood found there. Even today it is one of the greatest 'catch-grounds' in Orkney for driftwood and all kinds of flotsam.

Volunes: "En þeir Sveinn gengu upp á eyna ok kómu á bœ þann er heitir á Vǫlunesi. (O. Saga.) The island was Sanday and at that farm lived a relative of Sweyn called Bárðr. No farm or other place of that name can be traced in Sanday today, and its meaning is also obscure.

Walbroch: 1500; Waldibrek 1586 (R.E.O. 308); Waldbreck 1595. A 9d. land or half-urisland which at the date of the 1500 R. was in the hands of Sir Wm. Sinclair of Warsetter, brother of Lord Henry Sinclair, who had a tack of the Earldom at that time and for whom the Rental was prepared.

From the order in which it comes in the Rental immediately north of Garth, it would appear that its lands were incorporated with others to form the Warsetter estate. Though this 1500 R. of Sanday was "maid at Worsetter" there is no entry of that place as a skatted unit, and it would appear to have been created as a composite Sinclair manor, or 'bu,' out of various adjacent lands.

The approximate location of the old Walbroch or Waldbreck is probably to be traced to a field on the present farm of Warsetter called

Voldibrae where until comparatively recently there was also a cottar's house of that name.

Origin of name uncertain : the 1500 spelling is doubtless wrong as there is no evidence of an old broch anywhere near.

Waldgarth : 1595. A 1½d. land which according to the Rental was included in the area (see *sub* Stove) from which the bishop drew skat. In later sasines one frequently finds record of "Walgarth in the town of Woxatter," or Voxetter, and it can fairly safely be identified in the present Volyar—a field on farm of Stove. The first element of name is of the same obscure origin as that in the preceding name Waldbreck, but the suffix needs no explanation.

Voxetter, however, the tunship in which Walgarth lay, is a complete puzzle as it does not seem to appear in any other context and is quite unknown today. The same name is found in Shetland, and is as Jakobsen indicates, pretty certainly O.N. *vág-setr*, ' bay-homestead.' The bay in question must have been the Bay of Stove, but *vág-setr* is a curiously inadequate or weak name for a tunship that must have embraced the important 2½ urislands around Stove.

Walls, Bull of : 1595. A former urisland and another of the private earldom farms and as such unskatted bordland. Situated in the north end of the island it is no longer a separate farm-unit, but the name survives in that of the district adjacent—viz. Nort-Waa. It has been so named to distinguish it from Southwall (q.v.) in the south of the island.

O.N. *vágar*, ' bays.'

Warsetter : Worsetter 1500. This was the seat of the Sinclairs of Warsetter, the senior branch of that family in Orkney. Sir Wm. Sinclair of Warsetter was a grandson of Earl William Sinclair who was the last earl of Orkney under Norse rule. That Earl's grandson Lord Henry Sinclair was tacksman and ruler of Orkney in the early years of the 16th century, and after his death at Flodden in 1513 his brother Sir William of Warsetter was appointed " Justice " in Orkney.

Warsetter does not appear in either of the early rentals as a skatted unit itself, and as suggested above (*sub* Waldbreck) there is reason to believe it is a composite manor-farm built up by the Sinclairs out of neighbouring lands which they acquired. Even today after the loss of some of its best ground for the enlargement of crofter holdings, it embraces the old lands of Neebister, Garth, Waldbreck and Whiteclett.

The farm-buildings stand near The Wart, the highest point of the island and so named from having been one of the old Norse beacon hills : O.N. *varða* (or *varði*), a beacon; hence *varð-setr*, ' beacon-homestead.'

OTHER NAMES

Airan (ɛ·rən) : in Burness. O.N. *eyrar-endi*, 'ayre-end'; it is situated at the end of an ' ayre ' or gravelly beach.

Appiehouse : in Evirbist; near by is the complementary Nearhouse. These two names respectively mean the Upper-house and the Nether-house of some old unit—probably that part of Evirbist which was termed Northgarth in the Uthel Buik and other old records.

Backaskaill (bak·ə-skil) : the main farm in the old Southerbie tunship; situated in a picturesque setting at the north-west corner of the Bay of Backaskaill, at the west end of the Bea Sand. O.N. *bakka-skáli*, ' skaill or hall at the *bakki*.' What bakki refers to here is uncertain; the word means a slope, and the house is at the foot of a very steep slope. But *bakki* is also used in Orkney of the shore-bank, and the house is at the very edge of the beach.

Beafield (be·fild) : a farm in Evirbist. The termination is of course English and the name thus relatively modern. But the first element points to O.N. *bœr*, one of the most dignified of old farm names in Orkney. Near by in Evirbist is a farm called The Bu, but no old Bea can now be traced in this part of the island (apart from this hybrid name).

Benziecot (bɛn·ji–kɔt) : a farm in Evirbist. The last element is O.N. *kot*, a cottage or hut; for first element see Binscarth (Firth).

Boloquoy (bɔl·o–kwi, –kwai) : a farm at beach. Termination O.N. *kvi*, enclosure or yard for animals; prefix doubtful—name may be either *Bolla-kvi*, ' quoy of a man Bolli,' or *bola-kvi*, ' bull quoy.'

Braeswick (brez·wik) : a bay in southwest of Sanday and the ground around its shores. From time to time this bay seems to have been referred to as a *vik*, a *vágr* and a *hóp[r]* : see Southwell *supra*. Origin of Braes- quite uncertain. *Breis-* occurs as first element in several Norw- names but its source is obscure; see G.N. XII, 119.

Breckan: croft in Burness. O.N. *brekkan*, ' the slope.'

Canker: a croft in Evirbist district, and the name occurs again in Sellibistertun, and in Birsay.

The earliest recorded instance of the name I have come across is in the trial of Anie Tailzeour who was accused of witchcraft in 1624. (See *Abbotsford Misc*. Vol. I, p. 145). She was stated to have been one night " in Thomas Mure of Quoykankeris hous."

The origin is no doubt obscure, but the name might possibly represent O.N. *Kanu(n)k(a)r*, ' canon(s),' and indicate a quoy of land bestowed or acquired for the maintenance of cathedral canons. Parallel examples can be cited from Norway, e.g. *Kankerud*, etc., and it is perhaps significant that the 1500 R. indicates that " the bischop and the vicar " took the whole skats of 5¼d. lands in this Everbist district of Sanday. Cf. Quoycanker (Deerness).

Crotrive (krɔt·rəv) : Crowthreve, Crothrive, Croutheve, 1805; a small croft in Evirbist district. The termination is almost certainly the pre-Norse *trave* which is found as a farm-name repeatedly in Sanday : see **Treb**, N. Ry. The first element is evidently Icel. *kró*, No. *kru*, a pen or cattlefold, and if so the whole compound would probably signify a cattle-fold at a treb-dyke.

Cruddy (krʌd·i) : a farm in old Cleat-tunship. An obscure name. In Rousay Croddy (krɔd·i) is the name of a cottage and also of a small patch of ground in a larger stretch of pasture-ground. In N.G. XVII, p. 32, a farm-name in Norway—Krøttøen—is recorded, but the origin is also regarded as obscure.

Curcasetter: U.B. 1601. Entered as a tumail of cultivated ground in Burness. Curc(a)- (kark·ə) is a rather common but puzzling first element in Ork. place-names. In general it probably represents O.N. *kirkju-*, and indicates church land of some sort. But the fact that this Curcasetter is entered in the *Uthell Buik* as udal ground would seem to imply that the name could hardly mean ' church-setter.'

The only other apparent alternative would be O.N. *korki*, oats, a borrowing from Celtic *coirce*, id. In Foula that term was in use in the dialectal *korka-cost* for oatbread. But 'oats-setter' would be a rather unlikely compound, and we are thus left to choose between two possible sources—each of which presents difficulty.

Erraby (ɛr·əbi) : The Sand of Erraby is the name of the long narrow isthmus connecting Tresness with the rest of Sanday. No early records of the name are available, but it indicates fairly definitely an O.N. *eyrar-bœr*, 'ayre-, or beach-farm.' In the absence of further data speculation is risky, but sand-blowing and erosion have taken place here on a large scale, and it is probable that in this name we have a suggestion of another of those large early settlements referred to already, one that may have included the present Tresness and a good deal of additional arable- or pasture-land now buried under masses of sand dunes.

Garbo (gɛrˈbo) : one of the Sellibister farms. In a sasine of Sellibister lands in 1624 one of the witnesses was a John Linkletter in *Gairbak*. See Gerbo, N. Ry. A treb-dyke runs close past this farm-house.

Geramont (gɛrˈəmʌnt) : now a house and park only. Entered in U.B. 1601 as an udal tumail. In a disposition dated 25/3/1568 John Hall, 'indweller in Sanday,' sold to David Scollay, burgess of Kirkwall, "my towmall callit Garamont with the pertinents which I conquist fra Jonet Garamont, heritable possessor thereof."

This is a curious example of a place-name which has lost its termination, whatever that was—e.g. *quoy*, *garth*, etc. Only the personal name of an old owner is represented, viz. O.N. *Geirmundr*, and from the fact that Jonet had her surname obviously from this property it must have gone under that presumably truncated name for some time before the above sale.

Gleat (glɛt) : farm in Northwaa. Origin obscure; a farm-name *Glette* is on record in N.G. X, 439, but the origin there is also uncertain.

Hammerbrake : a large but relatively modern farm. So named probably from a field name—O.N. *hamra-brekka*, slope with 'hammars' or projecting rocks in it.

Howar : (houˈər) : there are several small farms or houses of this name. O.N. *haugar*, plur. 'mounds.'

Howbell (houbel·) : a small croft in Burness. The first element is no doubt O.N. *haugr*, a mound, but the second is quite obscure. *Bell* appears frequently in Norw. farm-names, but there is no general agreement as to its significance : see N.G. XII, p. 111.

Howland : small farm in Simbister-tun, Burness. O.N. *hauga-land*, 'land of mounds'; there are many old mounds near here.

Hynegreenie (hain-grin·i) : Howing Greinay 1633 (Trial of a witch : *Abbotsford Misc.* Vol. I, p. 150). A house in Burness. O.N. *haugrinn grœni*, the green mound.'

Isgarth (aizˈgər) : Ysgairthe U.B. 1601. First element rather uncertain. This farm lies at the north corner of the Bay of Kettletoft where a low narrow spit of land projects across the bay towards Elsness. Inside that point is a large sheet of shallow water which ebbs almost dry at low tides and is known as The Oyce (øs), i.e. O.N. *óss*, a term used of a rivermouth or estuary-like inlet. Isgarth might thus be O.N. *óss-garðr*, 'farm at the Oyce,' but if so the initial vowel change would present difficulty.

An alternative source may be suggested. The narrow spit of land referred to above, though it does not actually provide a dry path all the way across to Elsness even at low water, does suggest an isthmus, and may have been deemed such in olden times : hence one might assume an O.N. *eiðs-garðr* (isthmus-farm), which would present no phonetic difficulty.

Kettletoft : name of present village and island pier, but the name was in use long before either village or pier came into existence. It is uncertain what exact place was the original Kettletoft; it seems never to appear on record as a farm of any kind, but "the shore of Kettletoft" was the main port of the island for centuries back. It is situated in the important old tunship of Housgarth or Bae, and indicates the house-site of a man Ketill : O.N. *Ketils-topt*

Langskaill : there were two houses of this name in Sanday—one in Burness (in Noltland tunship?), the other in Marykirk or Gardemeles. The name is still remembered in Burness, but in the other case it is quite forgotten. It appears however in a list of field names on the Elsness estate dating from the year 1700 : "the tumelt of Loongskeall" which was one of the Marykirk fields.
O.N. *langi-skáli*, 'long hall.'

Langomay (laŋ·əme, -ome) : there are two small crofts of this name, East and West L——, in the Northwaa district on the neck of land between Lopness and the rest of Sanday. In the absence of early forms of the name it is impossible to decide what the -*may* signifies. It is a rather common element (me) in Ork. place-names, e.g. Maeshow, Maesquoy, Maizer, Howmae, Ossamae, etc., and so far no satisfactory derivation has been suggested.

Lettan : (1) a small district in extreme north end of island; several small crofts there go under the name of "The Lettan" houses; (2) a field on links of Warsetter—The Lettan—which pretty certainly denotes the location of a farm which in old records appears as Leata or Leatan (see Southwall *supra*); (3) in Burness several fields bear the name The Leti (let·i), one being known as Tammy's Leti.
All the above have one feature in common—the flat or level nature of the ground, and they must all derive from O.N. *flǫt* or one of its sideforms *flati*, m., or *flata*, f.—all of which indicate a flat expanse of land. (See Rygh's *Indl*. 50). In the 'Lettan' forms the def. art. is suffixed.

Confirmation of that derivation is to be found in the record of a sale of Sanday lands in 1649 which included the 4½d. land of Tofts and houses thereof called *Fletta*.

In Ork. place-names initial 'f' before 'l' tends to be lost. Skerry names e.g. Less, Lesh, Lashy, Tara-lash, etc., certainly derive from O.N. *fles*, a skerry; The Lother Rock from O.N. *flæðar-sker*, etc. Cf. also Lobust and Lythe in my *Place-Names of Rousay*.

Maizer (mez·ər) : Measetter U.B. 1601; Meazar 1734. A farm in Evirbist. Evidently an old -*setr* name, but the prefix is puzzling; see Langomay *supra*.

Myres : there are two farms of this name—one in Evirbist, the other in the old Southerbie. O.N.*mýrar*, 'mires' or boggy grounds.

Neigarth (ni·gər, ni·djər) : Neagir in Stangasetter U.B. 1601. Besides this Burness farm there are others of the same name in Lady parish. O.N. *ný-garðr*, 'new-garth.'

Newark: a Scottish name = 'New-work,' a common name for a new building in the 17th cent. particularly. Here in Sanday the building was probably due to the Stewarts of Brough who acquired much of the former tunships of Arstas and Langtas and made Newark their residence for a century or so.

Nouster (noust·ər): farm at beach in Burness where there is an old landing-place for boats. O.N. *naustar*, 'boat nousts,' i.e. berths for boats drawn up at beach.

Northgarth: U.B. 1601. This was a 4½d. land in Evirbist, the memory of which is quite forgotten locally.

Okamby (ok·əmbi): old house-site close to Lopness. This is another name which seems to preserve the memory of an otherwise lost *bae* (O.N. *bœr*): prefix obscure.

Ortie (ort·i): there are several small crofts in Burness which go under this name—all lying adjacent to each other and spoken of as the Ortie houses or simply The Orties. In Shetland the name Orter occurs in several districts, and Jakobsen tentatively suggested it might derive from a Celtic source—viz. Gaelic *oirthir*, an edge or border. That, however, would seem very doubtful, and the origin of the Sanday Ortie must be regarded as obscure.

Over the Water (ou·ər-): id. 1682 (in a legal process). A small farm at edge of a loch. This is an interesting example of a prepositional name, and must be a direct translation of an O.N. *yfir vatnit*, 'over or across the loch.' In Orkney *vatn* (a lake or loch) is sometimes found untranslated in the form Watten, but it is also often translated to Water, e.g. The Muckle Water and The Little Water (Rousay).

Oyce (øs, but also with a narrower vowel—es): a Burness farm adjacent to a shallow bight in Otterswick from which it has its name: O.N. *óss*, a river-mouth or estuary-like inlet.

Quivals (kwe·vəls, kwi·vəls): a farm near southwest shore of Otterswick. Evidently the same name as recorded from Iceland by Finnur Jónsson (*Bœjanöfn*, p. 471)—O.N. Icel. *kvia-vellir*, 'quoy fields'; for Quoy names in Orkney see Part III.

Rusness (rʌs·nez): a district adjacent to Sellibister: O.N. *hrossa-nes*, 'horse-ness.'

Savil (sa:vəl): Savell 1668. A farm in Burness which was the manor-place or 'laird's house' of the Westove estate in the 18th century. On Mackenzie's chart (1750) the mansion-house appears near the southern border of Noltland tunship, and near the shore of Otterswick.

There is another farm of this name in Firth parish (Savale 1595), and in Lady parish in Sanday is a third which has the adjective *-green* attached—Savil-green. That adjective would seem to confirm that the second element in Savil is O.N. *vǫllr*, a field, and that being so the first element can hardly be other than O.N. *sáð*, seed, the compound **Sáðvǫllr* thus meaning 'seed-field' or 'corn field.' Cf. O.N. *sáðjǫrð* and *sáðland*, though oddly enough **sáðvǫllr* is not recorded in Norse dictionaries.

Scarrigar: id. 1711 and 1762. A cot-house of Warsetter situated on top of a ridge which is traversed by a pronounced gap or 'col' immediately north of the house. O.N. *skarð-garðr*, 'farm at a skarth or notch on a hill-ridge.'

Scofferland: a croft near Lopness. First element obscure.

Seater (set·ər): there were at least two farms of this name, one still in being in Sellibister, the other near to (and now incorporated in) Boloquoy. O.N. *setr*, a homestead.

Skaill (skil, skjil): there were formerly several farms of this name of which Skaill in Evirbist and North Skaill in Burness still exist. But there were formerly also South Skaill and Langskaill in Burness and another Langskaill in Marykirk.
O.N. *skáli*, a hall.

Skedgibist (skidȝ·ibəst): a croft in Braeswick. Same name occurs in Orphir. Origin uncertain; the first element is possibly a man's name—*Skeggi-*, hence O.N. *Skeggja-bólstaðr*, 'Skeggi's farm'? Cf. however Skidge (Birsay).

Skitho (skɪð·o): small farm in Burness. First element probably O.N. *skeið*, which was used *inter alia* of (1) a race-course; (2) a track through fields. The final '-o' is obscure, but may perhaps represent O.N. *haugr*, a mound.

Stiglister (stɪg·ləstər): a croft in Hobbister tunship. The *-ster* is no doubt O.N. *setr*, a homestead, but the first element is more obscure. There is an Ork. dial. word *stiggle* used of a thin worthless straw or grass stalk (e.g. "nothing growing there but stiggles"; "a poor stiggly crop," etc.). But whether that is the term represented in this farm-name is extremely doubtful.

Stretigar: Strettagarth U.B. 1601. A farm in Burness. Origin of first element quite uncertain. In N.G. certain Strett-, Stræt- names are tentatively related to O.N. *strœti*, a roadway (a term borrowed from Latin), and in Shetland Jakobsen found the name *stred* (applied in the sense of a field or piece of ground) which he derived also from O.N. *strœti*. Such ascriptions would seem a little bold, and for the origin of this Sanday name one might perhaps compare the Norw. *Strituland* or *Stretuland* which Falk (N.G. V, 12) relates with No. *streit*, trouble, strife, etc. In that case one might cf. the Rousay Tratland.

Thrave (trév): there are three farms in Sanday so-named—East and West Thrave in Burness, and another Thrave in the Lopness area. Cf. also Crotrave, *supra*. See Treb, N. Ry.

Treb: id. U.B. 1601. A vanished farm in Southerbie. See Treb, N. Ry.

Whistlebare: a small farm in Burness. The name occurs again in Eday, Egilsay and South Ronaldsay. Origin quite obscure, and all the more as the main stress is on the final element *bare*.

Whiteclett (hwait·let): Quhytclett U.B. 1601. An old farm now represented only by a field of this name on Warsetter. O.N. *hvit-klettr*, 'white-rock'; see Cleat *supra*.

Westove: an old 17th—20th-century estate or lairdship in Burness: not a normal farm-name. In the 1500 R. is the entry "Item be west O twa towmalis"; in the 1595 R. Pro Rege is "The 2 Tumails Be-west ow," and in the 1595 R. Pro Episcopo—"Westow in Burness, 12d. terre." A stream runs down here, and a house near its mouth is still known as Woo (u:), i.e. O.N. *á*, a stream. It would thus appear that Westove is an erroneous name, manufactured by mistake from 'west o,' i.e. 'on the west side of the stream'

STRONSAY

O.N. **Strjónsey**, *O. Saga.*

Though the scribe of the Flateybook MS. of the *Orkneyinga Saga* (puzzled no doubt by an unfamiliar name) 'emended' it to Straumsey and Streaumsey, the normal spelling of the MSS. is Strjonsey—a form that is confirmed by present-day pronunciation—Stron·se or Stron·se. The final element of that name is of course O.N. *ey*, island, but the preceding element is even to this day somewhat obscure.

O.N. *strjón* is found only in place-names, and the term occurs in several Norwegian names. Its cognate, however, occurs in the O. English *gestreon*, which had the meaning of gain, profit, wealth, etc., a sense that is borne out by another cognate—O.H. German *strinnen*, to gain or win something. The O.E. term occurred also in a few place-names, one of which was Streoneshal[c]h, an older name for Whitby; and once at least it is on record as a personal ekename (in Eadric Streona).

In place-names it is thought that the term must have conveyed that idea of gain or profit, and in most of the Norwegian examples Magnus Olsen, the greatest authority on the subject, has suggested that a good fishing place was the underlying reason. It may very well be that such was the reason for the application of the term in the Orkney *Strjónsey* also; for centuries the island has been famed for its herring-fishing (though now unhappily at an end), and fishing-vessels from the continent are known to have frequented the island as far back as the 16th century at least.

Though Stronsay is for the most part low-lying and very fertile, its early skatting was on a considerably lower scale than that of Sanday. The island is missing from the surviving copies of both the earlier Rentals of 1492 and 1500, and appears first in that of 1595. It is thus impossible to ascertain exactly the number of urislands on which it was skatted, but along with the 12d. land of Papa Stronsay there were approximately 13 urislands or ouncelands altogether.

RENTAL NAMES

Airy (ɛ·ri) : Air, Are, 1595; Airie, 1664. A 4d. land in Evirby. A Celtic survival : see Airy, Sanday.

Airafea (ɛr·əfi) : Arefea, 1595 and 1632 &c.; Airifea, 1679. A farm on the southern border of Evirby. The same name occurs in Sanday, q.v.

Aith (e:ð) : A 24d. land. O.N. *eið*, isthmus; the isthmus here indicated is that joining Roithisholm with the main body of the isle.

Auskerry (ɔu·skɛri) : Ouskery, (Fordun) c. 1380; Owskare. 1535 (R.E.O., 219); Owskerrie 1595.

Though bearing a 'skerry' name this is really a grassy islet where a considerable number of sheep can be pastured, and it was entered in the 1595 R. as a skatted unit. It lies about a mile S.E. of Stronsay.

From the time of the earliest recorded form of the name its pronunciation seems to have remained unaltered, but its origin is indefinite. Munch suggested that it represents O.N. *aust[r]sker*, 'east skerry,' or perhaps *útsker*, 'out-skerry,' but in N.G. XIII, 205, it is noted (*sub* Ausnes) that O.N. *ausa* (a ladle or baling-scoop) is very often used as a skerry name, and oddly enough in the Shetland dialect *auskerry* is the term applied to such a baling-scoop. In that case, however, the second element is a different term altogether, viz. O.N. *ker*, a tub, bucket or goblet, etc.

Of the different possibilities—*Aust[r]sker*, 'east skerry' is the most probable.

Brecks: Now vanished, but the name appears frequently in sasines as that of 'a mansion-house' in North Strenzie (the present Whitehall farm). It was pretty certainly the site of the residence of the Rikarðr who, the *O. Saga* states, lived "at Breckum," i.e. dat. pl. of O.N. *brekka*, a slope.

Cleat: Clet 1595. An urisland farm in south of island. See Cleat, Sanday.

Clestran (klɛs·trən): Clestram and Clestrian 1595—both forms mistranscripts of Clestrain. A 6d. land adjoining Strenzie. This was the estate from which the Feas of Clestran, one of the best known old Orkney families, took their territorial designation. The first James Fea of Clestran on record obtained a feu charter of the estate in 1591, and the family held it until about 1770, when it was acquired by Robert Laing.

Origin of name obscure. In Orphir there is another farm of the same name which in the 1492 R. is spelt Clatestrand and Cletestrand, in the 1500 R. Claistrand, and in that of 1595 Clestrain. The origin would thus seem to be O.N. *klett-strǫnd*, 'klett- or rock-strand'—a somewhat curious name as most Orkney strands are rocky. See further *sub*. Strenzie.

Dishes: id. 1595. "Ane Skatland" (i.e. a 4½d. land). A small tunship on east side of Bay of Roithisholm. At the beach here the house of Lower Dishes is built on a prehistoric mound which has doubtless given the name—O.N. *dys*, a cairn or mound. It is uncertain whether we have here the plural *dysjar* (with the termination Englished as so often happens) or a compound with *hús—dysjar-hús*, 'mound house.'

Dritness: Dirtness 1595. A vanished farm near a point on east side of island still called Dritness. O.N. *drit-nes*, 'dirt-ness'; reason for name unknown.

Gart: Garth 1595. An old 4d. land now represented only by a field on Cleat called The Gart. The same name occurs as a field-name on Whitehall; this in the 1595 R. figured as Quoygarth. O.N. *garðr*, a farm or enclosure.

Grobister: Grobuster, Grobustar 1595; Groabuster and Grogabuster (Vik. Club Sas. 114) 1620. This was a two skatland or 9d. land, and is still a district name today. Despite the appearance of an internal 'g' in the above Sasine form the name must represent an O.N. *Gróu-bólstaðr*, i.e. the settlement of a woman *Gróa*. That name is on record in the O. Saga as that of a woman living on the Mainland of Orkney who went insane and was cured by going to the shrine of St. Magnus. An earlier Gróa (grand-daughter of Aud the Deepminded) was married in Orkney, and had a daughter who married Earl Thorfinn Skull-cleaver.

Hescombe (hɛsk·əm,·ʌm): Hescum 1595; Haskum 1624 (K.R.R.). A small farm near shore on south side of Mill Bay; skatted as ¼d. land. An obscure name; in Norway Hesk(e)- is a rare element in place-names, and for the most part obscure there also.

Holland: id. 1595. An old bordland urisland, and still one of the largest and most fertile farms in Orkney. It was the only bordland recorded in

Stronsay, and thus probably the earl's 'bu' which Sweyn Asleifson's brother Valþiófr held (see *O. Saga*, p. 172). See Holland, N. Ry.

Houseby (hus·be, ·bi): Housbie 1595. Adjacent to Holland and like it an old urisland. Still a large and most excellent farm. O.N. *hús[a]-bœr*, 'house[s]-farm,' i.e. a farm or settlement notable in some respect for its houses.

Huip (the true native pronunciation was høp, but through the influence of outsiders, and natives imitating them, it has now become generally perverted to hjup.): Hupe 1422 (R.E.O.); Hupe and Huip 1595.
An old urisland settlement on northern extremity of island, and still today a large farm. The name undoubtedly comes from an adjacent large, and almost landlocked, bight or 'hope': O.N. *hóp* (or *hópr*), a sheltered bay or 'hope.' This headland whereon Huip lies was certainly the ness referred to in the O. Saga as *Hofsnes* (a mistake for O.N. *Hópsnes*).

Hunton (hʌnt·ən): Hunto and Huntoune 1577 (REO); Hunto 1627 (P.R. p. 95). A large farm now apparently coincident with the former South Strenzie. On McKenzie's chart (1750) the two houses of Hunton and S. Strenzie are marked close together. Though Hunton does not appear in the 1595 R. at all (it was kirkland) its 6½d. lands are constantly in evidence in old records. In the earliest, a disposition of the lands by David Sinclair to his son, the lands are designated " of Hunto," but the signature is by " David Sinclair off Huntoune wt my hand." (See REO p. 296.) Subsequent record forms vary between Hunto and Hunton, but the name is never pronounced Hunto today, and probably never was so locally. The dropping of the final -n in records is probably due to a scribal error in failing to note the usual mark of abbreviation for a final -n.
It should be observed, however, that in Birsay is a farm Hunto—so pronounced today and thus spelt in records also. The origin of both names is uncertain.

Kirbuster (kɪr·bəstər): Kirbustar 1595; Kirkbister 1624. A 3d. land in Evirby tunship. This example of a 'bister' farm in a 'bae' tunship is noteworthy: see Part III. O.N. *Kirkju-bólstaðr*, 'kirk-bister or farm.'

Linga: Meikle Linga 1595; Linga-meickle Sas. 1664. This uninhabited holm lying off the west side of Stronsay and now known as the Holm of Midgarth, was entered in the 1595 R. as a skatted-unit. A smaller (unskatted) holm lying to the north is known as Little Linga (in a 1669 lease entered as Lingalittle). O.N. *Lyngey-in-mikla* and *Lyngey-in-litla* respectively—'Big and Little heath-isle.'

Midgarth (mɪd·jər): id. 1595. A 3d. land quoyland (and as such unskatted). O.N. *mið-garðr*, 'mid-farm.' It lies between Huip and Whitehall (formerly North Strenzie) tunships, and was apparently not broken out or cultivated when skat was imposed.

Musbuster: id. 1595. An old 4¼d. land lying between Huip and Whitehall, and now apparently incorporated with one or other of these farms. Probably O.N. *mós-bólstaðr*, 'farm or settlement on the moor.'

Mussater (mʌs·ətər): id. 1595. An old 1½d. land now remembered only as a field-name on Airy. Probably O.N. *mós-setr*, 'moorland house or settlement.

Odness: id. 1595. A 4¼d. land at the point of Odness, a precipitous rocky headland on the east side of the island.

A very puzzling name. The headland is frequently spelt Odin Ness, and curiously enough the most southerly point of the island is called Torness. O.N. *oddi*, ' a point,' is applied as a name for a farm—Oddie— near the point on the north-west of this island now known as Linksness, but one would hardly expect an O.N. *odda-nes*, though the two elements in inverted order are actually found in the name Nesodden which occurs in Papa Westray as well as in Norway. That name, however—the Ness-point—is understandable : whereas ' point-ness ' = ' the ness of the point ' makes no sense.

But at Odness there is a further complication. Close to the beach below the house of Odness is a large mound (about 70 yards in diameter by 12 feet or more in height) which bears the astonishing name of God-Odina (ɔd·əna). That name was practically extinct some years ago when I was collecting Stronsay place-names, and my informant told me that he made many inquiries before he found an old man who remembered the name. It would be most unlikely to encounter the two words *god* and *Odin* conjoined in one place-name, but it is difficult to disassociate the name of this mound from that of the farm and headland. The origin of Odness must therefore be left at present obscure.

Papa Stronsay (paːpi, and in Stronsay itself sometimes The Pap) : Papey in litla (*O. Saga*); Papay 1535 (REO 219); Pappay Stronsay 1566 (REO 377); Papa and Papa Stronsay 1595. A small island lying in the mouth of Whitehall harbour; was skatted as a 12d. land. It was here that Earl Rognvald Brusason was slain in 1046. O.N. *Papey in litla*, ' the little isle of the Papae ' (early Celtic clergy) as distinguished from *Papey in meiri*, ' the larger Papey,' i.e. Papa Westray.

Quoybora : Quoybana 1595; an evident mistake as in a 1630 Retour and later charters it is spelt Quoybora. A tumail in Aith. Origin obscure.

Quoygarth : 1595. A quoy in North Strenzie. Now a field of Whitehall called The Gart.

Quoygrana : 1595. A quoy in North Strenzie which in later charters appears as Quoygrain. O.N. *kvi-grœna*, ' green quoy.'

Quoykindness : 1595. A tumail in Aith. Reason for name obscure.

Quoyolie : near farmhouse of Roithisholm; evidently same as appears in the 1595 R. as Quoyolassa—an evident error for Quoyolefea which is the usual record form, i.e. ' Oliver Fea's quoy ' : Fea was a well-known local family.

This and the four preceding names were not farm names at all, but merely cultivated areas which for some reason were detailed in the Rental.

Rothisholm (rouˑsəm) : Rothisholme 1595 and 1632; Rothisholm 1600 (REO); Rowsholme 1633; Rowsum 1676; Rousum 1693 (Wallace); Rousholm 1750 (Mackenzie's charts).

This is a large peninsula which juts out from the west side of the island, with which it is connected by a narrow isthmus from which the neighbouring tunship on the ' mainland ' side has its name—Aith (O.N. *eið*, ' isthmus '). Between the peninsula and the ' mainland ' of Stronsay is a large bay—2 miles in length by 1½ miles in breadth—known as the Bay of Rothisholm. The peninsula is now dotted with a number of small crofts, but it consists for the most part of rough peaty moorland, and it terminates in the spectacular red sandstone cliffs of Rothisholm Head. The whole area was skatted as a 36d. land or two-urisland.

The name offers peculiar difficulty. The written form of today denotes a pronunciation that has evidently been obsolete for three centuries at least. The first element is no doubt O.N. *rauðr*, ' red '; a former alternative name for the headland was Rodneip Head—a name which is found in records as well as on Blæuw's Map of Orkney (c. 1650). Rodneip represents O.N. *rauða-gnipa*, ' red promontory ' (from the reddish cliffs), and ' Head ' conjoined therewith is obvious tautology. But the headland was probably also known as Red Head (O.N. *rauða-hǫfuð*).

The farm-house and rich old cultivated lands lie at the head of Rothisholm Bay, and it is possible that the final element of tthe name indicates O.N. *hǫmn*, ' a haven or harbour,'—the whole compound thus having been O.N. **Rauð(a)hǫfuðshǫmn*, ' Red Head haven.' Against such a source, however, one has to record that Ham (the local form of *hǫmn*) is never found elsewhere in Orkney applied to such a large bay, though in Rousay Sound, which is also a large bay, one corner of it where there is specially good anchorage is called Ham.

If it were not for the fact that there is no certain example of a genuinely old *heimr* (home) place-name in Orkney one might have been tempted to suggest an O.N. **Rauðs-heimr* (home of a man Rauðr) as the origin of Rothisholme, but, even so, that would not have accounted for the name Rodneip Head.

In view of the absence of sufficiently early forms of the name its origin must be regarded as uncertain.

Spurquoy: 1595 R. An old 4d. land quoyland in Strenzie : it was the post-Reformation church glebe. The first element is probably a personal name; Sporr, as an ekename, is on record in the Icelandic *Landnámabók*.

Strenzie or **Strynie** : " north Stryme in Stronsay" 1500, where the 'm' must certainly be a mistake for ' ni ' ; (v. sub Brusgarth, Sanday); North Strenzie, South Strenzie and South Strynzie 1595; Strynzie 1653 Val.; Stranzie 1706 lease; very often on record from 17th and 18th cent.—usually in form Strynie.

The name is entirely obsolete, but it denoted a large area of the island, most of which is now represented by the two large farms of Whitehall and Hunton. North Strenzie, which coincided roughly with Whitehall, was a whole urisland, and in view of the fact that large tunships such as this were usually skatted as whole or half-urislands, it is fairly certain that South Strenzie was an urisland also, though the rental is not explicit on the point. Here the absence of data from earlier rentals is especially disappointing. The farm of Hunton, which we know from charters to have been in South Strenzie, is never mentioned in the 1595 R. at all, but from charters we know it to have been a 6½d. land. Along with it we find associated another farm—Sanger—which was a 5½d. land. But both together only amounted to 12d. land; where was the other 6d. land required to complete a hypothetical 18d. land or urisland ?

To answer that question a glance at the topography of the place is necessary. Here in Strenzie the narrow peninsula of Griceness juts out eastwards at right angles from the main body of the island for about a mile. In the mouth of the bight thus formed on the north side of Griceness lies the island of Papa Stronsay, making a safe and sheltered haven between the two islands known as Papa Sound. Here for centuries—how many none can tell—was the chief centre of the herring-fishery in Orkney. On the south side of Griceness is the much wider but less protected Mill Bay.

From the main island ridge of Stronsay (here less than two miles across) North Strenzie sloped down north-eastwards towards Papa Sound, while South Strenzie sloped down south-eastwards towards the waters of Mill Bay. Over the island ridge to the west, behind the two Strenzies as it were, the only skatted unit was Clestran, and as that was a 6d. land it would seem probable that it was the missing unit required to complete the South Strenzie urisland. It must indeed be noted that in the 1595 R. are specified two other units which lay more or less between Clestran and Hunton, viz. Spurquoy and Yearnsetter, but as each was quoyland and thus unskatted they cannot be included among the original skat lands: their situation, indeed, near the ridge of the island, would imply that they were later out-breaks.

If the foregoing diagnosis is correct the original undivided Strenzie must have stretched right across the island and been skatted as a 2-urisland unit. It has been pointed out above how for centuries Stronsay has been famous for its herring-fishing, and it was here in Strenzie that the industry was centred. We have records of Dutch and other fishing vessels frequenting the place as far back as the 16th century at least, and the "Dogger Beach" of Strenzie was for long a source of considerable revenue to its owners.

About the end of the 17th century Patrick Fea, who had acquired the property of North Strenzie, built a new house there (at the head of the present pier) and called it Whitehall. Thereafter Patrick styled himself "of Whitehall" and the older name of Strenzie or Strynie gradually passed out of use.

The old name, however, is one of the most interesting in Orkney, though its origin is by no means clear. Its consonantal sequence—s, t, r, n—together with the fishing associations and what has been said above in regard to the island name itself, might tempt one to regard Strenzie and Stronsay as cognate names. But the probabilities are against any common derivation from O.N. *strjón*. Though spellings vary, the true local pronunciation of this tunship name was pretty clearly Stren·ji (the z = palatal j), and that being so it would appear to be the same word that we find in Strenyie-water (stren·ji-), the name of a small loch at the beach near Rothisholm Head. That name can hardly be other than O.N. *strandar-vatn*, 'beach loch,' and on that analogy Strenzie would mean simply the "beaches" (O.N. *strandir*, nom. plur.). The beaches in question would be the fishing beaches there which were so important in former days (for drying of nets and probably also fish); in other words the "Dogger Beaches of Strenzie."

On that theory we may have the more confidence also in regarding Clestran as originally part of the old Strenzie, for that name also appears to be a compound of *strand*—(*Klett-strǫnd*) and must have been applied to the beach on the west side of the island, on Linga Sound, which was in former days a rival fishing-station of Papa Sound itself.

The phonetic development *strandir*>strenzie offers no serious difficulty. The inflexional *r* drops, *nd* is assimilated to *nn* which then becomes palatalized : strandir>stranni>stra(e)n·ji ; cf. the Shetlandic Strandadelds which Jakobsen renders sträṇ·adelds.

Stensy (stɛn·si) : Stainsay 1627); not included in 1595 R. as it was Kirkland, but it was a 4d. land—probably part of the urisland of Houseby. There used to be a house there, but now it is merely one of the Houseby fields. Origin of second element obscure.

Yernasetter: Yernsetter 1595. A farm on ridge of island west of Hunton; former quoyland. The first element of name is almost certainly a personal name—probably O.N. *Jǫrundr*; hence—*Jǫrundar-setr*, the homestead of J.

OTHER FARM NAMES

Banks: id. 1630. Farm at beach in Grobister tunship. O.N. *bakkar*, 'banks', i.e. of the shore.

The Bay (be:): the chief farm in tunship of Aith; not mentioned at all in any Rental, but the pre-Reformation parish church of St. Mary was situated here, and its kirkyard remains in use today.
O.N. *bœr*, 'a farm settlement." In Stronsay there were certainly three, and probably four, early settlements bearing this name, and they were spread in such a manner that they must have embraced most of the cultivated ground in the large southern part of the island: Everby on the east side, Houseby in the south-east, and this—The Bay—in the north-west; the fourth, which is at least probable, was in the south-west on the large farm of Holland where there is a field near the shore called Erraby, near an old boat-noust. See further *sub* these names, and also Part III. It should be noted that this farm is always referred to with the def. article, as *The* Bay.

Boondatoon (bøn·dətun): a farm in Grobister tunship. It lies inland from the coast, and is obviously the Scots "Abune (above) the toon," a name that exactly fits the location.

Cataquoy: Cattaquoy U.B. 1601. Now a field on Whitehall, north of farm-house. First element doubtful, but most probably a personal name—e.g. O.N. *Kati;* hence ? *Kata-kvi,* 'Katis quoy or enclosure.'

Cruanna (krua·nə): Cruianay 1709. A house near the Little Water, and formerly part of the common hill pasture. Airafea, Fellsquoy, Milkquoy and Cruanna are all near together, and each is reminiscent of the old farming economy. The name is some form of O.N. (Icel.) *kró*, (Norse) *krú*—a loan-word from O. Celtic *cró*, a sheepfold, enclosure for animals, etc., and here we have probably the plural with def. article attached—*krúarnar,* 'the enclosures.'

Erraby (ɛr·əbi, -e): now a field-name on Holland; near to the old Inya-noust. O.N. *eyrar-bœr,* farm at the 'ayre' or beach. Cf. same name in Sanday.

Errigarth (ɛr·igər): Airigair 1664 Sas.; a farm near beach in Aith tunship. O.N. *eyrar-garðr,* 'ayre-, or beach-garth.'

Everby (ɛv·ərbe, -bi): id. 1625; Ovirbie 1632; Everby Blæuw's map c. 1650. A large tunship on east side of island. Not mentioned in 1595 R., but the farms included in it—Odness, Airy, Hescum, Musseter, Dritness and Kirbuster—total up to exactly 18d. land in a 1794 Rental. On Blæuw's map is marked a house of this name and it is also on record elsewhere, but the site is now unknown. O.N. *efri-bœr,* 'upper "bae" or farm.' Cf. The Bay, Houseby and Erraby. See further, Part III.

Holin (ho·lin): a farm near S.W. corner of Mill Bay. A curious name as there is no noticeable hill or hollow here.

Hunday (hʌnd·i): an old house and park near Airy. The last element cannot be O.N. *eið* as there is no isthmus there. First element also obscure; cf. Hunton.

Leaquoy (lɛk·wi): house on west slope of Stebb Hill. O.N. *leik-kvi* or *leika-kvi,* 'sports-quoy'; from *leika,* to play. This name occurs in several of the isles, and is actually on record from 1329 in South Ronaldsay where among farms sold was a Leika kwi: (*Dipl. Norv.* II, No. 170). The nature of the games played at such places is uncertain, but they probably included the old Norse sports of horse-racing and ball-games.

Linksness: farm at N.W. tip of island opposite Little Linga. The name probably represents Lingey's Ness, dating from a time when the English 's' had replaced the O.N. genitive *-jar*.

Linnabreck: a small farm on slope facing Muckle Linga. The second element is O.N. *brekka*, a slope, but the former is uncertain.

Quoyangry (kwi-aŋ·ri): "the cot called Q―――― a pendicle of the said room of Housbie." (Document of 1761.) Now a field-name, on Houseby, but it occurs as a field-name at Hescum where a neighbouring creek is called Quoyangry Geo. The name is found also in Westray, South Ronaldsay, Stromness (Quoyanger) and even in Kirkwall where an old park was so named. Origin of last element quite obscure.

Sangar: a 5½d. land now lost sight of, probably swallowed up by Hunton. O.N. *sand-garðr*, 'sandy-farm.'

Scoulters (skult·ers): a small farm on south slope of Stebb Hill. O.N. *skoltar*, plur. of *skoltr*, 1. a skull; 2. a hill-knoll. The final -s is a tautological plural, unless it represents a curtailed word, e.g. *-hús*, 'house.'

Staves (stevz): id. and Staffs 1735. A house north of Midgarth. O.N. *stafir*, plur. of *stafr*, a staff or stick. The term is not infrequent as a farm-name in Norway, but the precise reason for its application here is uncertain.

Stensy (sten·si): Stainsay, a 4d. land, 1627 (P.R. 96); Stainsay 1738. Now only a field-name on Houseby. It was old Kirkland. Origin obscure.

Stokaquoy: a quoy in N. Strenzie mentioned in a lease of 1656. O.N. *stokkr*, a pole or beam enters into many Norwegian place-names, but its exact significance there is often obscure, as it is here.

Whitehall: name given to a new house built by Patrick Fea in North Strenzie in 1676. It has now for long replaced the old Strenzie name, and is applied as a village-name as well as a farm name.

WESTRAY

Vestrey (O. Saga); Westra 1492; Westra, Westraw, Westray 1500. O.N. *Vestr-ey,* 'west isle.'

In regard to this island the early Orkney Rentals are not a little tantalising. They reveal certain curious facts of quite unique interest, but on the other hand large parts of the island are scarcely referred to at all. Much of it of course was in the hands of churchmen, and of the remaining earldom lands some had been granted to Sir William Sinclair of Warsetter, brother of Henry Lord Sinclair who was tacksman of the earldom at the time. The resultant picture of the island farms is thus extremely vague and confusing.

So far as can be ascertained, however, the island was skatted as containing approximately 13 urislands. A noteworthy feature was the relatively large size of the constituent tunships—two being 2-urisland units, and three 1½ urislands each. Only about a quarter of the island was udal property, the remainder being either bordland or bishopric or prebend lands.

RENTAL NAMES

Aikerness (ɛk·ərnes[z]) : Akirnes 1492; Aikarnes 1500; Aikernes, Akarnes 1595. A 1½ urisland (27d. land) tunship in which the chief farm was Skaill. Probably O.N. *akra-nes,* 'ness of cultivated fields,' though the first element may represent O.N. *ekra* which was used in much the same sense as *akr* (of arable land).

Are: Are 1492; Air 1500. Now obsolete and absent even from the 1595 R. In the two previous rentals it was linked with Tuquoy : "Tuquoy and Air ane uris terre . . . ." 1500 R. O.N. *eyrr,* "a bank of shingle or sand thrown up by water, especially rivers." (Rygh's *Indl.*)

In Orkney the term is specially applied to a low spit of beach composed of shingle and stones cast up by the sea. At the inner end of the Bay of Tuquoy is a narrow spit of land of that nature which has probably given rise to the name of the farm here.

Brebuster (bre·bəstər) : Brabustir 1492; Braebustar 1500; Brabustar 1595. A former 3 meils coppis or ½d. land in Rapness.

The name suggests a much larger unit originally. O.N. *breiði-bólstaðr,* broad-bister or farm.

Brough (broχ): Burgh 1492; Brugh 1500; Brught 1595. A former half-urisland, now the largest farm in the island. O.N. *borg,* a stronghold : the name has been applied from an old broch which has now vanished but traces of which were found in a field north of the farmhouses.

Bu of Rapness, The (bu :) : The Bull of Rapness 1492, 1595; The Bull of Ropness 1500. A 9d. land or half-urisland bordland estate. See Bu— Part III.

Burrogarth : 1492; Burogaith 1500. A 3 meils coppis in Wasbister bordland which is absent from the 1595 R. Obviously a *-garth* name, but as no broch is known to have existed in this area the reason for the first element is obscure. For 'meils coppis' see Part II.

Cleat (klet): Clait 1492 and 1500; Clett 1595. An old 6d. land. See Cleat, Sanday.

Cot: 1500. 1 meils cop (= one-sixth pennyland) in Rapness. Now perhaps represented by a croft named Cotorochan. O.N. *kot*, id. There are two or three other Cotts in the island.

Dykeside: Dyksyde (R.M.S.) 1565; Dikeside 1595. A name applied to a curving hill slope north-west of Noltland Castle on which are situated a number of small farms. The district was skatted as 6d. lands. The name arises from the 'hill-dike' which ran along the hill side, separating the arable lands of the tunship from the 'hill' or common pasture-grounds outside. If, as is almost certain, this is a translated name, the original would have been O.N. *garðs-siða*. At Stove in Sanday a field bears much the same name—Gornside.

Fribo (frɪb·o): Furbou? 1492 (final letter indistinct in MS.; might be 'u', 'w,' or 'n'); Forbo-bewest (R.M.S.) 1565; Quoys of Firbo 1740 and 1794 R. Entered in 1492 R. as 6d. land, but in the 1565 Charter to Gilbert Balfour only 3d. land was conveyed. When it reappears in 18th century Rentals there is no substantive farm, but merely a half-dozen 'quoys of Firbo.' The name is puzzling; if the second element is O.N. *bú* it would be a unique appearance in Orkney with a prefix attached. First element equally obscure as it might represent any one of several Norse words.

Garth (gɛrt): Garth 1492, 1500 and 1595. An old 3d. land. O.N. *garðr*, farm. A fairly large farm still today.

Glenna: Glennay 1492; Glenna 1500 and 1595. A 6 uris coppis or 1d. land in Skelwick. Though non-existent as an independent farm today its approximate location is indicated by the Point of Glen, a little to the south of Surrigar. The name is of special interest because it is not on record in O.N. literature, and when it occurs as a farm-name in Norway it seems to be confined mainly to the two sides of Oslo Fjord. Rygh defines *glenna* as an open space in a forest, or a grassy patch between cliffs. For 'uris coppis' see Part II.

Grassquoy: 1595. A quoy entered under Glenna. Self-explanatory.

Grimbust: Granabustir 1492; Grambustar 1500; Grenebustar 1595. A 5 meils coppis (= five-sixths pennyland) in Rapness. Still surviving as a small farm today. O.N. *grœni-bólstaðr*, green-bister or farm.

Grother: 1595. Small arable unit entered under Brabustar. Name seems extinct. Origin obscure.

How: 1492 and 1500. Small 1 meils cop forming one of the "umbesettis" that were attached to the Bu of Rapness, q.v. O.N. *haugr*, a mound.

Kirbist: Kirkbustir 1492; Kirkbister-bewest (R.M.S.) 1565; Kirbustar 1595. Entered as 4½d. land in 1492 R. O.N. *kirkju-bólstaðr*, kirk-bister or farm. There was an old chapel in the tunship.

Kirkhouse: the present farm of this name in Skelwick was represented by two farms in the old rentals: Overkirk and Netherkirk—each 4 uris coppis (= ⅔d. land). "Upper" and "Lower Kirk."

Langskaill: Langscale 1492; Langscaile 1500; Langskale 1595. A 4 uris coppis = ⅔d. land. O.N. *langi-skáli*, 'long hall.'

Today there are no fewer than four small neighbouring farms bearing this name—viz. Langskaill, North L——, South L——, and West L——. These are situated in the heart of a district known as Skelwick, which takes its name from the adjacent bay on the east side of the island, a bay which in turn almost certainly had its name (O.N. ? *Skála-vík*) from the original ' Long Skaill,' which (to judge from the situation of the present farms) stood overlooking the bay.

The fact that such a large continuous stretch of this island was bordland (personal earldom estate) tends to obscure the nature of the original settlements. The bordland area in question stretched a distance of about five miles in length, varying from one to two miles in breadth, and embraced the whole long southern arm of the island. By the date of the earliest Rental (1492) all the arable lands of the area, with the exception of the Bu of Rapness, were parcelled out in about 30 small units—none exceeding a pennyland in value. The entry of each of these in the Rentals helps, no doubt, to give a clearer idea than we can have in other areas of the actual distribution of population at the time, and their economic basis, but it unfortunately obscures the larger early units of settlement.

The northern part of the area (referred to in the 1500 R. as the Swartmeil bordland—from the first-named farm therein) was valued in uriscoppis; the remainder—from Wasbust southwards (with the exception of the Bu of Rapness)—was termed the bordland of Wasbustar, and was valued in meils-coppis. But among the small units entered in those areas there are a few which by their very names bear a stamp of higher dignity than others, and may thus be regarded as having originally been much larger units, which were later broken up by some earl in an early unrecorded " small holdings scheme." Of such names Grimbust (*Grœnibólstaðr*) on the bay at the south end of Rapness might be one; Brabuster (*breiði-bólstaðr*) near the present Clifton might be another, Wasbuster (*vaz-bólstaðr*) a third; and this Langskaill would have been the head-house of a fourth early settlement—whatever name it may have gone by prior to the erection of the ' long-hall ' therein.

It may be noted that there is another Langskaill in this island, not mentioned in the early Rentals, but situated in what was formerly called Noltland-Bewest.

Milgarth : Mylgarth 1492; Milgarth 1500. A one meils cop. Self-explanatory; it was probably located near the present Mill of Rapness.

Mo : Mow 1492 and 1500; Mo 1595. A 2 meils coppis unit. Still a small farm today. O.N. *mór*, poor sandy or gravelly soil; but in Iceland this word is also used of peaty moorland.

Mousland : Mobisland 1492 and 1500; Mousland 1595. A former 6 uris coppis (= 1d. land), and still a farm today.

An interesting name which recurs in Stromness parish, and was (though now obsolete) a Sandwick name also. In each of the three early rentals the Stromness name was spelt Mousland, but the Sandwick name was (as here) Mobisland 1492 and 1595; Mobisyord 1500; Moabsland 1739 R. -Yord is of course, O.N. *jǫrð*, = -land.

There can be no reasonable doubt that all three names are of the same origin, and the prefix is almost certainly a personal name, though most obscure. In Lind's *Dopnamn* the only one that might be considered is the woman's name *Móbil* or *Mabil*, but it would be most unlikely for such a rare name to appear thrice in Orkney farm-names. A like reason *inter alia* would seem to exclude the possibility of the Early Celtic (Irish) name *Mobi*. Origin therefore quite uncertain, but it may be noted that among the Westray men recorded from 1492 was a Mavius (? Magnus) Maibsoun (probably an error for Moibsoun).

Neagarth: Neagarth 1492; Negarth 1500; absent from 1595 R. A small 1 meils cop. unit in the Wasbust bordland: site now unknown. O.N. *nȳ-garðr*, 'new garth or farm.'

Noltland (nɔut·lən[d]): Lady Kirk of Noltland 1500; le Bow de Notland (R.M.S.) 1565; Noltclet (an obvious error) 1595. This tunship, in which stands Noltland Castle, was an old 12d. land which together with the adjacent 6d. land of Dykeside made up a whole urisland. Noltland itself is still a substantial farm today. O.N. *naut(a)-land*, 'nowt-, or cattle-land.'

Noltland Bewest: Nouteland be Wast Hay, and Wester-Nouteland 1492; Noltland be Westhay 1500; Noltland Be-west 1595.

Strangely enough this name is now forgotten and quite obsolete locally but the tunship occupied the southern slopes of Fitty Hill on the west side of the island, and the adjunct Bewest was applied to distinguish it from the other Noltland. The curious term 'hay' in Wast Hay, Westhay *supra* is O.N. *ey*, 'island,' as is proved by a similar entry under Graemsay where a certain 4d. land is named in the 1500 R. 'Outohoy,' but in the 1492 R. 'Oute apoune the Ile.'

It is well-nigh impossible to unravel the facts about this tunship as the Rental entries are so confusing. In the 1492 R. it is recorded that the earldom portion amounted to 23¼d. lands, and the bishopric portion 12½d. lands, i.e. 35¾ pennylands in all. We may thus safely conclude that originally it was a 2-urisland tunship (36d. lands) of which a farthing land had somehow been lost sight of. The information to be gleaned from the 1500 and 1595 Rentals is of little help, but in the next extant Rental—that of 1727—Noltland Bewest is entered merely as 27d. lands, while its adjacent tunship on the east—Midbea—is entered at the same figure. Midbea does not appear at all in any of the earlier rentals, having evidently been bishopric land.

From the still later Rentals of 1740 and 1794, however, we have lists of the different lands in those two tunships as follows:—

Noltland Bewest.			Mid-bea.		
Roveland	... ...	3d. land	Midbea	... ...	1½d. land
Pow	... ...	4d. land	Quina	... ...	1d. land
Trebland	... ...	3½d. land	Crowland	... ...	3d. land
Howan?	... ...	1d. land	Quinagrina	... ...	1d. land
Langskaill	... ...	7d. land	Midhouse	... ...	4½d. land
Powstink	... ...	2½d. land	Cott	... ...	3d. land
Gerry	... ...	2d. land	Gorn	... ...	5d. land
Brecks	... ...	4d. land	Pow	... ...	3d. land
			Velzie	... ...	5d. land
			Bakka—a quoy		
Total	...	27d. lands	Total	...	27d. lands

Now since 1614, when the excambion took place between earldom and bishopric, all Westray lands had been 'earldom' (in the sense that all island skats were payable to the earldom). Hence we know that by the 18th century there were no skattable lands in Noltland beyond the 27d. lands indicated. What had become of the odd 9d. lands?

Some light can be had from Bishop Bothwell's charter of church lands to Gilbert Balfour in 1560. *Inter aliá* there were conveyed 27d. lands in Mydbe-bewest, 4d. lands in Notland-bewest, and 9d. lands in *Bakka*. There is no form of Bakka today, but below the arable lands of Noltland

NAME MATERIAL 35

and Midbea is a wide expanse of sandy links stretching down for half a mile or thereby to one of the loveliest beaches in Orkney which still goes by the name of Bakkie. It may be regarded as certain that some part of that area represents the site of the old arable lands of Bakka which by reason of sand-blowing has long been rendered useless for cultivation. Even in the 1492 Rental we read of Noltland Bewest: "jam thrid Part blawn till Issland (Iceland) and mansworne down," i.e. uncultivable and thus not skattable.

In the later rentals it may be noted that Bakka was entered merely as a 'quoy' with no pennyland valuation, and the only skat charged for it was a barrel of butter, in respect no doubt of its value still as pasturage. And though it was entered under Midbea there can be little doubt that it represented the missing 9d. land of Noltland Bewest. And, incidentally, we may note that in conveying to Balfour 9d. lands in Bakka and 4d. lands in Noltland-bewest the bishop was making sure that he was conveying *all* his Noltland lands, which the 1492 R. estimated, however, only at 12½d. lands!

Before leaving those two old tunships of Noltland and Midbea it may be pointed out how here again we seem to be face to face with another of those large original settlements of *bœr* type. As indicated above the two are adjacent, but the very name Midbea implies the existence of a third tunship on its other side from Noltland. As to what that was there can be little doubt; it must have been the urisland of Tuquoy and Are which stretches from Midbea down to the Bay of Tuquoy on the east. This suggestion of the original unity of those three tunships is strongly supported by another interesting fact. It has frequently been pointed out that practically every urisland in Orkney had within its bounds an old chapel which was the property of the owner of the land whereon it stood. Moreover, in every case that can be checked it was the chapel of the leading man or local chieftain that ultimately became the parish church of the area. Now the only church or chapel known of in these three tunships is the old Cross Parish Kirk, the remains of which stand at the edge of the beach at the east end of Bakkie. It is a dignified little 12th century structure with Romanesque vaulted chancel, one of the very few ecclesiastical buildings from that period left in Orkney. And the fact that it is by no means central for the parish merely clinches the argument that here was the seat of the most powerful family in the parish at the time parish churches were set up. Everything thus points to the conclusion that those three tunships, amounting to no less than 4½ urislands in value, represent the 'land-take' of one of the greatest of the early Norse settlers in Orkney. And a more attractive settlement he would not readily have found.

Noust: 1492 and 1500; absent from 1595 R. No longer in existence but it must have been near the landing-beach at Helzie at the south end of Rapness. O.N. *naust*, a berth for boats when drawn up at beach; this is the regular word for such a place in the Orkney dialect of today.

Perth (pert): Pretty 1492; Precti 1500 (when 'c' is certainly a misreading of 't'); Prettie 1595. A 2 meils coppis in Rapness, and still a farm today. Origin quite obscure; the present form of the name is no doubt due to an assimilation to the name of the Scottish town (cf. Sanquhar). The English *pretty* can hardly come into consideration.

Quyskega: 1492; Quiskega 1500; absent from 1595 R. A quoy-name in Swartmeil bordland. Second element quite uncertain; Jakobsen cites a Shet. *skega*, a tabu-name for a boat's sail, and compares Fær. *skeki*, a clout or rag. He refers also to O.N. *skiki*, a strip or corner, small piece of something, but whether such a term appears here is more than doubtful.

Quoycurris: Quytrs [sic] 1492; Quoy curris 1500. A quoy in Glenna, but origin quite obscure.

Rackwick: Rekavik and Recavic (*O. Saga*); Rakwic 1492; Rakweik 1500; Rackwik 1595.

The *O. Saga* Rekavik was the residence of Þorliotr, a 12th century chieftain, but whether this Westray Rackwick or Rackwick in Hoy was meant is uncertain; Mr. Clouston argued for this Rackwick, though in his *History of Orkney* he refers to a third Rackwick near Stromness which has claims to be considered. There is actually a fourth Rackwick which is also in this island; see below.

As for the origin of the name there is no doubt: O.N. *reka-vik*, ' bay of jetsam,' i.e. a bay where driftwood, etc., is washed ashore. This Rackwick was a large 2 urisland tunship lying to the north of Pierowall, and it is noteworthy that in it is an old *bœr* farm—Trenaby, and a skaill-farm also—Breckaskaill.

At the other Westray Rackwick (in Rapness) the old Rentals cite a small farm—Rakwic 1492; Rekink (mistake for Rekvik) 1500, which was a 3 meils coppis or ½d. land. Now probably included in Clifton.

Ramsay: Ramyshow 1492; Rammishow 1500; Ramshow 1595. A 6 uris coppis (= 1d. land) in Skelwick; a small farm still today. Almost certainly O.N. *Hramns-haugr*, Hramn's (Hrafn's) mound. Hramn was a fairly common personal name.

Ranisgarth? Ran?[m]ysgarth (reading of MS. doubtful) 1492; Ranisgarth 1500. Absent from 1595 R. A small quoy in Skelwick—now vanished. The prefix is no doubt a personal name, but exactly what is uncertain.

Rapness: the Westray pronunciation of 'a' in such a word as this is almost identical with South English ' a ' in *fat*, etc.—(ræp·nez, rɛp·nez). Hreppisnes (*O. Saga*); (Bull of) Rapness 1492; Ropnes 1500; Rapness 1595.

This is not a farm-name, but is applied to the whole peninsula (containing many farms) at the south end of the island. It was the residence of Kugi, a notable Saga figure of the 12th century, whose hall was probably situated at the Bu (of Rapness). Here it was also that Earl Rognvald was on a visit when John Wing arrived bringing along with him Sweyn Asleifson's young son Olaf whom he had kidnapped at Eynhallow. John got sound advice from the earl to take him back without delay if he wished to avoid trouble from Sweyn and Kolbein Hruga—the boy's foster-father.

If the Saga form of the name is correct the name must represent O.N. *Hreppis-nes*, the ness of a man Hreppir. If by chance, however, the ' i ' is an intrusion, the name would indicate ' district-ness,' from O.N. *hreppr*, a small district containing at least 20 farms with a " thing " of its own : (the English *Rape*).

Sangar: this Rapness farm—now absurdly spelt Sanquhar in imitation of the Scottish town—was likewise represented in the early rentals by two names: Over Sandgarth 1492 and 1500; Oversanger 1595; and Nether Sandgarth 1492 and 1500; Nethersanger 1595. Each was 3 meils coppis (= ½d. land). O.N. *sand-garðr*, ' sandy-garth or farm.'

Skagarth: 1595. A small quoy in Glenna. Origin uncertain; see Quoyskega.

NAME MATERIAL 37

Spannysquy: 1492; Spannisquoy 1500. Absent from 1595 R. A small quoy somewhere in Swartmeil bordland. A rather interesting name if, as would seem certain, it represents O.N. *spanns-kví*. A *spann* was a measure which, though probably obsolete in Orkney by the time of the early Rentals, was regularly employed as a unit in the assessment of skat butter. It was reckoned as = 5 lispunds. *Spanns-kví* might thus indicate a quoy of a certain size or quality, e.g. one that would be rented at a spann of grain, or which could be sown with a spann of seed.

Surrigarth (sʌr·igər): North Sour-garth and South Sour-garth 1492; id. (minus hyphen) 1500; Northserigair and Southsoregair 1595. Each a 4 uris coppis = ⅔d. land.
    Apparently O.N. *saur-garðr; saurr*, mud, filth, etc., appears as an element in several Norw. farm-names. Cf. Sorquoy (Orphir).

Sulland: Suyirland 1492; Sutherland 1500 and 1595. A 3 meils coppis farm in south of Rapness. O.N. *suðr-land*, 'south-land.'

Swartmeil: Swarthmale 1492; Swartmeill 1500; Swartmell 1595. A 6 uris coppis (= 1d. land) farm. Still a farm today. O.N. *svart-melr*, 'black-sand.' The farm is at a sandy beach where seaweed lies and rots.

Todds Zir: 1492; Todis Yor 1500. Absent from 1595 R. A small quoy somewhere in Swartmeil bordland. Name now extinct. Rygh (*Gamle Personnavne*) cites a name Toddi which appears in Norw. farm-names, and regards it as possibly a pet form of *þórðr* (Thord): hence = Toddis-jǫrð, or 'ground.' By 1500 the Scots genitive 's' was replacing the older Norse form; cf. Hermisgarth (Sanday).

Tafts: Tofta 1492; Cofta (misreading of 'c' for 't') 1500; Tofta 1595. A 4 meils coppis. A farm on the Bay of Tafts at the isthmus between Rapness and the rest of the island. There is a huge mound at the site which kitchen midden remains indicate to be an important prehistoric structure.
    O.N. *toptir*, old house-sites or remains of old buildings.

Tuquoy (tu·kwi): Tukquy 1492; Tuquoy and Tuquy 1500; Towquoy 1595. In the 1492 and 1500 Rentals Tuquoy and Are (Air) are said to make one whole urisland (18d. land). See further *sub* Are. The first element of this name is too indefinite to justify any suggestion as to its origin.

Twiness (twɪn·jez): Tonzeis 1492; Cunyenes (where 'c' is undoubtedly a misreading of 't') 1500; Twynzes 1595. A 3 meils coppis; still a farm today situated on the coast at a point (ness) of the same name.
    The name is also found as a coast name (with same pronunciation) in N. Ronaldsay, Shapinsay and St. Andrews parish. At none of those four places is there any sort of twin-ness or two nesses, and Jakobsen was disposed to see in the word a hybrid containing *-ness* with a Celtic prefix e.g. Welsh *twyn*, a headland. Such an assumption would seem quite unnecessary, as the early forms might very well indicate O.N. *tungu-nes*, 'tongue-ness,' a name that actually occurs in Norway; see G.N.X, 206.

Wa: Waw 1500; Wa (R.M.S.) 1565; Waw and Wall 1595. The name is now obsolete (except in compounds Pierowall and Breckowall), but it was a 2-urisland tunship on the western shores of the Bay of Pierowall. The name Pierowall was formerly the Pier o' Wa, i.e. a pier built in this tunship of Wa. The bay is one of the most sheltered and secure harbours in the North Isles, and is certainly the Hǫfn (haven) where in the 12th century, according to the *O. Saga*, there was a 'thorpe' where a rich

and wise man Helgi lived. That 'thorpe' was no doubt the beginning of the present village of Pierowall.
O.N. *vágr*, ' bay.'

Wasbust (waz·bəst) : Wasbuster 1492; Wasbustar 1500; Wosbustar 1595. A 6 meils coppis which gave its name to one portion of the bordland; a farm still today.
O.N. *vaz-bólstaðr*, ' loch-bister, or farm '—so-named from a small loch (O.N. *vatn*, genitive *vatns* or *vaz*) now called the Loch of Wasbust.

Wattyn : 1492; Wattin 1500. Absent from 1595 R. A small half meils cop unit in Wasbuster bordland. The name has disappeared, but the site was probably near the Loch of Wasbust. O.N. *vatn*, lake or loch. Watten is a fairly common place-name in Orkney.

Weatherness: Wedderness 1492 and 1500. A 2 meils coppis unit — now vanished. It was no doubt somewhere on the point of Weatherness—the most southerly tip of the island. It is impossible to say whether the name represents O.N. *veðr-nes*, ' weather ness ' or *veðra-nes*, ' wedder-nes,' i.e. a ness where wedders (sheep) were wont to graze. In Norway it is equally difficult to decide which of those two words appears in certain names, but ' weather ' is certainly the meaning in some names, and may probably be the meaning here also, as the ness is a particularly bare and windswept headland. But Veðranes is an actual cape-name in Færoe.

Whiteclett: Quhiteclet 1492 and 1500; Quhytclet 1500. A 2 meils coppis satellite of the Bu of Rapness. Still a farm today. ' White rock '; see Cleat, Sanday.

Woo : (u:) : Owe 1492; Ow 1500 and 1595. A 2 meils coppis, and still a farm today, situated by a small stream. A fairly common Orkney name; O.N. *á*, a stream or river.

OTHER FARM NAMES

Airy: A small farm in Cleat-tun area. Celtic name; see Airy, Sanday.

Backarass: high-lying croft near Noup. Probably O.N. *bakka-ras*, ' slope-sliding,' or skree; the hillside here is very steep.

Benziecot (bɛn·ji–kot) : small croft in Rapness. See *sub* Binscarth (Firth).

Berriedale: a farm in Rackwick. Probably a relatively modern name, as there is neither cliff (*berg*) nor valley here to suggest an O.N. origin.

Bigging : a cluster of small houses at the beach of Stave (a landing beach) in Rackwick Bay. Probably of Scots origin like last name; Scots *bigging*, a building. O.N. *bygging* is used in a rather different sense.

Breckaskaill : id (1664 charter). A farm in Rackwick tunship. O.N. *brekku-skáli*, ' hall or skaill on a slope.'

Breckowall: Breck 1740 R. O.N. *brekka*, a slope. This compound name denotes the farm of Breck in the tunship of Wa[ll]. The medial ' o ' may be either Scots *o*' = ' of,' or O.N. *á*, ' in ' or ' on ' or ' at.'

Brecks : 1740 R. Name now obsolete, but this was the minister's glebe in the tunship of Noltland Bewest. O.N. *brekkur*, ' slopes.'

NAME MATERIAL 39

Burness: small farm at Loch of Burness, near Noltland Castle. It does not appear in old rentals, having probably been part of the 12d. land of Noltland. There is a small ness projecting into the loch and on that is a mound which A. M. Inventory describes merely as a 'burnt stones' mound. The name, however, implies an O.N. *borg(ar)-nes*, 'fortressness,' and suggests that this was the site of an old broch. Probably the stones thereof were utilized in the building of Noltland Castle.

Clifton: large farm in Rapness; a modern creation.

Clouster: in Aikerness tunship—now vanished. In a 1664 charter of feu-farm to Geo. Balfour of Pharay there was conveyed *inter alia* "6 penny land of Clouster in Aikerness of old called Bishop's Land." And on Blæuw's map of Orkney (c. 1650) a house in that neighbourhood is marked "Clastre." O.N. *klaustr*, cloister; the name does not necessarily imply that any religious house or cloister ever existed there, but the land at some period had probably been bequeathed to some such religious house, though no record survives to indicate which. A considerable amount of land in Shetland, for example, belonged to Munkeliv Kloster at Bergen, to which the famous Orkney Bishop Bjarni also bequeathed some of his lands in Norway.

Another farm of the same name is to be found in the parish of Stromness today.

Cooan: small farm near Skelwick School. Origin obscure; in N.G. XV a farm Koen is recorded from N. Trondhjem's Amt which may be the same name, but in that case the meaning is also obscure.

Crowald: a 3d. land in Midbea (1740 Rental). Both farm and name seem extinct. Origin uncertain.

Cubbygeo (kʌb·i–gjo) : small croft at beach on west side of Skelwick district. O.N. *kobba-gjá*, seal-geo; the geo or inlet was evidently a haunt of seals. On the Holm of Scockness near Rousay a low shingly point at the side of a tideway is called Cuppa-taing, and is a favourite haunt of seals still today. The Faroese name for a seal is *kópur*.

Curquoy: part of old tunship of Wa (1740 R.). The same name occurs as a farm name in Evie, q.v.

Damsay: a small farm in Rapness—not recorded as an old bordland unit and thus a name probably introduced by an immigrant into Rapness from the small isle of Damsay in the Bay of Firth.

Faraval (far·əvəl) : Fervel 1740 R.; Tervail (mistake of 'T' for 'F') 1794 R. See Faraclett (Rousay).

Fiold (fjold) : Feold 1794 R. A farm near top of ridge in Rackwick. O.N. *fjall*, a hill. In this case local pronunciation must have followed the old Orkney spelling of the name; cf. Low's spelling: "Of other hills are Cringlefiold, Bailiefiold and other such names." (*Old Stat. Acct.*)

Gairy: Gerray, 1665 Charter; Gerey 1740 R. A 2d. land in Noltland Bewest. A small farm still today. Origin of name almost certainly O.N. *geiri*, a gore or triangular patch of land; cf. Gyre, Orphir. Close behind the farmhouse is a large mound called The Knowe of Gairy, but at no great distance is a bit of rising ground called The Brae of Gowrie. Cf. Gowrie, Papa Westray.

Gill: a farm in Rackwick adjacent to pier in Pierowall. "Geo. Rendall of Gill" was a lawrightman attending a circuit court in Westray in 1683. O.N. *gil*, a valley or ravine; here the name may have been applied to the low neck of land between Pierowall Bay and The Bay of Rackwick. The bay below the house of Gill ebbs dry for a long distance out, leaving exposed a great expanse of sandy beach known as the Sand of Gill. It is noteworthy that a sandy beach in the Bay of Berstane near Kirkwall is known as the Sand of Gilles, and a house of that name formerly stood near that beach. But in neither case is there anything in the nature of a deep ravine among hills such as one naturally associates with the word 'gil.'

Gorn: id. 1740 R.; a 5d. land in Midbea. See Goir, Sanday.

Grindally: Grindelay 1740 R.; a 3d. land in Wa tunship, and a small farm still today. See this name—Sanday.

Haabreck (ha:brɛk): (1) Habreck, a 2d. land in Rackwick (1664 Charter). (2) Halbreck and Kirkhouse formed 5d. land in tunship of Wa. (1665 Charter). Second element of name O.N. *brekka*, a slope; first element doubtful: ?*haga-brekka*, 'pasture slope,' from O.N. *hagi*, outpasture.

Hammar: a 3 meils coppis in Dykeside (1664 Charter). O.N. *hamarr*, projecting rock on hillside.

Hildival (hɪld·əvəl): Holdwail and Haldwall (1664 Charter); Hildavel 1740 R.; Heldaveld 1794 R. A 3 meils coppis (= ½d. land) on hillside in Dykeside, and still a farm today. Second element no doubt O.N. *fjall*, hill, but the first is uncertain. It may possibly represent O.N. *huldu-fjall*, 'fairyhill.'

Howan (hu·ən): small farm in Noltland-Bewest. A mound nearby is called the Knowe o' Hooan. O.N. *haugr*, a mound; here with def. article suffixed = 'the mound.'

Iphs (ɪfz): Iffs 1740 R.; Ifs 1794 R. A 1 meils cop in Dykeside, and a small croft still today. O.N. *ups*, eaves of a house; frequent in place-names, applied to a rock wall, or 'a perpendicular or very steep side of a cliff.' In Norway the name takes various forms—Ups, Ufs, Øfs, etc. This farm is situated at a very steep, craggy, hill-slope north of Noltland Castle.

Knugdale (nʌg·del): Knokdale 1740 R. Seems to have been formerly a quoy of Wa tunship. It lies on the slopes of Knucker Hill which derives its name from O.N. *knjúkr*, a steep roundish hill. Knugdale thus indicates O.N. *knjúks-dal*, 'Knucker (Hill) valley.'

Lakequoy (lɛk·wi): small croft at Pt. of Cott between Bays of Brough and Cleat. O.N. *leik-kvi*, sports- or games-quoy'; see Leaquoy, Stronsay.

Lcean (li·ən): Lian 1740 R. and 1794 R. A 3 meils coppis in Dykeside. O.N. *hlið-in*, 'the' slope. It was situated on the steep hill-slope near Farafield.

Midbea (mɪd·bi): see Noltland-Bewest. O.N. *mið-bœr*, mid-farm or -settlement.

Newark: a farm in Skelwick. Scots term = 'New work.' In 1725 Wm. Balfour of Pharay, in arranging a better jointure for his wife Jean Elphinstone, made over to her "the manor-place and dwelling-house of New-wark and offices, etc., built by himself upon the grounds and lands of Skelwick in Westray" where they were then living. He had apparently created a new 'manor' for himself out of some of the Skelwick lands.

Nistiger: In the 1740 R. this was one of the quoys of Fribo. O.N. *neðstigarðr*, 'lowest farm' of some tunship—probably Fribo.

Noup (nup): id 1740 R. A 3d. land. An outlying farm in a deep valley near the Noup Head of Westray. O.N. *gnúpr* (No. *nup*), a mountain or (as here) a high headland with steep overhanging face.

Ouseness (ouz·nes): there are three or four small farms today of this name, e.g. North-, Mid-, Nether-Ouseness, and on the O.S. map the name is applied to a point half a mile or more to the south of those farms. From their situation, however, it would seem clear that the 'ness' referred to in Ouseness is that running out to end in what is now termed the Point of Vere. This ness forms the southern side of a large oyce (ø*s*), or shallow inlet which ebbs dry at low tide, and which has certainly given its name to the adjacent ness.

O.N. *óss-nes*, 'oyce-ness,' lit. 'mouth-ness,' but in Orkney the term oyce (ø*s*) is frequently found applied as here to a shallow bight or lagoon having a narrow spit of land or 'ayre' running across its mouth and leaving only a narrow entrance from the sea outside.

Pole: Pool 1740 R. A 3d. land in Wa tunship. Now called Lochend; situated near Loch of Saintear. O.N. *pollr*, which was used of a small loch, etc.

Pow (pou): there were two farms of this name—one in Midbea and another in Noltland Bewest. The name is no doubt of the same origin as the last—O.N. *pollr*—translated into Scots. Cf. Pow (Sandwick and Stromness).

Quoy. In the early rentals *quoyland* is a specific term for land which was unskatted, the obvious explanation being that such lands were not broken out or in cultivation at the date when skat was imposed. But some farms throughout Orkney actually named quoys, as e.g. Tuquoy *supra*, were skatted in the normal way, and such farms must be regarded as very early outbreaks which had been made from parent tunships even before the imposition of skat.

Many quoylands (and as such exempt from skat) do not bear *quoy* names and need not therefore be discussed here further. But of quoy-names as such there are a quite unusual number on record in Westray, and there were no doubt many others which have not chanced to appear on record, or been unnoted by the present writer.

In Geo. Balfour of Pharay's feu-charter of 1664, for example, we may note the following: in Rackwick—Quoykilling, Quoyburrigar and Vassaquoy; in Dykeside or Wa—Quoyer (or Quear), Quoybirse, Coupersquoy, Ringsquoy, Michael Ringsquoy, Pallie- or Palace-quoy, Quoyangrie (otherwise Angerquoy), Olicorsquoy, Chalmersquoy, Quoybreckan; from other sources may be noted Quinamuckinshay and Curquoy in Wa, Brabnersquoy and Simblestersquoy in Fribo, Quiña and Quinagrina in Midbea, Quinabreckan in Aikerness, etc.

In that list it will be noted that several have personal-name prefixes —some even of Scottish origin, e.g. Couper and Chalmers. Such, as also Palace-quoy, must obviously be of relatively late origin. Some e.g. Quoybreckan have *quoy* prefixed, others e.g. Vassaquoy—suffixed, but from such placing no valid deduction seems warranted. Ring is a farm in Dykeside, and Ringsquoy evidently a quoy originally dependent thereon.

Michael Ringsquoy must have been one belonging to a Michael who had his surname from the farm. Such surnames may mostly be assigned to the 15th or 16th centuries. As to the age of the others nothing can be said.

Quoy itself derives from O.N. *kví*, a cattle-fold or place where animals congregate (for the night chiefly); Quoyer or Quear is the plural *kvíar*; Quina probably represents the accus. sing.—*kvína*.

Ring : id. 1664, 1740 R. A 3 meils coppis (1½d. land) in Dykeside, and still a farm today. O.N. *ring*, a ring or circle. In Norway this name occurs also uncompounded as a farm-name, but more frequently in combination with another element; here in Westray the name seems to have been applied with reference to the semi-circular contour of hill on which it is situated (above Noltland Castle).

Roveland : 1740 R. and 1794 R. 3d. land in Noltland Bewest—skatted in three portions in 1740—1½d., 1d. and ½d. lands. Today there is no farm of this name, but on the Glebe farm in that tunship there is still a Brae of Rovi, and a field called Rovi. O.N. *rófu-land*, ' (animal) tail-land '; the term *rófa* (tail of an animal) enters into many Norse place-names, and (like *hali*, id.) may, according to Rygh, have reference to a long, narrow and perhaps crooked roadway through fields, or may refer to some long tongue of land projecting from some higher ground.

Rusland (rʌs·lənd) : a farm near Ness of Rapness; not recorded in Rentals, and thus apparently of later origin. A common Orkney name; O.N. *hross(a)-land*, ' horse-land.'

Saintear : Santier 1740 R., where it is entered as a 3d. land in tunship of Wa. There is no farm of this name today, but there is a Loch of Saintear to perpetuate the name. Origin obscure. The same 1740 R. records a Kirkhouse, a 2d. land somewhere in Wa also, and that name implies the presence here of a kirk or chapel at some time, though no memory of it seems to survive today.

The dedication of that chapel is also shrouded in mystery. In Caithness there is a St Tears chapel near Castle Sinclair in Wick parish, of which the minister of Wick, writing in 1726, said it was " ane old chappel called by the common people St. Tears, but thought to be in remembrance of Innocent day, the commons frequenting that Chappel having their recreation and pastime on the third day of Christmass " (Dec. 28). (Macfarlane's *Geog. Coll.* 1, p. 159.) Prof. W. J. Watson in his *Celtic Place-Names of Scotland* (from which the above was quoted) says : " The name appears to be a translation of Gaelic *Cill nan Deur*, ' church of tears.' But in his *Pictish Nation*, etc., p. 136, the Rev. Dr. A. B. Scott, referring to this Caithness chapel, asserts that " S. Drostan's foundations are Kirk o' ' Tear,' that is the Caithness pronunciation of ' Deer '." And in a footnote he adds that " A popular legend turned the name into ' Kirk of Tears,' and connected it with a celebration of Innocents' Day, which was really a celebration of S. Drostan's Day, old Style."

The presence here in Westray of a ' Saintear ' only increases the mystery of the name.

Saverton : Savertoun 1740 R. This was a 3d. land in Wa tunship, and the name is still applied today to a house at the shore off Pierowall Bay. The same Rental also records a 3d. land in Wa called Overtoun. Those two names respectively indicate O.N. *sævar-tún*, and *øfra-tún*, the (part of the tunship) by the sea, and that higher up.

Skaill (skil, skjil) : the main farm or head-house in Aikerness tunship. O.N. *skáli*, a hall.

Smerskeal : 1740 R. A small unskatted unit in Wa. Name now apparently extinct. The second element is O.N. *skáli*, a hall, but the first is uncertain. *Smer-* occurs in other Orkney names, e.g. Smerquoy in St. Ola, and probably denotes NO. *smære*, clover; Mod. Icel. *smæra* and *smári*. But 'clover-hall' would be an improbable formation.

Smittaldy : a small farm on slope above the upper end of Bay of Tuquoy. The absence of old forms of the name is regrettable, because one may suspect that this name should have to be placed alongside Spittalsquoy in Evie, q.v. 'Smit' is the Ork. verb used instead of the Eng. 'infect' —in connection with disease, and confusion may have crept in between Spittal- and Smittal-.

Snukesbrae (snøks-bre) : small farm at point in Skelwick called The Sneuk (*snøk*). Jakobsen cites various Snjug-, Snook- names from Shetland, deriving them from O.N. *knjúkr* with changes of initial consonant in Shet. dialect. It is doubtful if that is correct, and one may think rather of No. *snók* (O.N. *snókr*) a snout. The Snook is a name for a cape also on Holy Island (Northumberland).

Sponess : small farm at point of same name in Skelwick — a low narrow promontory on south side of Bay of Swartmeil. Probably O.N. *spóa-nes*, 'ness of curlews.' O.N. *spánn* or *spónn*, a spoon, which appears in several Norse names would seem here to be quite inapplicable from the nature of the topography.

Tifter : a small farm in Midbea; not recorded in old Rentals. O.N. *toptir* or *toftir*, pl. of *topt* (*toft*), a house site, or site where building has stood of old. Vowel change rather odd, but Rygh notes that in mediæval Norse records various spellings appear, *inter alia*—typt.

Tirlot : a farm in Midbea area. In the absence of early forms of the name no suggestion can be offered as to its origin.

Trebland : 1740 R. records 3½d. lands of Trebland in Noltland-Bewest. In this tunship there used to be a house called Gairbacks which stood a little way above the present Netherhouse. It is thus of especial interest to note here in Westray the presence once again of the Celtic *treb*, in association moreover with the old 'gairsties' or 'gerbacks' as in N. Ry., Sanday, and Stronsay. See Treb, N. Ry.

Trenaby (tren·əbi) : Trennibie 1664 Ch.; the main farm in Rackwick tunship, and the place from which the Balfours of Trenaby took their title. It is situated near a large and almost land-locked salt-water 'Oyce,' or shallow bay, where probably the old Norsemen laid up their vessels in winter. This is another O.N. *bœr* name, but the prefix is uncertain. It is most probably a personal name, and may represent O.N. *Þrándr* or *þróndr*, one of the commoner names in the north : hence ? *Þrándar-bœr*, 'Thrand's farm or settlement.'

Here, again, one may suggest there has been one of the large original settlements or 'land-takes' by a Norseman. It would probably have embraced the tunship of Aikerness to the north-east as well as all Rackwick—a large 3½ urisland area—with the parent *bœr* of Trenaby in the middle, and a safe haven for ships down below. And if *Þrándr* was the chieftain who made the conjectured settlement, he would be the only one of those hypothetical 'landnamsmen' whose name has come down to us, as a personal name compounded with bœr is unknown elsewhere in Orkney. It is just possible, indeed, that the first element represents an O.N. *prœndir* or *prœndr*, denoting settlers from the Trondheim area of Norway.

Trumland: In Balfour's feu-charter of 1664 he had granted to him 1d. land Bishopland and ½d. land Kingsland of Trumland in Rackwick. The name is almost forgotten locally, but it survived, until recently at least, applied to a piece of land near Fiold. Origin here uncertain, but cf. Trumland, Rousay.

Vell: Velz 1740 R.; Velzie 1794 R. A 5d. land in Midbea. O.N. *velli*, dat. sing. of *vǫllr*, a field.

Vere (ver, vir): small farm at point of Vere, the outer end of the ness that was formerly called Ouseness, q.v. O.N. *ver*, "a place where one stays at certain seasons of the year to carry on fishing, etc." Rygh's *Indl*. Here the adjacent Oyce (a land-locked bay) would have provided shelter for boats.

PAPA WESTRAY

'til Papeyiar hinnar meiri'; 'i Papey.' (*O. Saga.*)

The name *Papey hin meiri*, 'the larger isle of Papae' (Celtic clergy), was applied to distinguish this island, which lies near Westray, from *Papey in litla*, 'the little Papey' or Papa Stronsay, which lies adjacent to Stronsay. It was to this isle that the body of Earl Rognvald Brusason was taken for burial after his death in Papa Stronsay.

Papa Westray, or Papey (pa:pi) as it is called locally, is rather over four miles in length by about a mile in average breadth, and running across the middle, dividing the isle into two roughly equal portions, was one of those prehistoric earthen ramparts or massive dikes known as gäirsties or treb-dikes. Here in Papey, however, that latter term seems to be unknown, but a portion of the dividing dike still goes by the name of The Cunnan (rabbit) Gairsty. (For that term see further *sub* Treb, N. Ry.)

The two portions of the isle are still known as North Yard and South Yard (cf. the three divisions in N. Ronaldsay), and in the three early rentals they were designated Benorth the Zarde and Besouth the Zarde, though on one occasion Garth is substituted for Zarde. Each half was skatted as 2 urislands, but unfortunately those early rentals mention no individual farm names. From the 17th century, however, down to the late 19th practically the whole island belonged to one family—the Traills of Holland—and from old rentals of that estate, etc., some older forms of names are accessible.

T.R. TRAILL RENTALS, c. 1740.

Backaskaill: Backiskill, T.R. A small farm at main landing place on west side of island. O.N. *bakka-skáli*, 'skaill or hall at the bank (of the shore).'

Breck: 'Old Breck' is now a ruinous site, and the old farm has lost its identity altogether, its lands having been shared among other farms. In T.R., however, appears Underbreck, i.e. O.N. *undir brekku*, 'under the slope,' an interesting example of a prepositional name.

Breckaskaill: Breakiskill, T.R. O.N. *brekku-skáli*, 'slope skaill or hall.' This farm is adjacent to Backaskaill—the two farmsteads being only a few hundred yards apart. The existence of two *skaill* farms so close together would seem to imply the division of an early udal property between two heirs such as (in the *O. Saga*) we read of as happening at Langskaill in Gairsay.

Cuppin: a small farm near St. Tredwell's Loch. O.N. *kopp[r]inn*, 'the hollow' or cup. *Cuppo* is an old Ork. dialect term for a basin- or cup-like hollow on the landscape.

Gairbolls (ger·bəls): Gerbow, Gerboues, Gerbous, T.R. A farm near Backaskaill. The older forms of this name suggest that it is a Scots plural of the name Gerbo, q.v. North Ronaldsay and Sanday; the remains of one old gairsty can still be traced running past the west end of the farmhouse. The 'll' in the present name can almost certainly be regarded as an English 'correction' of the '-bou-' on the false assumption that, like the Scots *bow* (bɔu) [of meal], it represented Eng. *boll*.

ORKNEY FARM-NAMES

Gowrie (gou·ri) : Gaurie, T.R.—almost certainly this farm. Near the farmstead is a patch of land known as the Gerry Toomal.
It is interesting to note here the close parallel with the Westray farm Gairy, and its neighbouring Brae of Gowrie. It would seem certain that in both cases Gowrie is a corruption or Scotticising of Gairy or Gerry (both names sound alike though spelt differently), and that the origin in each case was O.N. *geiri*, a 'gore' or triangular piece of ground.

Holland : O.N. *há(hó)-land*, 'high-land.' This was the 'laird's house' or manorplace of the island for about 250 years down to the latter part of last century. The lairds were the Traills of Holland, the senior branch of the Orkney family of that name. The house occupies a dominating position on the ridge of the island.

Hookin (huk·in) : Howquyne, U.B. 1601; Vquoy 1618, Howquoy 1634, Howkin 1636 (Traill Charters); Ouquin (Blæuw's Map); Uquinland, 1658; Ouking T.R. 1740.
A small farm on east side of island where a small stream runs out to the sea and used to drive tthe island's grinding mill. A difficult name, occurring also in Sanday and North Ronaldsay (*vide* Hookin, N. Ry.). Many years ago when dealing with the Place-Names of North Ronaldsay (P.O.A.S. Vol. I) I suggested that the name represented O.N. *ár-kvern*, 'river-mill,' i.e. a mill driven by water, taking the place of the old handquern. Later, when discussing the Place-Names of Papa Westray (P.O.A.S. Vol. III) I was impressed by the absence of final '-n' in some of the early forms of Hookin in that island, and suggested it might rather be O.N. *ár-kvín*, 'the burn quoy.' Against that idea, however, is the actual presence of a mill at those places, and I now revert to my original suggestion of O.N. *á(r)-kvern*.

Maeback (me·bo) : Maback (Blaeuw's Map 1654); Maiback 1671 (Traill Charter); Meback, T.R. This farm-house stands at the shore about midway along a sandy beach, and the name might well represent O.N. *miðbakkar*, 'mid-banks'; but the syllable *may* (me) occurs in some Ork. names where it cannot be explained as 'mid,' and consequently the origin here must be regarded as somewhat uncertain.

Moclett : farm near Head of Moclett. O.N. *mó-klettr*, 'moor-clett' or rock.

Nesodden : Nesshodden T.R. Not a farm now, but only a field of Holland, on north side of St. Tredwell's Loch. There is a small point of land here running out into the loch, but no 'ness' to warrant such a name—O.N. *nes-oddinn*, 'the point of ness.' It would thus seem probable that the name may be due to an immigrant from Nesodden, a place near Oslo in Norway.

Nouster (noust·ər) : id. T.R. A small farm at a landing-beach on east side of island. O.N. *naustar*, plur. of *naust*, a berth for boat when drawn up at beach.

Quoyolie : Quoyollie, Quoyole, T.R. 'Olie's Quoy'; Olie is the usual Orkney contraction of Oliver (O.N. *Óláfr*).

Skennist (skɛn·əst) : Skeanes 1658; Skaenes 1664; Skenest T.R. 1740. This farm lies adjacent to one of the main boat-nousts in the island where there is no ness of any kind. Moreover, local pronunciation confirms the

NAME MATERIAL 47

final 't.' The name undoubtedly represents O.N. *skeiða-naust*, 'shipnoust,' a berth where sailing vessels of the *skeið* class were drawn up. A *skeið* was a fast-sailing type of longship.

The same name occurs in Shapinsay in the compound Skennestoft.

Tifter: T.R. 1740. The site of this old farm is now unknown. **See Tifter,** Westray.

Via (vi·o): Via, T.R. There are two small farms now—North and South Via. At the latter is a mound which may account for the *-o* of the final syllable, and it is just possible that the name may represent O.N. *vé-haug(r)*, 'holy mound,' but in the absence of further evidence that must be regarded as very uncertain.

EDAY

Etheay, c. 1475 (Fordoun); Ethay, 1560 Sasine; Ethay 1595.

Though not on record in the O. Saga, the origin of this name is not in doubt. Eday is an elongated isle—over seven miles in length from north to south but less than two miles in average breadth. Near the middle, however, the width shrinks to little over a quarter of a mile, and from the isthmus so formed the island undoubtedly took its name.

From some obscure reason Eday was not included in detail in any of the early rentals. But from the fact that Bishop Adam Bothwell in 1560 made a grant of the isle to his stepfather Oliver Sinclair of Whitekirk (the one-time favourite of James V of Scotland) it would appear to have been part of tthe bishopric estate, though whether all of it was really bishopric is rather uncertain: the skats at least were no doubt all payable to the bishop. From the Sinclairs the island soon passed into the hands of the Stewarts—the first of whom, John Lord Kincleven, brother of Earl Patrick Stewart, was created Earl of Carrick and obtained a royal charter in 1632 by which the northern portion of the island bordering on Calfsound was erected into a Burgh of Barony with the usual burghal privileges.

For an island of such size its skattable value was very low, probably only two urislands all told, though the exact amount is uncertain. That low valuation was merely a reflection of the small extent of arable land in the island, which is for the most part hilly and covered with some of the finest peat-moss in Orkney. In former days peats used to be boated from Eday or the Calf of Eday, a small adjacent isle, to most of the neighbouring islands. The arable land lies mostly in pockets around the coast, and much of that would seem to have been brought into cultivation in comparatively recent centuries. It is impossible, however, to know exactly how much was cultivated at the time skat was imposed, but the number of really old farms is certainly small, and as already stated there are no early rentals to indicate their names or size.

Backaland: a 5½d. land (1733 Charter). Today there is no farm of this name, but it survives as a district name. This old tunship lay along the east side of the island from about the present pier of Backaland to Skaill, including the present farm of Stenaquoy, etc.

O.N. *bakka-land*, 'slope-land.'

Bredakirk: a small croft in Cusvie district. Probably the site of an old chapel dedicated to St. Bride.

Carpaquoy: a ½d. land (1733 Ch.). A small croft on west side of island a little to the north of Seal Skerry. Origin of first element obscure; *karp*- occurs as a rare element in Norw. place-names but (acc. to N.G. XII) no definite source can be suggested.

Carrick: the island mansion-house and a farm in the north of the island. It may be confidently asserted that the name is foreign to Orkney, and almost certainly due to John Stewart, Lord Kincleven, afterwards Earl of Carrick though its actual origin is enshrouded in some obscurity. In Tudor's *The Orkneys and Shetland* it is stated that Lord Kincleven, having obtained in 1616 charters of certain lands which had once formed part of the ancient earldom of Carrick, desired to procure a grant of the Carrick title as well, and did actually get a patent from Charles I in 1628.

NAME MATERIAL 49

When the patent was presented to the Privy Council, however, the Lord Advocate took objection on the ground that the title of Earl of Carrick was one always borne by the heir-apparent to the Scottish crown. "After some delay apparently caused by Lord Kincleven changing his flank and stating that it was the Orcadian and not the Ayrshire Carrick that was meant, the patent was finally confirmed."

The probability therefore is that the Eday name Carrick dates from about that time, and replaced an old local name.

Cusvie (køs·vi) : not a separate farm-name but a name applied to a small district or tiny tunship in the north-west of the island, and containing a few small crofts. Origin obscure.

Fersness : a 1d. land (1733 Charter). Fairisnes (1623 Oblig.). This farm is situated on a narrow ness or headland of the same name which juts out towards the neighbouring island of Faray which is only half a mile distant. It is tempting to regard the name as comparatively modern and representing 'Faray's Ness,' or even Ferry-Ness, as it is the point of nearest access to Faray. But in each case the 's' would seem to forbid such an origin. In Norw. place-names Fers- and Færs- appear as first elements in a few cases also, but even there the origin is doubtful.

Furrowend : a small croft near shore in Calfsound district. A fairly common Ork. place-name. The first element appears again in Furrowdike or Forrowdike, a name applied to an old ridge or dike at the west end of Calfsound beyond Carrick. Forrowdike, indeed, is practically a generic term in Eday for such a dike, which is built on the beach and down into the sea to prevent animals from straying along below the cliffs. Icel. *forvað*, "shoal water between the cliffs and the flowing tide." O.N. *forvaði*, "a cliff projecting into the *forvað*, where the rider has to wade through water." (C. & V.)

In Furrowend it is uncertain whether *-end* refers merely to the upper end of such a dike, or whether it indicates a place where the *forvað* ends, i.e. where further passage along the beach under the cliffs becomes impracticable; most probably the latter.

Garth : an old farm or tunship which was situated immediately north of Backaland, as appears from evidence in an inquest held in 1722. O.N. *garðr*, a farm.

Greentoft : the largest farm in the south end of Eday. O.N. *græn[a]-topt*, 'green-steethe,' or site of old buildings.

Guithe (gjøð) : not a farm-name, and it is doubtful if it was a really old tunship at all. It is a name applied to a small district immediately south of Cusvie, and contains only a few small crofts. It may be regarded as a 'neighbourhood' unit like the Norw. *grend*.

The origin of the name is extremely obscure, but it seems to occur again in Geoth, Harray, q.v.

Holland : a vanished house, of which the name only survives as a field name on Skaill. See Holland, Papa Westray.

London : the ruins of two old houses—East and West London—may still be seen on what is now the sheeprun of Skaill, near the Bay of London on the east side of the isthmus in the middle of the island.

There can be little doubt that this name represents O.N. *lund[r]-inn*, 'the grove, or woodland,' and its occurrence here is of special interest in regard to the presence or absence of trees in Orkney a millenium ago. The name London was also applied to an old house in Rousay situated in a sheltered southward-facing valley in Frotoft, half a mile or so up from the beach. From the evidence of peatbogs we know that there has been no significant forestation in Orkney since the Boreal Period, i.e. since roughly 5000 B.C. But scrubwood probably persisted, and that may have been what gave rise to such names as London, Skogar (Birsay), Skooan (Rousay), etc.

In the case of this London, however, there may be another explanation. At several places in Orkney, below high-water mark, forest remains are to be found buried under the sand and clay of beaches, and are occasionally revealed after heavy storms. In the Orkney Geological Survey handbook it is stated that such remains are to be found in the sandy beach of the Bay of Musseter which is just across the narrow isthmus from the Bay of London, and it is quite possible that such remains may have given rise to the name London.

But there exists a curious old legend which can hardly be dismissed offhand. It is current in several of the North Isles of Orkney, and tells of an Orkney skipper over in Norway who met there an old Orkney woman who had not re-visited Orkney since her youth. She inquired regarding various places in Orkney, but these vary in each island version. In the Eday version the legend runs that the skipper was over there for a cargo of timber, whereupon the old woman asked him in surprise: "What's become o' the bonnie woods o' Skaill?"

Musseter: an old farm on mossy soil near the bay of the same name. O.N. *mó-setr*, 'moor-settlement or farm.'

Noup (nup): an old farm now incorporated with Carnick. It was the nearest farm to the precipitous headland now called the Red Head, which must have been known formerly by the term Noup, like the Noup Head in Westray, where it is of interest to note that the adjacent farm is also called Noup. See that name.

Papley-hoose: a small croft in Cusvie district. I know of no tradition attached to this place, and the name may have arisen from that of a man who built it or lived there in comparatively recent times, but of whom no record survives. On the other hand it might possibly indicate the site of an early Celtic mission-station like Papley in Holm and in South Ronaldsay, though that is less likely. N.B., however, the neighbouring name Bredakirk.

Quoy-faulds: a small fertile farm in south end of island. The name is almost certainly a corruption, as the Norse *kvi* and Scots *fauld* (fold) are practically synonymous. It is probably of the same origin as the Sanday Quivals, q.v.

Seal-skerry: an old 1½d. land adjacent to the skerry of this name on the west side of the island. Self-explanatory.

Skaill (skil, skjil): in early charters Skaill and Puckquoy go together as a 5½d. land. It is situated adjacent to the parish church on the east side of the island, and from that fact must have been the residence of the island chieftain in early Norse days. O.N. *skáli*, a hall.

NAME MATERIAL 51

Stenaquoy: a farm in the old tunship of Backaland. O.N. *stein[a]-kví*, 'stone-quoy.'

Whistle-bare: formerly a small farm or cot, but now merely a field name on Carrick. See this name—Sanday.

Windy-wa: a small farm in Southend district.

This name—either in the sing. as here, or in the plural as Windywaas—occurs in many other districts of Orkney (e.g. Westray, N. Faray, Rousay, Firth, Graemsay, etc.) as the name of a small farm or cot. It is usually regarded as a Scots term = Windy walls, applied in a semi-derisive way to a poorly built house with draughty walls. But Jakobsen disagreed with such an interpretation, and regarded it as a corruption of O.N. **vinjar-válir*. In an article in *Danske Studier* (1919) "Om Orknøernes Historie og Sprog" he wrote as follows: "But partly because the name Windiwas occurs repeatedly, partly because it cannot be shown to be new (or recent) as a place-name, and finally because a corresponding name Winniwals occurs in several places (e.g. as name for a wide stretch of pasture in N. Roe) it is far more reasonable to think of an O.N. *vinjar-válir*. A combination *vinjar-váll* occurs repeatedly as a place-name in Norway. *Váll* seems to signify ground which is cleared by burning. See N.G. Indl. p. 85."

Vinjar is the genit. case of O.N. *vin* (pasture) a term which was going out of use in Viking times, but it is found as an element in various Orkney place-names, e.g. Vinquoy, Vin Hammers, etc.

Windywa, however, is never found as the name of a large old farm or tunship, and its real origin must be regarded as uncertain.

In concluding this section on Eday, attention may be drawn to two classes of small croft-names, or cot-names, that are of special interest. Quoy-names, as shown in Part III, are among the youngest of the farm-names that date from Norse times. But here in Eday are to be found names such as Essonquoy, Harcusquoy, Hourstonquoy, in which the first element is a personal surname, and such quoy-names cannot therefore be many centuries old.

The other class of names is that in which the suffix is *-ha'* = Scots-'hall.'

Such names occur all over Orkney to some extent, e.g. Whiteha', Blackha' etc., and have sometimes a derisive character such as Tarryha'. But in Eday there is an astonishing concentration—Benstonha', Brandyha', Cooperha', Galtyha', Gowkha', Groatha', Hungryha', Noneyha', Ploverha', Purmha', Runtha', Shoeha', Shuttleha', Sillerha', Warrenha', Waterha'. It is doubtful if any of these is over 300 years old, and the number occurring in this one small island would seem to point to some social factor which is as yet obscure, and which increases the regret one feels for the paucity of Eday records.

(NORTH) FARAY
[fɛːre]

Fareay, c. 1380 (Fordun); Faray, c. 1590 (Jo. Ben); Farray, Faray, 1595 R.; Pharay, 1627 (Peterkin, III, p. 95). O.N. *Fœrey*, 'sheep-isle.'

There are two small islands in Orkney bearing the name Faray, this one, adjacent to Eday, being sometimes distinguished as North Faray, the other in Scapa Flow as South Faray (or Fara). It is the same word as appears in the Færoe Isles, O.N. *Fœreyjar*, which the author of *Historia Norvegiae*, c. 1200 A.D., regarded as meaning 'sheep islands' (*insulae ovium*). From the time of the Norwegian writer Peder Claussen (c. 1600) down to our own day various writers have queried the correctness of such an interpretation on the ground that while **fœr* was used of sheep in Old Swedish and Danish it was not strictly speaking a Norse word at all. A recent sceptic is Professor Brøgger who in his *Løgtingssøga Føroya* (1937) discusses the name at length, and seems disposed to regard it as possibly a Norsification of some old Celtic name such as *fearann*, 'land.'

The occurrence of the name, however, in Orkney for a small island, not once but twice, would seem to nullify such a suggestion. Nor is the above-noted philological objection of any real weight. Prior to and during the early Viking Age the language spoken throughout all Scandinavian lands was one and the same, and though after that common language—the *Dǫnsk tunga*—split up, *fœr* may not have been retained in Old Norse, there is no reason to conclude that it was not a term in the vocabulary of the men who first settled in Orkney or Færoe. Hence we can with confidence accept the traditional interpretation of Faray as 'sheep isle.'

This small island, only about 1½ miles in length by half a mile in average breadth, is never mentioned in the Orkneyinga Saga. Nor does it figure in detail in any of the early Rentals, and its old skattable value is unknown. Like its larger neighbour Eday it formed part of the Bishopric estate, but prior to 1600 it had become the property of the Westray Balfours, in whose possession it remained until, through the marriage of a Balfour heiress, it passed to the Stewarts of Brough in 1734. Today it is in the hands of trustees of a Stewart Mortification for religious purposes of the Church of Scotland.

Until recently there were eight small farms or crofts on the island, and in the end of last century it maintained a population of 70 persons. Modern social and economic conditions, however, combined to render life on the island less and less attractive, and the last few remaining families deserted the island in 1946. Today it is an uninhabited sheepwalk.

RENTAL NAMES

Cott: O.N. *kot*, a cot or small house.

Doggerboat: a quaint name for a farm. Possibly a dogger-boat or fishing vessel may have been utilized in some way as a temporary dwelling. In any case the name can hardly go back beyond the 16th century.

Hammer: O.N. *hamarr*, a rock projecting from a cliff or hillside.

Holland: the largest farm on the island. See Holland, North Ronaldsay.

Leaquoy (lɛk·wi): see this name—Stronsay.

Ness: O.N. *nes*, a headland. This is the most southerly farm in the island.

Quoy: the most northerly farm in the island. See Part III.

Windy-wa: see this name—Eday.

* In the East Scand. form—*får*.

SHAPANSAY

(According to the spelling on present-day O.S. maps, etc., the middle letter of this name is 'i,' not 'a' as above, but I prefer the latter which is more in accord with the oldest spelling and local pronunciation.)

Scalpandisay (Fordun, c. 1375); Schalpandsay 1492; Schapinscha and Shapinscha 1595; Gjalpandisöi (Peder Claussön c. 1600); Schapinschaw 1601; Schapinsay 1614; Schapinshay 1642; Shapinshay 1739.

This island, strangely enough, is not mentioned either in the Ork. Saga or in any other old Norse saga, and the most nearly satisfactory derivation of the name is that suggested by Munch a century ago—O.N. *Hjálpandisey*, an origin which he considered "not merely probable but as good as certain." *Hjálpandi* is a present participle meaning 'helping,' and Munch suggested that such a name had perhaps been applied to the island "from its convenient situation and good harbour for ships coming in from the ocean."

As the genitive of *hjálpandi* would not normally end in '-s' I had considerable doubt as to Munch's derivation, and consulted Professor Magnus Olsen as to whether the first element might not rather be a personal name—*Hjálpandr*. He was kind enough to reply as follows:—

"Jeg er tilbøielig til å tro at Munch har funnet frem til den gamle form for Shapansay:

"*Hjálpandisey*. Denne sammensetning synes, formelt betraktet, å ha paralleller i *Liðandisnes* (Lindesnes) og i *bindandismaðr* (avholdenhetsmann) og *b—stími*. Ikke så å forstå at der nødvendig må forutsettes et *hjálpandi* (subst.) som ønavnet er sammensatt med. Forholdet kan godt være det at der til infinitiv fritt kan dannes et sammensetnings-ledd på *-andis-* efter analogien *bindandis-*, *Liðandis-* (med flere andre tenkelige dannelser). Det må merkes at mange hunkjønnsord på *-i* kan ha gen. på *-is*; *atgervismaðr* ... etc. Endog ord som *hjálp* og ønavnet *Hising*, f. kan som første ledd ha gen. på -s: *hjálpsmaðr* (=*hjálparmaðr*), *Hísingsbúar*. Like ovenfor vil *Hjálpands-ey* nærer også jeg betenkeligheter, kanskje endog sterkere enn De. *Hjarrandr* og *Stígandr* (se Lind) gjenspeiler neppe en analogi som kunde fjerne disse betenkeligheter.

"Det skulde vel være mulig å finne en realforklaring av *Hjálpandisey*, 'hjelpe-øen' (*hjálparey*, *hjálpsey*)."

In view of the above opinion from such an authority there is evidently no formal objection to Munch's *Hjálpandisey*, though the precise reason for the application of such a name to this island is still rather uncertain.

The early settlement-units of this, the nearest of the North Isles to Kirkwall, are extremely difficult to unravel, partly owing to the fact that by the time of our early rentals the *pro rege*, bishopric and udal lands therein were so inextricably intermingled. The island is included in the 1492 R. but with curious omissions; it is absent altogether from the 1500 R., while in the 1595 R. and subsequent rentals the entries are so mutually incoherent that it is difficult or impossible to reconcile them. Furthermore, in the 1492 R. some names of quite substantial units occur which are never to be found later, and it would thus appear that from a quite early date considerable regrouping and renaming of lands had taken place.

The skattable valuation is also quite indefinite. Peterkin prints a Taxt Roll made up by the Bishop of Orkney in 1617 in which the parish of Shapansay is specifically stated to contain 108 pennylands. That is of course exactly six urislands, but from other evidence it would appear to have been an under-estimate, and that there was at least half an urisland more.

The arable lands lay around the shores, and beginning at the south-east corner and going clockwise round the island we may enumerate the main townships as follows, omitting two or three smaller units interjected along the route: Sands, Kirbuster, Elwick, Sound, Waltness, Weland, Holland, Meaness, and Burrowston. These will be discussed in further detail below.

RENTAL NAMES.

Archdeansquoy: 1739 R. A quoy which apparently pertained at one time to the Cathedral archdeanery. Site unknown. Cf. Canker (Sanday).

Burrowston: id. 1492, 1614 and 1642; Borrowstoun 1595; Burrestoun 1739. A 3d. land (1739 R.) in north-east of island adjacent to an old broch ruin known as The Broch of Burrowston. This name occurs again in Sanday and in Westray in the metathesised form Broughston—in each case applied to a small area or tunship adjacent to the site of an old broch. The term *broch* derives from O.N. *borg*, a fortification, the genitive of which is *borgar;* hence originally the compound form must have been *borgar-tún.* The ' s ' in those names can date therefore only from a time after the O.N. genitive *-ar* had been superseded by the Scots or English *-s*.

Brawel: 1595. A cot in Weland-tun. Origin obscure; perhaps same name as Breval—a small hill-croft in Rousay.

Bigging: A farm recorded as in Meanes-tun (1739). See this name—Westray.

Caskald (kaskald·): Castcauld, Sas. 1618; a small farm in Meanes-tun (1739); still existing today. As the second syllable bears the stress there seems little doubt that it represents the O.N. adjective *kaldr*, cold; the first element is however obscure.

Clerksquoy: 1739. Apparently another small piece of church-land; cf. Archdeansquoy. Site also unknown.

Cultisgew: 1492. This appears as a 3¾ pennyland, situated apparently in the Hollandstun area. No later trace of the name can be found. Origin obscure; cf. however Culdigo (Egilsay).

Elliar Holm (ɛl·jər or hɛl·jər): Heleneholme (Fordun c. 1375); Eloerholme (Jo. Ben, c. 1590); Elgerholme (Monteith, 1633); Elginholme (1642); Heiller holm (Blaeuw's Map, c. 1650); Elgine Holm (1739); Eller Holm (Mackenzie's Chart 1750). See Elwick.

Elwick: *Elliðarvik* (Hakon Saga); Ellandwik and Ellenwick 1492; Ellavick and Elwick 1595; Elwik 1642; Ellwick 1739.

A 9 pennyland or half-urisland tunship on the shores of an almost semi-circular bay of the same name. In the mouth of this bay is a small island known as Elliar Holm or Helliar Holm which protects the bay from the south and makes it one of the most sheltered anchorages in Orkney. This was the harbour where, as narrated in his saga, King Hakon Hakonsson's fleet lay for some days at anchor on his ill-fated expedition to the west in 1263 A.D.

Elliðarvik, the Saga form, signifies the bay or 'wick' of a man Elliðr, and the present-day pronunciation of the island-name leaves no doubt that it was named *Elliðar-hólmr* after the same man.

The 'n,' however, which appears in the 1492 record of the tunship name, as well as in several recorded forms of the island-name, is

NAME MATERIAL

undoubtedly puzzling. Whether that has been due to early scribal error, or confusion with the woman's name Helen, or to an early alternative name for the tunship, e.g. *Elliðarland*, it is now impossible to tell.

Frustigar (frʌst·igər): Frustager 1614; Frustigair 1642; Frustiger 1739. A 1½d. land in Nether Holland still surviving as a farm today and still occupied by people named Meason as in 1739.

Probably O.N. *Frosta-garðr*, 'farm of a man Frosti.' That was a somewhat rare personal name, but it is found as the first element in two Icelandic farm-names, and probably (acc. to Rygh) in a few Norwegian names also.

Fuxtown: 1739; Fixtown 1764. A 1½d. land in Nether Holland. The name has now been corrupted to Feastown. Origin obscure; quite close is a small house called Fuag (pronounced fjog) which may represent the first element in Fuxtown.

Gairsty (gɛrst·i): Gairstay 1642; Gairsty 1739. A once notable farm in Kirbuster tunship. Though no longer in existence it was the residence of Wm. Irving of Gairstay, bailie of the island in the 1630s.
See under Treb, N. Ronaldsay.

Garth: a farm north of Waltness. In 1739 Gairth was entered as a 1½d. land in Waltness tunship. O.N. *garðr*, a farm.

Girnigo: 1739, 1764. A small farm near shore in the old Meanes-tun; still surviving. The same name appears in Girnigo Castle, Caithness, where there is a marked cleft or 'geo' in the shore line that would seem to confirm an O.N. *grœna-gjá* as the origin of the name. Here in Shapansay, however, there does not seem to be any noticeable geo or creek in the vicinity, though the original shore line may have changed.

Grassquoy: Girsquoy 1642; Grassquoy 1739. A small farm in Kirbuster-tun.

Greenwall: id. 1595; there entered as a cot in Weland-tun. Evidently O.N. *grœn-vǫllr*, 'green field.'

Haquoy (ha:kwi): id. 1739. A small farm in Meanes-tun. Origin of first element uncertain; the name may have been O.N. *haga-kví*, 'quoy on the hagi' or outpasture, or perhaps simply *há-kví*, 'high quoy,' in contrast to others: there was another farm Quoys in this tunship.

Harroldsgarth: Thoraldisgarth 1492; Horraldisgair 1595; Horralsger 1614; Horrelsger 1614; Horreldsgairth 1642; Horraldsgirth and Horraldsgairth 1739. A farm in Sands-tun. O.N. *Þóraldz-garðr*, 'farm of a man Thoraldr.' For change of initial þ > h in Orkney place-names cf. Hurtiso (Holm) and Horraldsay (Firth).

Headgeo (hɛd·jo): Hogo 1739, where the first 'o' is probably a mistake for 'e'; Hego 1764. A small farm in Burrowston. Origin uncertain; the old spellings make one suspect the present name to be an assimilation to familiar words.

Holland: 1492 and all Rentals down to 1739. In the latest Rental, however, it is entered as Over Holland and Nether Holland; in previous Rentals Nether Holland appears as Hollandswick, a name which in one 17th cent. Rental is entered as "Hollandisweik alias Lintoun.' The original tunship of Holland seems to have stretched right across the island at the base of the Ness peninsula, and probably formed a whole urisland of itself. The name is now more or less obsolete. See Holland (N. Ron.).

Hollandswick: Hollandswic 1492; Hollandiswick 1595; Hollandisweik alias Lintoun c. 1642; Nether Holland 1739. See Holland; the termination no doubt represents O.N. *vik* (from the 'wick' or bay at Linton on the east side).

Holm (hwom): sic in 1739. Now spelt Quholm. O.N. *hvammr*, a small valley; cf. Whome, Rousay.

How: "How—a skatland" (i.e. 4½d. land) 1492. Still a considerable farm today. O.N. *haugr*, a mound.

Kirbuster: Kirkbustir 1492; Kirbuster 1595, 1601, 1739. A tunship on south side of island in which stands the parish church. O.N. *kirkju-bólstaðr*, 'kirk-bister,' i.e. a farm-settlement containing a church.

Lairo: in the former Welands-tun district there are today two small farms East and West Lairo, and down below at the beach is a loch or lagoon known as Lairo Water. In the old Rentals we find entered in Welandstun—" Ane Quoy callit Laverock " 1595; Laverock 1612; Quoylarok 1642; quoy Larock 1642. There can scarcely be any doubt that these names represent earlier forms of the present-day Lairo.
Strangely enough the only other known occurrence of this name in Orkney was in the town of Kirkwall, where the upper end of the old town (the present Victoria Street area) used in old days to be known as The Laverock, and was the early bishopric domain.
As yet no wholly satisfactory explanation of the name has been found.

Leaquoy: Leoquoy 1492; Leaquoy 1595. A quoy attached to Ness farm. See this name—Stronsay.

Laverock: See Lairo.

Lingrow: 1739. A 1d. land in Meanes-tun. See this name—St. Ola.

Linton: Lyntoun 1492; Hollandisweik alias Lintoun, c. 1642; Linton 1739. A farm on east side of island in old Holland tunship. Origin uncertain.

Meaness (mi'nez): Meones 1492; Meoness 1601; Mennes 1614; Menas 1642; Meanes 1739.
A half-urisland tunship between Hollandstun and Ness and west of Burrowston. Name now practically forgotten locally. O.N. *mjó-nes*, in sense of 'lesser ness'; the O.N. adj. *mjór*, slim, narrow, etc., is found used in Orkney for the smaller of two things, e.g. Mue-geo in Sanday as contrasted with a near-by Langi-geo. Here in Shapansay it would appear to have been applied to a small 'ness' jutting out from the west side of the main 'ness' (of Ork).

Midhouse: 1739. A small farm in Burrowston.

Morsetter: Morsettir 1492; Mossatter 1595; Mossetter 1612; Musetter 1614; Musseter 1642; Mussatter 1739. A 1½d. land in Sands-tun. Now vanished. Prob. O.N. *mór[ar]-setr*, 'moor-dwelling'; cf. Jakobsen's note in *Ordbog* on *mørsten*.

Neager: 1614 and 1642. A cot-house in Burrowston. Same name as Neigarth, Sanday, q.v.

NAME MATERIAL 57

Neares: 1739. Part of Meanes tunship. Contraction of Netherhus = **Lower House**. Cf. Nears, Rousay.

Ness: Nes 1595; Ness 1739. Farm or small tunship adjacent to headland termed Ness, and on O.S. map Ness of Ork. O.N. *nes*, headland or 'ness.'

Nisthouse: 1739. Part of Sands tunship. O.N. *neðsta-hús*, 'nethermost house.' There is another Nisthouse in Burrowston.

Ork: 1601, 1614, 1642, 1739; a ⅜ pennyland in Ness (1614 R.). An obsolete and puzzling old farm-name. The tip of the large promontory or 'ness' in the north-east of the island is marked on modern O.S. maps Ness of Ork, but on Mackenzie's Chart (1750) the only name appearing there is " Gio "—apparently indicating a small creek which is to be found there. In his accompanying " Directions for Sailors " Mackenzie refers to it again as " the Gio, or N.E. Point of Shapinsha." It would thus seem that the name Ness of Ork was not current at that date, and that it is a later formation from the adjacent farm-name.

Now Orknes is a farm -name which occurs in Norway, and is there regarded (G.N. xii, 205) as deriving from O.N. *ørkn(a)nes*, 'seal-ness,' from *ørkn* (*erkn*) a large kind of seal (*phoca barbata*)—a name which is found in Orkney dialect in the form *erkny* (N. Ry.) or *arkmae* (Sanday). Difficulty arises, however, in accepting such an origin for the Orkney name *Ork* owing to the absence of the letter ' n ' which is radical in the O.N. term. In the Norwegian Orknes the ' n ' runs together with the initial ' n ' of *nes*, but in Shapansay, though the 'Ness of Ork ' may perhaps be occasionally referred to, I am informed that ' Orkness " is an unknown form. It has been suggested (by myself among others) that in this Shapansay Ork we may have another example of the pre-Norse root which is seen in the name Orkney itself, and in the O. Saga name for the famous chambered mound of Maeshowe, viz. *Orcahaugr*, a name also found recorded as *Orkøuh* in 12th cent. runic characters in an inscription on the wall of the chamber itself. But the problem is further complicated. In his *Place-names of Shetland* Jakobsen states that a steep promontory in Unst is known to Unst fishermen as " de Orka," or " de Orki "; and in the south of Shetland there are two hills known as the Muckle and the Little Orka. He further recalls that Diodorus Siculus (a younger contemporary of Julius Caesar) records that one of the capes in the north of Scotland was called Orcas. Jakobsen comments : " there must be a connection between the old Pictish *Orc-* (*Orcas*) and the Shetland *Orka* as the name of a promontory, and Shetl. *Orka* as a hill-name can hardly be separated from them."

As yet, however, that whole problem is so speculative that the origin of this Shapansay name must still be deemed uncertain.

Ostoft: Ostoft, 1737 Retour; Ostit 1739. A farm in Kirbuster-tun. First element uncertain : perhaps O.N. *aust-topt*, ' east toft ' (site of former building). For *-topt* > *-tit* cf. Crantit, St. Ola.

Owendsatir: 1492. A 4½ pennyland or 'skatland' in tunship of Sands; never found on record afterwards. The first element is certainly a personal name : probably O.N. *Øyvinds-setr*, settlement of a man Øyvindr.

Rinnabout: 1739. A 1d. land in Waltness. Origin obscure

Sandgarth (saŋ·ər): Sandgarth 1492; Sanger 1601; Sangar 1739. Part of Sands tunship. Name self-explanatory.

Sands : Sandis 1601, 1614, 1642; Sands 1739. A half-urisland tunship in southeast of island. Though not mentioned itself in the 1492 R. the individual farms therein total up to exactly 9 pennylands (Morsettir with Thoraldisgarth 2⅝; Owendsatir 4½; Sandgarth 1⅞d. lands). The tunship name is still in use today for the district. The name must be a translation of O.N. *sandar* (pl. of *sandr*), sand.

Savaquoy : 1601, 1614. A quoy in Sound-tun. Perhaps O.N. *sœvar-kví*, 'seaquoy'; cf. Saviskaill (Rousay).

Shaltaquoy : 1614, 1642, 1739. A 2d. land in Meanes tunship. First element rather uncertain—probably a personal name. Cf. Sholtisquoy, N. Ry. which is probably of same origin.

Sound : 1492 and all succeeding rentals. A 2 urisland (36d. land) tunship of which seven-eighths was bishopric property. An interesting example of the use of the old technical term *skatland* is to be found in connection with this tunship in a Judicial Rental of Fea property which was compiled about 1761. In evidence it was stated by Magnus Williamson in Nearhouse in the town of Sound that "the town of Sound consists of eight Skattlands, each scatland being four and a half pennyland, and that each of the above eight scattlands has a Towmail belonging to it . . . ." (Kirkwall S.C. Record Room.)

The fine old name of this important tunship is now unfortunately obsolete, the lands thereof being mostly, if not wholly, incorporated in the large farm of Balfour Mains.

O.N. *sund*, a sound (strait between islands, etc.) : in this case the narrow passage between 'Sound' in Shapansay and Carness on the Mainland.

Skenstoft : Stangistoft 1492; Skennistoft 1612; Skennestoft 1614, 1642; Skeenstoft 1739. A farm at south-east corner of Veantro Bay, and probably included in the original Hollands-tun. The name is to be classed with Skennist, Papa Westray, q.v.; at each place there has evidently been a noust or shed for that type of longship known as a *skeið*; hence O.N. *skeiða(r)-naust-topt*, i.e. 'house-site or toft at a noust or shed for a *skeið*.'

This is an interesting type of name, with which may be compared Snaky Noust, a landing-place on the west shore of Rapness in Westray which perpetuates the memory of a station for another type of O.N. vessel—*snekkja* (a swift type of longship). Skippie Geo in Birsay and Skibbawick in St. Andrews recall berths for vessels of more indefinite type.

Sty : 1612. A 3d. land, now incorporated (at least partly) in the farm of Hilton, but the name probably survives in the present-day Steaquoy which is in the same neighbourhood. A very puzzling name; see further Stye (Deerness).

Swartaquoy : Muckle and Little S———, 1739. These two farms were in Over Holland, and a farm of this name exists still today. O.N. *svarta-kví*, 'black quoy' (perhaps so named from the colour of the soil as contrasted with neighbouring soil).

Trattletown : 1595; Trattletoun 1612. A cot in Welands-tun. The name survives today in The Banks of Traddleton which is applied to a stretch of land along the shore on the south side of Veantro Bay. An obscure name, but cf. Tratland (Rousay).

Iquiver (ek·wivər) : Vquever 1614; Oquiver 1627 (Peterkin, III, 28); Vquever 1642; Iquiver 1739. A 1½ pennyland in Ness; surviving as a field-name today. Origin obscure.

Vedesquoy (vɛd·ez—) : Vadasquoy 1595, (1642; Wadasquoy 1614. A small farm in Meanes-tun. First element uncertain, but probably a personal name; ? *Vaði*.

Weland (wi·lən[d]) : Weland 1492; Weyland 1595, 1612, 1642, 1739. The name is now practically obsolete. Formerly an 8 d. land tunship on south side of Veantro Bay.

This name is found twice again as a farm-name : (1) for an old 6 pennyland farm on the eastern border of the town of Kirkwall; (2) for a small croft in the north end of Egilsay. Origin doubtful; it most probably represents an O.N. *Víðiland*, from *víðir*, a kind of willow. All three of those farms are adjacent to wettish, low-lying ground where scrub-willow may well have been found at one time. In this Shapansay case, however, one might perhaps consider an alternative origin—O.N. *Véland*, from *vé*, a holy place, site of pagan worship, as on the beach here lies the so-called Stone of Odin.

Waltness (walt·nez) : Waltnes 1601; Weltnes 1642; Waltness 1739. An 8½d. land (1614), and probably an original half-urisland tunship. A rather large farm still today.

The *-ness* termination of the name is confusing and pretty certainly deceptive. Oddly enough the farm is near a headland, but that bears the name—not Waltness, but Saltness. A clue to the real origin of the name is to be found in connection with a North Ronaldsay farm—Ancum. In the 1653 Valuation of Orkney is an entry—" The Valtenes of Ankum," which in a rental of 1785 appears as—" Vaultnes of Anchum." There can be little doubt that this is the same word as the Shapansay Waltness. For the probable origin see Ancum (N. Ry.).

OTHER FARM NAMES.

Brecks : O.N. *brekkur*, pl. of *brekka*, a slope.

Bu-house : a farm in the former Welands-tun area. This name perhaps perpetuates that of the old head-house of the tunship, though no such name as The Bu of Weland is known to appear on record. For Bu see further Part III.

Erraby : near the point of Ness in the north of the island is a landing-place marked on the O.S. map—Noust of Erraby. No such farm exists there now, nor even appears in the old Rentals, but the name strongly suggests the existence of an original *-bae* settlement in this part of the island. The same name is to be found in Sanday and in Stronsay, and in neither case does it represent a present farm or a recorded farm. For discussion of *-bae* farms see Part III.

Gebro : a farm in Waltness area; origin obscure.

Gorn : or Gorn in Holland. Name now practically obsolete — the farm having been re-named Monquhanny by the proprietor of the island Col. Balfour about a century ago in memory of a place of that name in Fife from which the Balfour family sprang.

Hannatoft : a farm in the old Kirbuster tunship; not recorded in old Rentals. The termination is plain—O.N. *topt*, a house-site or steethe of old buildings. First element rather uncertain; *Hana-* is a common prefix in Norse names and is generally regarded as indicating a personal name or nickname—*Hani.*

Hestivald (hɛst·əval) : O.N. *hesta-vǫllr,* ' horse-field.'

Housebay (hus·be) : now a small farm on the fertile southern slope of the island; said to to be in the old tunship of Kirbuster. Though never mentioned in the old Rentals this probably represents the original main settlement in this part of the island. O.N. *húsa-bœr,* a farm-settlement somehow remarkable in its houses. Cf. Erraby. For *-bae* names see Part III.

Livaness (liv·ənez) : a farm near shore to east of Elwick farm and adjacent to a headland which juts out and shelters the famous old roadstead of Elwick from billows rolling in from the open eastern ocean. The name almost certainly represents O.N. *hlífar-nes,* ' ness of shelter,' from *hlíf,* shelter, protection, etc.

Odinstone : a relatively modern farm, so named after the famous Stone of Odin which lies on the beach below.

Quoymorhouse : a small farm in Sands district, perhaps a quoy-extension from the old Rental Mursetter which was in the same tunship.

Veantro (viəntro·) : a small farm on east shore of bay of same name, the origin of which is most obscure.

ROUSAY
(rou·ze)

Hrolfsey (O. Saga); Rollesay 1369 (R.E.O. 16); Rolusay or Rollisay (Fordun, c. 1375); Rolsay 1430 (Dip. Norv. xvii, 381); Rowsay 1500; Rousay 1595. O.N. *Hrólfs-ey*, 'Rolf's island.'

Rousay is a very hilly and compact island, roughly five miles in diameter, and its arable lands are stretched out around the coast-line in a not-quite-continuous belt, for the most part about half a mile in width. Its early skattable value was probably exactly 6½ urislands.

It is not included in the Rental of 1492, and for early forms of its farm-names we are mainly dependent on the Rentals of 1500 and 1595 together with a few sasine and other law deeds.

RENTAL NAMES

Avaldsay (av·əlse) : Awaldschaw 1500 and 1595. A 4d. land; part of tunship of Knarston. A considerable farm still today.
O.N. *Augvalds-haugr*, Augvald's-mound or cairn. For the phonetic development of *-haugr* in this name cf. Ramsay (Westray) and Horraldsay (Firth).

Banks : id. 1595. Two farms of this name; one an old 3d. land in Sourin, the other a small croft in Frotoft. O.N. *bakkar*, pl. 'steepish slopes.'

Bigland : id. 1595. A former 3d. land farm in Sourin; still a farm today. Most probably O.N. *bygg-land*, 'bere land,' from *bygg* = bigg, a kind of barley. The recent discovery on this farm of the very important prehistoric settlement of Rinyo, which indicates human occupation from a very early date, might suggest, however, that in the name Bigland we have not O.N. *bygg* (bere) but O.N. *bygð*, in the sense of a place already 'settled on.'

Breck : Brek 1500. An old ¾d. land in Whome; now extinct. O.N. *brekka*, a slope.

Brendale (brɛndil·) : Brindaill 1563 Ch.; Brendale 1595. A farm in Sourin; a former 3d. land. Pronunciation and topography combine in proving the termination of this name to be quite misleading. In Rousay true -dale endings (from O.N. *dalr*, valley) are pronounced -dəl, e.g. Swandale (swan·dəl), Quandale (kwan·dəl), whereas in this and in the adjacent farm name Ervadale the pronunciation is -dil. Nor is either farm situated in a dale at all : both lie side by side on the swelling breast of Kearfea Hill. Each was also a former 3d. land, as were five other farms in this Sourin area, a circumstance that plainly suggests constituent portions of a larger divided unit. One has therefore the more confidence in regard-the -dale ending in these two farm-names as O.N. *deild*, a share, portion, divisional part. Hence O.N. *brenn(u)-deild*, 'portion of land cleared by burning.'

Broland (bro·lən) : Brewland 1562; Browland 1595. A farm in Sourin; a former 1d. land. Origin rather doubtful; it lies on a pronounced 'brow' on a hill-slope, and may well be O.N. *brún-land*, from *brún*, which was applied to a 'brow' of exactly this kind. The disappearance of 'n' is however curious.

Brough (brɔχ): Brugh 1503, and so regularly in 16th and 17th century documents. The famous old homestead of the Craigies of Brugh, three generations of whom provided a lawman of Orkney. O.N. *borg*, a fortification. Though the old farm is now part of the huge farm of Westness, the ruinous old buildings still stand and are adjacent to the remains of a broch.

Cogar: Calgair 1500. A small farm in Wasbister—a former 1½d. land. Cf. Keigar (Deerness).

Corse: id. 1500 and 1595. A former 2d. land. O.N. *Kross*, cross, appears frequently in place-names and may have been applied for various reasons: (a) a wayside cross marking a praying site (e.g. at a point where a church came into view); (b) a burial site; (c) cross-roads; (d) ground pertaining to a church or chapel dedicated to the Holy Cross; (e) property boundary, etc. Reason for name here unknown.

Eastafea (ist·afi): id. 1595. A former 1¼d. land now incorporated with Faraclett. Self-explanatory—'east hill.'

Ervadale (ɛrv·ədil): Ovirdaill 1563; Overdale 1595. O.N. *efra-deild*, 'upper portion'; see further Brendɐ'

Essaquoy (es·əkwi): Ossaque 1563; Ossaquoy 1595. In each case initial 'O' is almost certainly due to a common scribal error in reading 'o' for 'e.' A former ½d. land in Sourin. First element obscure, but probably a personal name.

Faraclett (far·əklɛt): this, though the official form, is rarely if ever heard locally, the form currently used being Faraclee (far·əkli). Ferraclott 1563; Farraclet 1595. In a Births Register as early as 1739, however, is to be found the form Faraclay. A big farm still today and a former 3d. land. This is a difficult name which must be considered in relation to several other Orkney names.

1. Farraval (Westray): Ferval 1740; Fervail 1794. A farm at base of a hill rising about a mile west of Pierowall.
2. Farafield (far·əfil): a farm at base of Skelday Hill in Birsay.
3. Farabreck: an old house-site above Kirbest in Egilsay.
4. Faravill: one of the 'slaps' or gateways in the tunship dyke of Redland in Firth.
5. Farewell: a farm in South Ronaldsay at eastern base of a hill known as The Wart. This farm-name is now pronounced like the English word so spelt, but old people knew it as Farawel (far·əwəl).

In all these cases there is a presumption that the Fara- element represents O.N. *varði* or *varða*, a beacon. For the initial change cf. Fitty Hill in Westray where there can be no doubt that Fitty represents O.N. *viti* (a beacon also), and Fitful Head in Shetland where the Fitful undoubtedly represents O.N. *vita-fjall*, beacon-hill. The spelling of that name being assimilated to the Eng. word *fitful* has misled even Jakobsen to regard it as an O.N. *fitfugl*, web-footed bird, forgetful of the fact that every sea-cliff is the resort of such birds and that such a name would be meaningless. Fitful Head in fact is the first land to be sighted in Shetland when travelling from Orkney, and would be the obvious site on which to light a beacon to warn first Fair Isle and then Orkney.

As noted above, Farewell in South Ronaldsay is at the base of a Ward Hill—" The Wart." Faravill is in Redland tunship which lies at the eastern base of The Ward of Redland—another Ward (or beacon) Hill. Farabreck in Egilsay is within a few hundred yards of the farmhouse of Warset on the ridge of the island where—as the name shows—(*varð[a]-setr*)—there has also been a beacon site.

The hill at the base of which is the farm of Farraval in Westray is not known to have been a beacon hill, but, as it overlooks Pierowall (the old Hǫfn of the Saga) from which a beacon on Fitty Hill would be invisible by reason of two intervening hills, it is at least probable that it also was a beacon site. Nothing is known as to whether the hill above Farrafield in Birsay was a beacon hill.

This farm at Faraclett in Rousay lies on the steep southern slopes of an extensive elevation now called The Head of Faraclett, but formerly known as The North Hill, which rises to a height of over 300 feet and is the nearest of the Rousay hills to Westray. As the Ward Hill of Rousay is on the west side of the island, and thus invisible in the Wasbister district as well as most of the large Sourin district in which Faraclett is situated, it is very likely that this hill (visible in both the districts mentioned) would have been a beacon site also.

If then a *varði* origin can be assumed for the first element in all the above names, Farraval, Farafield, Faravill and Farewell might all represent O.N. *varða-fjall*, beacon-hill, or less probably *varða-vǫllr*, beacon[hill]-field. Farabreck would be *varða-brekka*, beacon-slope, and Faraclett *varða-klettr*, beacon-rock. The variant Faraclee might represent *varða-kleif;* O.N. *kleif* becomes klivvy in Orkney, applied to a steep path or track up the face of a crag or hill.

Against such an interpretation of the element Fara- it must be pointed out that the ' v ' of *varði* normally becomes ' w ' in Orkney as found in the numerous Ward Hills, Warsetter, etc. In view of that very serious objection the above-suggested interpretation of Fara- can as yet be offered only tentatively. It is noteworthy, however, that in Shetland the ' v ' of *varði* (or *varða*) is usually retained and not altered to ' w,' but, according to Jakobsen, when appearing as second element of a compound the ' v ' in *varða* often appears as ' f ' : e.g. Gamblavird or Gamblafirt, Nunsvird or Nunsfirt, etc.

It is interesting to note that Vørðuklettar occurs as a place-name in Færoe.

Frotoft (frot'et) : Frowtoft 1500; Frotoft 1595. A former half-urisland tunship; today the name is used for a district but not any farm. For the original name Munch suggested *Freys-tupt*, indicating the site of a temple to the Norse god Freyr. The phonetics, however, forbid such an explanation : more probably O.N. *Fróða-topt*, house site or homestead of a man Fróði.

Furse : Forse 1503. A farm in Wasbister; a former 1d. land. O.N. *fors*, a force or waterfall.

Grudwik : 1500. An old 1d. land in Wasbister. Both farm and name are now extinct, but it evidently lay near the Bay of Grithin, a narrow bight where the steep beach is composed of a mass of huge boulders. O.N. *grjót-vik*, ' stony wick or bay.'

Hammer : there were two places so-named—one a 4½d. land in Wasbister appearing in 1500 R. as Hammyir, the other a 1d. land in what is now

Westness—Hamer 1595. The latter is certainly O.N. *hamarr*, a projecting rock on a hillside or slope; but the early spelling of the former suggests a different origin which must be regarded as uncertain.

Hunclet (hʌnˑklɛt, haŋˑklɛt): this form, however, is never used locally unless one is trying to "speak proper," i.e. according to the spelling. The almost invariable local pronunciation is Hooklid (hukˑled) or Hooklet (hukˑlet). The earliest recorded spelling is Howclet, 1500 R., but the 1595 R. has Hunclet.

This name also occurs in Holm where a 3d. land (quoyland) is spelt in the 1492 R. Hundclett, in 1500 R. Hunclet, and 1595 R. Hunclet.

Though in the earliest recorded form of this Rousay name the 'n' is missing there can be little doubt that it was originally present, and that the first element of the name was O.N. *hund-*. The second element is of course O.N. *klettr*, a rock, but such a compound as 'dog-rock' for a farm-name is hardly likely.

Hund-, indeed, as an element in Norwegian place-names presents great difficulty to Norse scholars who have sought to explain it in various ways. And as the element appears also in another Orkney farm-name —Hundland, as well as in a Rousay crag-name—Humber, the actual source of all such names must be deemed uncertain.

In the early Rentals this tunship was divided into a 3d. land of Ovir Hunclet and a 4d. land of Nether Hunclet. It is probable that originally the 2d. land of Trumland was also included—the three making up a 9d. land or half-urisland.

Hurtiso (hʌrtˑəso): Hurteso 1595. A farm in Sourin formerly skatted as a ½d. land. See this name—Holm.

Knarston (narˑstən): Knarstane and Knerstane 1500; Knarstane 1595. Formerly a half-urisland tunship, and a moderate-sized farm today. This Rousay name appears to be the only example of an old *-staðir* in the North Isles of Orkney. Cf. Knarston (St. Ola) and Knarston (Harray).

Lairo (lɛˑəro, lɛrˑo): Liaro, 1627 Sas.; Lyaro, 1665 Sas. No longer a farm but only a house adjacent to Burn of Hullion in Frotoft. In 17th century it was a ¼d. udal land. Origin uncertain, but it seems to have no connection with the Shapansay Lairo. The banks of the adjacent burn are definitely clayey, and one might therefore think of an O.N. *leir-á*, 'clay burn.' But the old spellings would rather seem to suggest an O.N. *hlíðar-haugr*, 'slope mound,' the reference being to a large mound closeby which is one of the only three so-called "Horned Cairns" that have been noted in Orkney. The hill slopes up very steeply here.

Langskaill (laŋˑskil, –skjil): Langscale 1500; Langskail 1595. A large farm today and a former half-urisland by itself. O.N. *langi-skáli*, 'long hall.'

Mithvie (mɪþˑvi): Midfea 1595. An old 1¼d. land, now merely a field name on Faraclett farm. O.N. *mið-fjall*, 'mid-hill.'

Neo (nfˑo): now merely a field on farm on Faraclett, but it is the adjectival survivor of the name of a 1d. land farm (or at least quoyland) which appears in the 1595 R. as Quoynania, i.e. O.N. *kvín-nýja* (or some oblique case thereof), 'the new quoy.' For other interesting examples of similar adjectival relics cf. Mugly, and probably Digro, *infra*.

Pow (pɔu): id. 1595. A former ½d. land. Pow is a Scots form, but another Pow, in Sandwich, appears in the 1500 R. as Poll. No doubt this Rousay name is of similar origin—O.N. *pollr*, pool.

Quandal (kwan·dəl): Quendale 1500; Quendall 1595. An old 3d. land, now part of the sheepwalk of Westness, and uninhabited.

O.N. *kvern-dalr*, 'quern-valley'; a name applied to a scooped out valley or hollow bearing some imagined resemblance to the hollow of an old saddle-quern.

Quoyostray (kwi-ɔst·ri): Quy ostir 1500. Farm in Wasbister—a former 1d. land. O.N. *kví-aust*[a]*ri*, 'eastern quoy.'

Saviskaill (sɛv·əskil, —skjil): Savirscale 1500. A large farm in Wasbister—a former 3d. land. O.N. *sævar-skáli*, 'hall or house by the sea.' The houses stand at the brink of the shore.

Scockness (skɔk·nez): Skowkness 1549 (Craven); Skoknes 1576 (R.E.O.); Scoknes 1595. A peninsular farm in Sourin district; a former half-urisland or 9d. land.

First element of this name of doubtful origin, but most probably referable to O.N. *skógr*, forest (= Eng. *shaw*). The change of 'g' to 'k' is however puzzling, and the retention of the guttural at all even more so. The same element would seem to appear in the South Wales island —Skokholm. In Charles's *Non-Celtic Place-Names in Wales* early forms of that name are recorded, the earliest being Scogholm (1219-31). As early as 1275, however, Stokholm is found, and thereafter recurs along with Scugholm (1276) and Scokholm, etc. The problem is to decide which early form is the more authentic, and in view of the earliest recorded, viz. Scogholm, it would seem arbitrary not to regard the name as deriving from O.N. *skógr* as suggested above, though Mr. Charles does not do so. Other old Rousay place-names such as Skooan and Skooany tend to confirm the existence of brushwood at least in Rousay in early Norse days.

Skaill (skil, skjil): Skaile 1595. A former 5d. land in Outer Westness; now part of Westness farm. Here stood the old parish church of Rousay, and there can be no doubt that this was the seat of the Saga chieftain—Sigurd of Westness. O.N. *skáli*, a hall.

Swandale (swan·dəl): Swindale 1595. A small farm in Sourin; an old 3d. land. *Swan-* in Orkney place-names—e.g. Swanney, Swanbuster—is a somewhat puzzling element and is derived probably from different sources. Here, however, the old spelling indicates O.N. *svín-dalr*, 'swine-valley.'

Trumland (trʌm·lən[d]): Trymland 1500; Trumland 1595. A large farm today; an old 2d. land.

This is a rather difficult name which I discussed at length in *Place-Names of Rousay*. Its arable lands are not continuous with those of the neighbouring old Hunclett tunship, but its 2d. land would fit in with the 3d. and 4d. lands of Over and Nether Hunclett to make a half-urisland tunship. I cited record evidence also which suggested that all three were at one time parts of the same udal property.

Hence probably O.N. *trǫm-land*, 'fringe land,' or land on outskirts of tunship; from *trǫmr*, an edge, brim, verge, etc., the same word originally as Eng. *thrums*.

Wasbister (waz·bəstər) and colloquially Wazder (waz·dər). Wasbustar 1500 (and -er); Wosbustar, -er, 1595. A former urisland, but today Langskaill, a former half-urisland, is also reckoned in the Wasbister district. There is no *farm* so called today, nor apparently was there even in 1500: it was merely a tunship name. O.N. *vaz-bólstaðr*, 'loch-bister' or lake-b—— (farm settlement); so named from the *vatn* or lake in the middle of the district.

Westness: Vestnes, O. Saga. Vtterwestnes and Innerwestnes 1500. Under this name was included a wide sweep of country on the west side of Rousay—1½ urislands in all. Here was the estate of Sigurd of Westness in the 12th century, his home being at Skaill in Outer Westness as indicated above. This unit embraced such an extensive sweep of territory that it could hardly ever have been a single tunship. See further Part III.

Whome (hwom): Quham 1500 and 1595. An old 3d. land now incorporated in Westness farm. 2¼d. lands of this property, it is stated in the 1500 R., had belonged to a former claimant for the Earldom of Orkney—Sir Malise Sperra. See this name—Stromness.

Woo (u :) : Owe 1595. A farm on banks of the Suso Burn in Sourin; a former 3d. land. O.N. *á*, a stream or river. Woo is a fairly common name in Orkney.

OTHER FARM-NAMES

Breckan: farm in Wasbister. O.N. *brekkan*, 'the slope.'

Breek (brik): a vanished croft in Quandal. Origin obscure.

Breval (brɛˑvəl): small croft far up hillside in Sourin. Origin uncertain; cf. Brawel (Shapansay)

Classiquoy: small croft in hill above Avaldsay. A quoy name, but first element rather doubtful; most probably a personal name—No. *Klas*, a dimin. form of Nikolas which is recorded in N.G. as appearing in the names Klasmoen and Klasnes.

Clyver (klai·vər): old vanished cot at foot of a steep rocky hill slope on Westside. O.N. *kleifar*, pl. of *kleif*, a pathway running up a steep ascent.

Cott: a name formerly used of three small crofts or cot-houses. O.N. *kot*, cot-house.

Cruannie (krua·ni): small croft on hillside in Sourin. O.N. *kró* or *krú* (a loan-word from O. Celtic *cró*, a sheepfold or enclosure for animals); a very common element in Orkney place-names, and also a generic term (krø) for a small pen or enclosure. Here used probably in plural with suffixed def. article—*krúarnir*.

Cruar (kruˑər): small croft in old Knarston tunship. See previous word. Here in plural—*krúar*, pens, enclosures for animals, etc.

Curquoy (kʌr·kwi): Croft far up Sourin valley. Not recorded in early Rentals, and probably a later 'intake' from common. In that case it may well be a place named after another Curquoy (in neighbouring parish of Evie) q.v.

Cut-claws (kʌt-kla·z) : a vanished croft on Westside. Coatclaws in Birth Reg. of 1738. Despite the suggestion of a cat's claws this is most probably an old cot name with the personal name suffixed : hence *kot-klas*, cot of a man Klas (dimin. of Nikolas). Cf. Classiquoy; and for suffixing of owner's name another Rousay place-name may be cited—Cot-Mowat.

Dale (dil) : Deal, Birth Reg. 1816; an old croft in Quandal. O.N. *dalr*, a dale : cf. Rousay pronunciation of pail as peel, kale (keel), tale (teel), etc. This farm lies in a definite dale.

Digro (dɪgˑro) : small hill-croft in Sourin. Almost certainly an adjectival survivor of a compound name of which the substantive has been dropped; cf. Mugly and Neo. O.N. *digr*, stout, thick, substantial. What the lost noun may have been is difficult to suggest : ?*kvi*; O.N. *kvin digra*?

Falldown (fadun·) : small croft in Sourin lying at the foot of an abrupt and precipitous slope. One is disposed to regard this name as a parallel to a class of Norwegian place-names for which Rygh (*Indledning*, p. 19) suggests the designation " Imperative Names." Examples thereof are : *Vendom* (Turn round), *Kikut* (Peep out), *Sitpaa* (Sit upon), *Søkkned* (Sink down), etc.

Falquoy (fjal·kwi) : a farm in Wasbister. O.N. *fjall-kvi*, ' hill quoy '; a rather common Orkney name.

Finyo (fɪnˑjo) : a cot-house in a field of same name on the farm of Banks, Sourin. In Birth Reg. Housefinzie, 1738, 1745, 1819; Housefinian 1744, 1745, 1821; Housefinzean 1816, 1823, 1830, etc. The same name Finyo is found in Stronsay.

An interesting name—to be compared with another Rousay field-name —Brae-an-finyan—on farm of Faraclett. The upper part of that field is very steep—a pronounced ' brae.' Many years ago in P.S.A. Scot. ix., I suggested a possible Celtic origin for that name, but experience has made me more sceptical of Celtic names in Orkney. I would now suggest that the *Brae-* in this name is simply the Scots *brae* (a declivity) and that the whole name signifies a ' brae in (or on) Finyan.' Immediately south of this field the present public road separates it from a large field (on another farm) known as The South Meadows. If from that name we might deduce that there were also *north* meadows, these could be no other than the lower slopes of Brae-an-Finyan. One is thus tempted to think of O.N. *vin*, a term for pasture land, which already in Viking times is supposed to have become obsolete or at least obsolescent. It is however very common as a farm-name element in Norway, and, with the definite article attached, appears sometimes in the form Vinnan (Rygh's *Indledning*, 85), which may be the term occurring here as Finian. In that case the original ' V ' must have become unvoiced as it has e.g. in Fitty Hill (Westray)—an old *viti-* or beacon-hill.

It is probably the same term *vin* that appears in Finyo: cf. the Norwegian place-name *Vinju* (dat. sing.); see Rygh *op. cit*.

Garson : a vanished farm—now included in Westness farm. O.N. *garðs-endi*, ' dyke-end.'

Goarhouse (gjoˑər-hus, djoˑər-hus) : a croft in Sourin adjacent to the Goard gjoˑərd) of Banks. *Goard* is a name applied to at least four stretches of pasture in Rousay. No. *gjorde*, meadow-field, piece of enclosed pasture.

Grindally: Grindilla 1734 B.R.; Grindally 1832 B.R. A vanished house on farm of Knarston. See this name—Sanday.

Grinlaysbreck (grɪn·lez·brɛk) : small croft (now deserted) in Sourin. O.N. *grind-hliðs-brekka*, 'gate-slope'; the grind or gate in question would have been on the old hill-dike which passed close by.

Hamri-field : croft in Wasbister on face of hill. O.N. *hamra-fjall*, 'hammer-hill,' i.e. a hill where projecting rocks just out; O.N. *hamarr*.

Hanover (haino·vər) : croft in Sourin; spelling probably due to confusion with the German name. Origin uncertain; my former suggestion in *Place-Names of Rousay* being untenable.

Hestival : vanished croft in Westness. See this name—Shapansay.

Housteith : 1619 Sasine. A vanished name for some part of the present farm of Trumland. Evidently O.N. *hús-stœði*, 'house steethe,' site of house.

Howatoft : Charter of 1625 and 1634 etc. A vanished piece of udal land somewhere in Wasbister. O.N. *haug(a)-topt*, 'house-foundations at a mound or near mounds.'

Hullion (hʌl·jən) : a small farm in Frotoft, and from its situation pretty certainly the head-house of that old tunship. Phonetically it might well represent O.N. *hǫllin*, 'the hall,' but, if so, would be the only known example in Orkney of that word. It is interesting to note that this name is also found applied to a sunken rock on the coast at Aukengill in Caithness—" The Hullion " : (Old Lore Misc. ix, 155).

Husabae (hus·əbe) : a group of fields on the present farm of Hurtiso in Sourin. Near the highest point of this area of land is, or was until recently, a knoll called The Taft of Husabae, i.e. old house-site of H——. O.N. *húsa-bœr*, a farm-settlement or *bae* signalised somehow by its houses.

No farm or tunship of that name is recorded in the early Rentals of Rousay, but surrounding this area of very fertile ground lies a whole group of small farms, no fewer than seven of which appear in the 1595 R. and a charter of 1563—each skatted as a 3d. land ! Such a remarkable collocation can have only one reasonable explanation—namely that here we are face to face with an older, larger unit or tunship that had at some earlier date been divided up. And that Husabae, in view of its name and position, was the name of that large earlier settlement, is a conclusion hardly admitting of doubt. For Bae names, see further Part III.

Innister : not recorded in Rentals, but often in old charters : Ingisgarth 1606; Ingisgar 1624, 1627, etc.; Ingsgar 1633 to 1825; Inisgar 1816; Innesgair 1825. In a Birth Reg. of 1834 is recorded a birth of a daughter to James Inksater or Inisgar. Today the local surname has been stabilised as Inkster and the farm name as Innister.

First element a personal name—perhaps *Ingi* : though the old genitive was *Inga-*, in the later Middle Ages older genitives were sometimes assimilated to the -s form, e.g. the Sanday farm Hermisgarth though spelt Hardmundigarth in the 1595 R. was already in the 1500 R. Harmannisgarth; O.N. *Hermundar-*. Hence probably O.N. *Inga-garðr*, 'Ingi's farm.'

Mugly (mʌg·li) : 1. now merely a field-name on Swandale; 2. a field on Faraclett. An interesting example of an adjectival survival. In a sasine of 1665 one of the witnesses was a " Magnus Ketle in Queenamigle in

Swendaill.' That cot house has vanished, but the field name still testifies to the site. O.N. *kvín-mikla* (or some oblique case of same) 'the muckle or large quoy.' In a neighbouring case—Hammermugly—the noun has been retained. Cf. Neo and Digro.

Nears (njɛrs): a farm in Frotoft district—often spelt mistakenly as Nearhouse. It is a contraction of Nether-house (in this case of the old Hunclett tunship).

Outerdykes: a corruption of the older name Out-o-dykes, B. Reg. 1819, 1822 etc.; pronounced sometimes yet ut·ədaiks. A deserted croft in Sourin immediately outside the old hill-dike. O.N. *utan-garðs* was the regular term for any ground outside the wall or garth enclosing the cultivated lands of a farm settlement, and 'Out o' Dykes' is a Scotticising thereof. O.N. *garðs*, however, is a genitive sing. governed by *utan*, and the plural —dykes—of the translation is rather interesting as due to a misunderstanding of the significance of the '-s' in *-garðs*.

Quoys (kwaiz): a small farm in Wasbister, and a smaller croft in Sourin. O.N. *kvíar*, 'quoys'; see further Part III.

Skaebrae (ske·bre): a vanished farm, the lands of which are now included in Westness. An old 5d. land. A bold, picturesque headland adjacent to the old farm still goes by the name of Skaebrae Head. Same name as the Sanday Skelbrae, q.v.

Sketquoy (sket·wi): the most outlying farm in Wasbister, now incorporating the old farm of Stennisgorn, and probably the 1503 Grudwik also. In a charter of 1608 the name appears as Skataquoy, and in a 17th century charter as Skitquoy. The same name occurs in Tankerness. First element doubtful; as a rule-quoy farms were unskatted, but some *were* skatted. This may possibly have been a skat-paying quoy, but the 17th century form renders any connection with skat doubtful.

Skooan (sku·ən): Scowan 1738 B. Reg., 1820, 1822; Quoy Skows 1738 B. Reg. Both names certainly used, as the same witness at two baptisms in 1738 is described once as 'in Scowan,' and again as 'in Quoy Skows.' The name derives from O.N. *skógr*, forest, copse, etc., which in Orkney, as in Iceland, must have been applied to denote low scrub or copsewood. Cf. Shetland Skuen brenda, and Scockness in this island.

Sourin (sou·rin): name now used for a large district on east side of Rousay, and not strictly a farm or tunship name at all. The earliest record of the name known to me is in a Kirkwall merchant's account book where in 1724 it appears as Sourin, and in 1733 as Souring. But an earlier name for the district (or at least part of it) was Sorwick. In Gilbert Grote's Protocol Book there is a reference to the lands of *Sorweik* from the year 1563, and the 1595 R. refers to the Mill of *Sorwick* (misprinted by Peterkin *Serwick*). That name is frequently on record thereafter, but it became obsolete in the early 19th century. The lands of Sorwick included those already referred to under Husabae.

The two forms of the name are puzzling, but both are pretty certainly derived from O.N. *saurr*, mud, filth, mire, etc., and parallel forms occur in Norway. With Sourin may be compared Sauren (Sourenn 1567, Souren 1723) which in N.G. xvi, p. 27, is explained as O.N. *saurr*, with def. art.— *Saurr-inn*, and so applied by reason of the miry nature of the soil. And the exact equivalent of Sorwick may be found in *Survik* (N.G. xv, 330).

The lower part of the Sourin valley near where the Suso Burn issues into the bay is even yet very wet and boggy.

Stennisgorn : Stannesgair 1565; Stainsger 1570; Stennisgar 1578; Stennisgair U.B. 1601; Stennisgarth and Stennesgorne 1624; Stennisgoird 1629; Stenhousegorne 1664. Old farm in Wasbister now included in Sketquoy. O.N. *steins-garð[r]inn*, 'the stone garth,' i.e. 'the farm at a standing stone.' The same name (without def. art.) occurs in Birsay where a farm at a famous surviving standing stone is called Stanger (sten·ʒər), q.v. There is no trace of such a stone at Stennisgorn today. See further Goir (Sanday).

Swartifield : a small croft on hillside in Sourin. O.N. *svarta-fell* (or *-fjall*), 'black hill.'

Tafts : a deserted farm in Quandal. The old house of Tafts, once a house of real distinction, has been described in detail in P.O.A.S. II by Mr. Storer Clouston who regarded it as the oldest two-storeyed country dwelling-house in Orkney. See Tafts (Sanday).

Too (tu :) : small croft in Wasbister. Tow 1567, 1624, 1640; Quoytowe 1721. Prob. O.N. *tó*, a grassy plot, patch of grass among rocks or scrub, and a common Norwegian farm-name.

Tratland : farm immediately west of Hunclett. O.N. *præt(u)-land*, 'dispute-land,' from *præta*, a quarrel, litigation, etc. This farm is not mentioned in the old Rentals, but it is situated between the old tunships of Hunclett and Frotoft and probably formed a bone of contention between the two.

EGILSAY

Egilsey, *O. Saga.* On the face of it this seems to be plain enough—the isle of a man Egill. Munch, however (in *Annaler fer Nord. Oldk.* 1860) was strongly of opinion that though the name was no doubt so understood by the Saga writer and others of that time the first element was really an early Celtic term for a church—Gaelic *eaglais* (E. Irish *eclais*, from Latin *ecclesia*), and that the name was to be interpreted as 'Church Island.'

Munch, however, was under the impression that the famous old round-towered church in Egilsay was of pre-Norse date, and as that view is no longer tenable his suggested origin of the name loses much of its validity or plausibility. Furthermore, in Shetland there is a small island bearing the same name, but without any record of a church thereon. One fact, however, which apparently escaped Munch's eagle eye, may lend considerable support to his conjecture. At the north end of Egilsay, and separated from it by only a very narrow neck of water, lies a small holm called Kili Holm. There is no knowledge as to the possible existence of a chapel there at any time, but the name can hardly be explained from any Norse source, and the Gaelic *ceall* (genitive *cille*), a cell, at once springs to mind. And it is just the sort of *diseart* or 'solitude' that was sought out by early Celtic anchorites.

In these circumstances the origin of the name Egilsay must still be deemed an open question.

Egilsay is a small carrot-shaped island about three miles in length by rather over one mile in greatest breadth, and its early skattable value, though not precisely known, was probably two urislands. It does not figure in either of the earliest Rentals, having been church land apparently from the earliest days of the bishopric: the first Bishop—William the Old—is referred to repeatedly in the Saga as being in residence there. By 1595, however, it had been feued to Patrick Monteith, and from that family it passd by marriage in the following century to the Douglases, from whom in turn it passed (again by marriage) in the 18th century to the Baikies of Tankerness who own the island still today.

Culdigo (kul[d]·igjo) : a small deserted croft. Origin of name quite obscure; cf. however Cultisgew (Shapansay).

Cott : a small croft. O.N. *kot*, a cot.

Feelie-ha' : a small croft. Origin rather uncertain. The same name occurs in Rousay, and dates probably from a post-Norse period, the termination being Scots *ha'* (hall) which is found frequently applied in Orkney (especially in Eday) in a depreciatory or derisive sense to cots and similar humble abodes. The first element is thus probably = Scots *fail* in fail-dyke (an earthen or turf wall) which in Orkney is regularly—*feelie-dyke*. If such is the origin the name would seem to imply a house built, partly at least, of turf.

Howe : one of the larger farms. O.N. *haugr*, a mound. It is rather uncertain whether the name was applied in reference to an old burial-mound on the farm (near the beach), or to the near-by hillock or high ground of Howe Banks abutting on a narrow channel or sound known as Howa (hε·ui).

Kirbest (kɪr·bəst) : a fairly large and good old farm. O.N. *kirkju-bólstaðr*, 'kirk-bister,' kirk-farm. The 'kirk' in this name cannot be traced today; but it must refer to a kirk or chapel dating from a time previous to the date of the present St. Magnus Kirk, which moreover is not on the farm of Kirbest at all. See Part III for further remarks on Kirbuster names.

Onziebust (ɔn·je·bɪst) : Unzeabister, Unzebister (Commis. Rec. 1609). A biggish farm on the south point of the island. An O.N. *-bólstaðr* name, but the first element is of quite uncertain origin. This name occurs once again in Orkney, and by a very curious coincidence (if it be such) the other Onziebust is on the eastern point of the neighbouring island of Wyre immediately opposite the first, so that these two farms are so-to-speak adjacent except for the intervening mile of sea.

Cf. also Unye-tuo, the name of a hillock in Shapansay.

Skaill (skjil) : two adjacent small farms close by the main old landing-place and present pier are named respectively Midskaill and Netherskaill, indicating undoubtedly a once larger undivided unit—Skaill. O.N. *skáli*, a hall.

Sound (sund) : most northerly farm in island, and adjacent to the so-called Smithy Sound between Egilsay and Kili Holm. O.N. *sund*, a sound or strait.

Tofts : two small adjacent farms, one on each side of St Magnus Kirk, are named—North Tafts and South Tafts. See this name—Sanday.

Vaday (va:de) : a small seaside croft with a puzzling name. Close by the houses of Vaday is a narrow pebbly isthmus, very little if any above high water mark, connecting the island with an outlying little headland, inside of which a bay so formed is called the Bay of Vaday. Either of two O.N. words might account for the Va- in Vaday—*vágr*, a bay, and *vað*, a ford or wading-place (in Orkney dialect usually *waddy*). The most probable explanation is that the isthmus referred to had at times to be waded across, and that the name represents O.N. *vað-eið*, 'wading-isthmus.'

Warset (war·sɛt) : croft near highest spot on island—evidently the old beacon-site. O.N. *varó-setr*, beacon-settlement. Cf. Warsetter, Sanday.

Watten : croft in north end, close by a loch. O.N. *vatn*, lake or loch.

Weland : croft in north end of island. See this name—Shapansay.

Whistlebare : small croft near Skaill. See this name—Sanday.

Whiteclett : small croft on east side. See CLEAT, Sanday.

WYRE

Vigr, *O. Saga*; Were, c. 1375 (Fordun); Wyir, 1500 R.; Wyer, 1595 R. The Saga name—O.N. *vigr*—meant a spearhead, and from its resemblance in shape to that the island no doubt had its name. It is not impossible, of course (though rather unlikely) that the original name-giver had in mind an island of the same name off the west coast of Norway (in Sunnmøre), or even hailed from there himself. But there is a third island of this name off the west coast of Iceland.

It may be noted that from the time of Geo. Buchanan who Latinised the name to Viera, various writers and even map-makers adopted that corrupt form which it is to be hoped has now died an unlamented death.

This is a small island, only about two miles in length, and it was skatted as a 12d. land. But it is classic ground. Here was the residence of that famous 12th century personality Kolbein Hruga, and here no doubt grew up his even more famous son, the future Bishop Bjarni Kolbeinsson, poet and statesman as well as priest. Here too may still be seen the ruins of the "good stone castle" which Kolbein built, and which withstood a rather remarkable siege in 1231, as narrated in the Hakon Saga. Near by also still stand the walls of a roofless 12th century church.

RENTAL NAMES

Bu : "The Bu of Wyre": this farm is adjacent to the old church and castle (Cubbie Roo's Castle, as it is locally called), and here without any doubt was the homestead of Kolbein Hruga 800 years ago. For the name Bu see Part III.

Castle-hoan (—hoːn): a farm to south of castle; name wrongly entered on O.S. map as Castlehall. The second element of name is of uncertain origin.

Cavit (kev·ət, kjev·ət, -əd) : a small farm in south-west of island. Origin of name obscure.

Habreck : adjacent farm to last. First element uncertain; perhaps O.N. *haga-brekka*, 'pasture slope.'

Helyie (hɛl·ji) : this is the local form, but the more official name is Helziegitha (hɛl·ji-gīðə, -gīþə). A curiously interesting but puzzling name. The farm houses stand about 500 yards or so east of the old church, but one is strongly tempted to associate them therewith. The first element Helyie- can hardly be other than the O.N. *helgi(a)*, the weak form of *heilagr*, holy, and in the second—githa—one suspects some case of *garðr*, garth or wall or enclosure, etc., so that the compound would indicate 'holy garth'—whatever that might denote. The only parallel in Orkney place-names known to me of -githa is in the curious Rousay name Handimidgathy, which is applied to a very precipitous hill slope. In the *Place-Names of Rousay* I suggested that name indicated the 'hanging (slope) between the dykes,' i.e. between the Sourin and Wasbister hill-dykes (the first element representing an O.N. **hangð*, steep overhanging slope, a term assumed as the origin of several Norwegian names, e.g. Hande or Hanne; and the second element some case (probably dative sing.) of O.N. *miðgarðr*, 'mid-garth.').

Onziebust : see this name—Egilsay.

ORKNEY FARM-NAMES

Rusness (rʌsˑnez) : a small farm next to Onziebust. See this name—Sanday.

Testaquoy : small farm near west point of island. The same name is applied to a croft in Rapness, Westray. It is another -quoy (O.N. *kví*) *name*, but the first element is rather doubtful. In both Shetland and Faeroe *Test*- occurs as a first element in place-names, and is regularly explained as O.N. *þeist*, a bird (black guillemot). Here, however, that source has to be ruled out for two reasons : 1. Such a term would be quite irrelevant in a farm-name; 2. the island here is quite flat, with no cliffs such as these birds frequent. In Testaquoy the first element is almost certainly a personal name—probably O.N. *Teitr*, gen. *Teits*-; as various forms of that genitive Lind cites Teiz, -zt, -st.

GAIRSAY

Gareksey (O. Saga); the isle of a man *Garékr*, a name otherwise unrecorded.

This small island, about the same size as Wyre and lying a couple of miles south of it, is equally renowned from Saga times as it was the home of a chieftain Olaf Rolfson and his more famous or notorious son—Sweyn Asleifson. It was skatted as 13d. land. Today it is uncultivated and merely a large sheep-run.

Boray (bo:ri) : a farm on south side of island with the remains of an old broch. O.N. *borgi* (dat. case of *borg*), a broch.

Langskaill (laŋ·skil) : on the south-west point of island—the old home of Sweyn Asleifson. O.N. *langi-skáli*, 'long hall.' In the O. Saga reference is made to that long hall.

Rusness : a vanished farm-house on east point of island which is now called The Hen of Gairsay. See this name—Sanday.

Skelbist : probably the original settlement in Gairsay, situated above a sheltered bay known as The Millburn, one of the safest anchorages for small craft in Orkney. O.N. *skála-bólstaðr*, an old 'bister' farm with a *skáli* or hall.

DEERNESS
(de·ərnez)

Dýrnes (*O. Saga*), ' animal ness.' Durness (in the north of Sutherland) was also spelt *Dýrnes* in the Hakon Saga, and Duirinish in Skye is regarded as derived from the same root.

In view of the seeming irrelevance of such an origin in this Orkney name it has been suggested that the Norse form may be a corruption of some Celtic word (e.g. *daire*, originally an oak grove, later associated with a church building) having reference to the early Celtic church settlement on the Brough of Deerness. Such an hypothesis seems far-fetched and unnecessary as names were applied by Norsemen sometimes for quite trivial reasons. There is, however, a very curious coincidence (if it be such) that is worth noting in connection with this name. The precipitous headland (well-known to travellers by sea to and from Kirkwall), at the extreme outer tip of the Deerness peninsula, is called The Moul of Deerness. The O.N. word represented here—*muli*—was used of a bold headland such as this (Cf. also Mull of Galloway), but that was really a secondary meaning or metaphorical extension of the primary, which was—the snout of an animal!

Since pre-Reformation times Deerness has been linked with St. Andrews to form one parish, but they are really quite distinct units, each with an old parish church of its own. Deerness indeed is almost an island, being connected with the rest of the Mainland by a very narrow isthmus, only 50-100 yards wide. It was skatted as a six-urisland area, and is one of the most fertile and intensively cultivated districts in Orkney today.

RENTAL NAMES

Aikerskaill (ɛk·ɔr-sk[j]il): Akerskaill 1595. A farm in the old Skae-tun. O.N. *akra-skáli* or *ekra-skáli*, lit. ' acres-hall'; the precise interpretation is doubtful; it may have denoted a hall built amid cultivated fields (*akr* or *ekra*, a corn-field), or the farm on which it was built may itself have been named *Akrar* or *Ekrur*; cf. Aikers (South Ronaldsay).

Ayre (ɛ·ər): Air 1584 (R.E.O.); Air 1595. An old 6d. land in Skae-tun at south-east corner of Deerness: now represented by the farm Barns of Ayre. O.N. *eyrr*, a gravelly beach, but in Orkney the term is usually applied in a more specific sense to a narrow strip of beach between a salt-water lagoon and the open sea, e.g. The Ayre at Kirkwall. Here in Deerness at present there is no such lagoon, but there is a high ridge of stony beach with very low-lying ground immediately behind, and it is quite possible that at one time this may have formed a lagoon.

Braebuster (bre·bəstər): Brabustare 1500; Brabuster 1531 (R.E.O.); Brabustare and Brabustar 1595. A former urisland or 18d. land tunship, and the largest farm in Deerness today. O.N. *breiði-bólstaðr*, ' broadbister,' broad farm-settlement.

Brecks: Brek in Kirbuster 1500; Breck and Brecks 1595. Represented today by three units—Brecks, Hall of Brecks and Kitchen of Brecks. In the 1500 R. Brek was a 5½d. land, and apparently the chief farm in the urisland tunship of Kirbuster. O.N. *brekkur*, ' slopes.'

Calgarth: 1500; Calgart 1595; Kaygair 1595; Cager 1627 (Peterkin III, 93). A 2¼d. land in Kirbuster tunship, represented today by North and South

Keigar (ke·gər, ki·gər, -gjər). O.N. *kálgarðr*, ' cabbage garth.' This curious name for a farm is confirmed by the occurrence of the same name—*kaalgaardt*—in Færoe from the year 1587. (Information from Dr. Matras.) Cf. Cogar (Rousay).

Colster: Colsetter 1595. An old 1½d. land. An O.N. *setr* name, the first element being a personal name—Kollr or Kolli or Kolr: *Kolls-setr* or *Kolla-setr* or *Kols-setr*, K's settlement or farm.

Deldale: now spelt Delday (dɛl·di,-'de): Deldaill 1488 (R.E.O.); Deldell 1500 and 1595; Deldaill 1550 (R.E.O.).

An old 9d. land or half-urisland tunship. The second element is undoubtedly O.N. *dalr*, a dale or valley, but the first is quite obscure. *Del-* appears occasionally as a first element in Norwegian farm-names, and is sometimes derived from *deile, dele*, a division or boundary, but it is doubtful whether that would hold good here.

Flenstaith: 1500; Flenstaith alias Sands 1595 (misprinted in Peterkin Fleustaith). A vanished name—quite forgotten locally, having been displaced by Sands, q.v. which probably also preceded it. An interesting name, as the only Rental example in Deerness of an O.N. *staðir* name. Like the great majority of such names in Orkney as elsewhere it has a personal name as first element—O.N. *Fleinn*: hence *Fleins-staðir*, ' Flein's stead or farm.' For *staðir* names see Part III.

Gloupquoy: 1500. A small quoy evidently in Sandwick tunship, and situated near the famous Gloup of Deerness—a deep yawning chasm formed by the collapse of the roofing of a long sea-cave. O.N. **gloppa* or **gluppa*, a steep chasm.

Halley (ha·li): Hallay 1500; Halla 1595. An old 1d. land on a steep slope at the north end of the Kirbuster tunship. A deriv. of O.N. *holl*, which according to Rygh's *Indledning* is found in several old Norwegian farm-names, presumably in the sense in which it is still found in local speech —of a slope, e.g. Halle, Hallan, Hollan."

Hlaupandanes: (O. Saga). The residence of Amundi, father of Thorkel Fostri, and the scene of Earl Einar Sigurdsson's dramatic death in 1020 A.D. The Saga states that Amundi lived " i Hrossey i Sandvik a Hlaupandanese," and from the context it is clear that the Sandwick referred to was the Deerness tunship of that name. Today there is no place-name extant that can represent the old H———, but it can scarcely be doubted that the site was at or near the present farm-buildings of Skaill, adjacent to the old parish church and on the brink of the beach of the Bay of Sandside.

The name is puzzling in more than one way. The first element—*hlaupanda*—may be interpreted in two ways: (1) a participial adjective of weak declension qualifying *nes*, the compound thus meaning the ' leaping or springing ness,' from the O.N. verb *hlaupa*, to leap, spring, run, etc.

(2) The genitive case of a man's name—**Hlaupandi*—in which case the compound would signify the ' ness of H———.'

Both interpretations present difficulty. Taking the latter first we have to note that *Hlaupandi is not on record as a personal name, though its existence has been assumed by Norse scholars to explain a few place-names in Norway. Now in Sanday there is a farm-name Lopness which is presumably to be derived from the same original form as this Deerness name. It is then remarkable, to say the least, that such an unrecorded hypothetical personal name should appear twice in Orkney place-names.

On the other hand, '(out)leaping' or '(out)running' is a quite unusual, if not unique, qualification to find applied to a cape or ness. Moreover, there is no topographical feature at Skaill now to justify the use of the term 'ness' at all. On the O.S. map, however, one may observe that, though now covered at high tide, a ness of rocks runs out from Skaill for about half-a-mile. Much of Deerness is covered by boulder clay, and it is not improbable that a thousand years ago this ness had still its boulder clay mantle and was a real grassy ness. The beach here lies open to the billows of the North Sea, and a thousand years of erosion must have caused vast alterations on such a vulnerable coast.

Though in these circumstances the true interpretation of the name must still be deemed uncertain, nevertheless, taking all the above facts into consideration one is disposed to prefer the first alternative— '(out)leaping ness.'

Holland: id. 1595, 1627. An old 6d. land in Southend urisland of Deerness, and a moderate-sized good farm still today. See this name N. Ron.

Keigar: see **Calgarth**.

Kirbuster: 1500; Kirbustar 1595. An urisland tunship in northwest of Deerness. Name practically obsolete today.
O.N. *kirkju-bólstaðr*, 'church-farm.'

Meal: Maill, Maile, Male, 1595; Meall 1627. An old 3d. land farm at shore at eastern end of the Sand of Newark. The name was changed to Newark. in the latter end of the 16th century when a new house or mansion was erected there apparently by Lord Lindores, who in 1591 received a charter of lands in Deerness from Earl Patrick Stewart including Meal with "the new houses called the Newark." O.N. *melr*, sand.

Midhouse: 1595. An old 2½d. land in Kirbuster.

Newark: see **Meal**.

Newhouse: Newhows 1627. Part of Sands tunship.

Northquoy: 1595. A quoy in Sands tunship.

Oback (o·bək): Oabak 1531 (R.E.O.). A 2 merk land in Braebuster. Another small farm of this name still exists today in what was probably the old Sands tunship. O.N. *ár-bakki*, 'river- or burn-bank.'

Pool (pøl): Pwle, a 1½d. land in Skae tunship, 1595. The present farm lies on the east coast of Deerness, and the name is probably derived from O.N. (Icel.) *pollr*, No. *poll*, used (*inter alia*) of a small bay semi-enclosed or protected by an outlying reef of rocks.

Quoyburing: a quoy in Skae tunship, 1595. The name also appears in records as Boorin, but is pronounced Burrian (bʌr·jən). Evidently denotes site of an old broch—O.N. *borgin*, 'the b————.'

Quoycrusda: another quoy in Skae-tun, 1595. Origin of name obscure, but it would seem to appear again applied to a hill-croft in Rousay—Crusday (krøz·de).

Cuoyfea: a quoy in Deldale, 1595. Present-day farm simply Fea. O.N. *fjall*, hill.

Quoymos: another quoy in Deldale, 1595. Self-explanatory

Quoymyirland: a 1½d. land, 1595. Now simply Mirland. O.N. *mýr-lendi*, moorland.

Quoyrush, Quoyrushe: a quoy in Deldale, 1595. Origin of second element obscure.

Quoyturben: a quoy in Skae-tun, 1595. Origin uncertain.

Sanday: 1500, 1595; Sandy 1595; Sandakin, U.B. 1601.
The termination of this name tends to be misleading. Instead of (as usually) indicating an island it here represents O.N. *eið*, an isthmus, and has reference to that well-known, narrow, sandblown isthmus at Dingieshowe which connects Deerness with the rest of the Mainland, and which must formerly have been known as *Sand-eið*. The name, however, is of special interest for there were two tunship-districts so named—this on the Deerness side of the isthmus and another adjacent on the west side in the parish of St. Andrews, now generally known as Upper Sanday. This Deerness Sanday was a 3d. land, the other a 1½d. land, so that originally it is obvious that they went together to form one 4½d. tunship—a unit known in Orkney as a *skatland*. We thus have here one of the very rare examples of a parish division having been drawn in such a way as to cut a tunship in two.

This Deerness Sanday would seem to be unique in another way also, for from about 1600 it came to be known as Sandakin or Sandyaiken, Sandaitken, etc.—the termination obviously having arisen from the family name of Akin, or Aiken or latterly Aitken, who are known to have possessed some or all of the lands in the tunship from an early date. There is no separate farm of Sandaiken today, the lands now being incorporated in the adjacent farm of Stonehall.

Sands: Fleustaith [misreading of Flenstaith] alias Sandis, 1595. A 9d. land or ½ urisland, the alternative name of which—Flenstaith—disappears after the 1595 R. Sands remained in common use, and is represented today by three farms—North Sands, Wester Sands and Nether Sands. See further *sub* Flenstaith.
O.N. *Sandar* (pl. of *sand*) a term applied to a stretch of sandy ground. Cf. Sander, North Ronaldsay.

Sandside (san'sət, san·səd): Sandisend 1500, 1595; Sandsyd 1627 (Peterkin III, 93). A 4½d. land at north end of the sandy bay on east side of Deerness—the original Sand-wick. Sandside was part of the urisland of Sandwick, but it seems to have formed a little tunship by itself. (See R.E.O. 266). The Buchanans of Sandside were a leading family in Orkney in the 17th century.
The change of Sandsend to Sandside may be explained by the slurring over of the second 'n' in the local pronunciation of an unaccented final syllable—san·sεnd>san·sεd—which gave rise to the assumption that the second element was -*side*.

Sandwick: Sandvik, O. Saga; Sandvyk 1534 (R.E.O.); Sandwik 1561 (R.E.O.); Sandwick 1595.
This name seems to have been applied as a name for a whole urisland tunship, but not to any single farm in it. The area in question lies

round the bay from which it took its name, and in Saga times (11th cent.) a chieftain Amundi and his more famous son lived here "*i Sandvik a Hlaupandanese.*" See further under H——— *supra*. O.N. *sand-vík*, 'sandy bay.'

Skae (ske:) : Ska, 1500,1595; Skay 1595; Skae 1597 (R.E.O.); Skea 1627 (P.R. iii, 93). A 9d. land or half-urisland tunship, which with Boorin 1½d., Pool 1½d., and Aire 6d. made up an urisland generally referred to of old as Skae-tun urisland. A farm here is still called Skea.

This name occurs also in Rousay, in Westray and in Birsay, but not as a farm name. The origin is doubtless O.N. *skeið* which was the usual term for a place where horses ran, or could suitably run, races, but the modern Norse form (according to Rygh) is used also of a track through fields. Its precise significance in the Orkney examples must thus be left as uncertain.

af Skeggbiarnerstaupum : O. Saga. This farm, which cannot now be identified in Orkney, would appear from the Saga context to have been situated in Deerness. If so, it would be an interesting companion to Flenstaith (q.v.) for it is another of that class of *staðir* names which are rare in the East Mainland but plentiful in the West Mainland. Like Flenstaith this farm may very well have had an ' alias ' which has survived without any memory of its alternative *staðir* name. See further—*staðir*-names; Part III.

Stove : Stoif 1500, 1561 (R.E.O.), 1595; Stowffe 1627. One of the farms in Sandwick urisland now represented by Upper and Nether Stove. O.N. *stofa*, used generally of a room rather than of a house. In Orkney the name occurs five times (always uncompounded), but it would seem never to have been applied as an original-settlement name. The Sanday Stove has been discussed *supra;* the 9d. land in the *parish* of Sandwick which appears as Stove in the 1595 R. figured in the earlier 1492 R. as Scorowell, and in the 1500 R. as Scrowell alias **Fermandikis**; the fourth Stove was in the tunship of Outer Stromness, and the fifth in the tunship of Swanbister in Orphir. As in the case of the Skaills it is fairly certain that the use of this word as a farm-name must have been due to the presence of a house of some kind, but the special nature thereof is obscure.

Wattin : 1595. A 9d. land between Sands and Newark, which evidently originally included the 3d. land of Meal or Newark, and which together with the 9d. land of Sands made up an urisland formerly known as the Watland urisland. In the valley here in olden days there must have stood a loch which gave rise to the name : O.N. *vatn*, a lake.

Yarpha : 1595. A 2½d. land in Kirbuster urisland. This farm-name occurs again in Orphir, and as a generic term *yarpha* or *yarfie* is still in use in Orkney for peat-moss which is only one peat deep. When such peats are cut they normally have gravel attached to them. See *O. Norn.* Fritzner defines O.N. *jörfi* as sand, gravel or bank formed thereof; in Mod. Icel. *jörfi* is used also of a ridge or elevation, an earthen bank, etc. It is thus difficult to determine exactly the nature of the ground that gave rise to the farm-name here.

OTHER FARM NAMES

Brandyquoy : the *kví* or quoy of a man Brandi or Brandr (or one of its numerous compounds, e.g. Guðbrandr, etc.).

Cattaby (kat·əbi) : Cattibie 1601 (U.B.); name now applied to a portion of lands in Skae-tun still remembered as the site of an old farm of that name. Probably O.N. *Káta-bœr*, or *Kattar-b*——, the farm of a man *Káti* or *Kǫttr*.

In this same tunship is another farm Netherby. Those two presumably -*bœr* names are specially noteworthy as they seem to suggest that here again may be detected a large original -*bae* settlement which later was divided up, as indeed the prefix in Netherby would seem to imply. It is of course impossible now to determine with certainty the actual course of events, but one may suggest that after the settlement had been divided the farm of Skae became the main farm in the area, and gave its name to the whole tunship. This south-east corner of Deerness is one of the most fertile districts in Orkney, and in it also is a farm Aikerskaill, a -*skáli* name that suggests a family of distinction had its seat here at one time.

Creya: a small croft on hilly ground. This name is found in several districts in Orkney applied as the name of a small farm or croft on hilly ground or common. See Crya, Orphir.

Denwick: a small farm adjacent to bay of same name; never referred to in old Rentals, and probably a comparatively recent outbreak from the hill or common. The bay is close to the Moul Head, and being very exposed can never have afforded more than a temporary shelter or landing-place for ships. The name is curious and puzzling, as what is probably the same name—Dannvik, Danviken, etc.—occurs repeatedly in Norway and Sweden where the meaning of the first element is likewise in doubt.

Garth: small farm below Holland. Not one of the old settlement 'garths,' but one which has taken its name probably in more recent times from a field so called. O.N. *garðr*, an enclosure, field, farm, etc.

Grind (grɪnd) : a farm which must have its name from its proximity to a 'grind' or gate in an old hill- or tunship-dyke. O.N. *grind*, a gate.

Grindigar (grɪnd·igər): a farm or 'garth' near a 'grind': see last word.

Gritlay: Gruthill 1488 (R.E.O. 197), Gruthailay, Gruthlay 1505 (R.E.O. 204). Formerly part of Deldale tunship. See Gruttill, Sanday.

Horries: origin quite uncertain, but the final 's' may possibly imply this was a farm pertaining at one time to a family Horrie; cf. Horrie, St. Andrews.

Mirkady (mɪrk·ədi, -e) : a farm on west side of Deerness (running out to a long narrow point) below the old Kirbuster tunship. Not included in old Rentals, and thus evidently of later origin.

The first settlement can hardly be other than O.N. *myrkr*, dark, dusky, 'mirk,' but the termination is uncertain. Many Orkney place-names end in -*dy* (di, de)—a termination always more or less obscure. Sometimes, as in Delday *supra*, it must represent O.N. *dalr*, a valley, dale (probably in dative case—*dali*), and that might possibly be the case here. Another possible source would be O.N. *dý*, bog or marshy ground. The point of land or 'taing' is so prominent, however, that one is tempted to think of O.N. *tá*, a toe (now pronounced te in Orkney), a term that occurred in O.N. place-names, in which case the adj. 'dark' might refer to the long ridge of seaweed showing up at low tide. Origin thus quite uncertain.

Netherby: see Cattaby.

Noltland (nɔut·lənd) : O.N. *nauta-land*, cattle-land.

Quoybelloch (kwaibɛl·ək, –bil·ək) : Quoy Belock 1643. A small farm in Deldale tunship. Origin of second element uncertain, but probably a personal name. In Deerness dialect an unaccented final vowel is often followed by a 'k' sound, e.g. Jimmy>Jimmick, Mary>Merrick, etc. That may probably have occurred here in the case of a Billie or a Bella.

Quoycanker: a small farm now apparently conjoined with Daystar. See Canker (Sanday).

Quoylanks: farm in Deldale tunship. Second element obscure.

Skaill (skil, skjil) : old farm house in Sandwick adjacent to parish church. This has very obviously been the head-house of the tunship, and may well represent the site of the Saga *Hlaupandanes*, q.v.
    O.N. *skáli*, a hall.

Snippigar (snɪp·igər) : a small farm in what was Deldale tunship. Not one of the early *garth* class, but (like Garth *supra*) a later formation from a larger unit. The name originally may have been a field name. The first element is probably No. *snipp* (found in several Norwegian place-names) signifying a 'snip' or corner of something; cf. *Eng.* to snip off.

Stye (stai) : a farm in the old Sands tunship. A reference is found in a charter of 1625 to lands "under the house of Sty." A puzzling name, as it is hardly credible that a farm should bear the name of a pig-stye Nor is it likely that the *-staith* of Flenstaith (the old alternative name for Sands tunship) would have survived in such a form as Stye. The same name is on record in Shapansay.

Tiffyha' (tiv·i·ha) : a small farm in valley between Sands and Skae-tun. The second element is doubtless Scots *ha'* (hall) used in a semi-derisive sense as so often in Orkney (especially Eday); the first, however, must be Teeve, a place-name occurring in several parts of Orkney, and applied to boggy or marshy ground—The Teeve. Origin of that term quite obscure.

ST ANDREWS PARISH

Unlike Deerness, with which it is now conjoined, and which has a name that had been in use for that distinct geographical unit long before the erection of parishes, St. Andrews comprises a number of tunships or small districts which do not seem to have had any previous unity. The name it bears is that of the parish church which was situated at the Hall of Tankerness where the 12th century chieftain Erling of Tankerness presumably had his seat, and no doubt his urisland chapel. For the selection of such a chieftain's chapel to be the parish church see further—Part II.

The skattable valuation of this parish is difficult to ascertain with accuracy from the data in the early rentals, but it probably included 5½ urislands of skatted land, together with some unskatted quoyland in addition.

RENTAL NAMES

Campston (kam·stən) : Camstaith 1492; Campstane 1500, 1595; Campsta 1546 (R.E.O.). A former 9d. land or half-urisland, and a large farm still. This is one of the few *-staðir* names in the East Mainland. The first element is probably the O. Norse personal name Kampi, as in the Norwegian farm-name Kampestad (G.N. v, 416).

Carabrek : 1583 (R.E.O.); Carrabrek (1581 deed). This was a small 'onsett' or holding attached to Sabay. Termination—O.N. *brekka*, a slope; first element doubtful.

Essonquoy : Essinquye, Esinquye 1550 (R.E.O.); Estinquoy 1601 (U.B.). A 3d. land forming the north-western extremity of St. Andrews. There used to be a Hall of E——, and a farm The Barns of E—— (now swallowed up by Grimsetter Aerodrome), and in a sale of E—— in 1550 the 'Chapel of E——' was included. These facts all point to this having been an estate of some dignity, which its *-quoy* name seems to belie, and the following may be regarded as a tentative explanation.

The north side of the Tankerness peninsula embraced four old tunships—Linksness a 4d. land, Yinsta an 8d. land, Veda a 3d. land, and Essonquoy a 3d. land, and it is probable that they all went together to form a single original settlement which was skatted as an 18d. land or whole urisland. Yinsta is a *-staðir* name and thus presumably not an original settlement name; Links- in Linksness is a Scots form and thus late; the original form of Veda is obscure, but Essonquoy again denotes a quoy-settlement on the outskirts (as it indeed is). It may therefore be suggested that the original name of the urisland has been entirely forgotten, and that only names of constituent parts have survived. But the presence of a chapel in Essonquoy—the only one in the whole urisland— would seem to imply that the original head-house was in that end of the urisland.

The prefix in Essonquoy is perhaps the family name Esson.

Fea : Feaw 1492, 1595; Feall 1500. An old 1½d. land quoyland. O.N. *fjall*, hill.

Foubister (fu·bəstər) : Fowbustir 1492; Fowbustare 1500; Foubistar, Foubustar 1595. An old 3d. land. The second element is O.N. *bólstaðr*, a farm; the first probably O.N. *fúll*, stinking, an epithet often appearing in Norwegian names of shallow bays from the beaches of which a stench arises at low water. The beach below Foubister is just such a place.

Gart: 1583 (R.E.O.). A small 'onsett' of Sebay. O.N. *garðr*, a farm or enclosed piece of land.

Grotsetter: id. and Grottsetter 1536 (R.E.O.) and 1595. A 3½d. land of old forming part apparently of Tolhop 'Above the Yard'. The second element is O.N. *setr;* the first rather doubtful; a personal surname Grot would be unlikely to appear compounded with a -*setter* name, and the element here is therefore probably O.N. *grjót,* stone, or stony ground.

Grutquoy: Gruttquhy 1509; Gruitquhoy 1519; Grutquoy 1583 (R.E.O.). O.N. *grjót-kví,* 'stone-quoy.'

Hawell (ha·wəl): Havell 1578; Hawell 1584 (R.E.O.). A former 3d. land in Tolhop; still a farm today. The second element is doubtless O.N. *vǫllr,* a field, but the first is uncertain; the topography would seem to rule out any possibility of O.N. *hár,* high, but it is possible it may represent O.N. *hagi,* pasture.

Horrie: Hurre 1510; Hwrre, Horre 1516; Horrie 1567; Horry 1568; Horrye 1569 (R.E.O.); Horrie 1595. A farm of today and an old 1½d. land.

Origin rather doubtful; but it is probably the same name as the Norwegian farm name Horr (on record also as '*pá Hore*') which in N.G. x is stated to be "probably the same (or essentially the same) word as O.N. *horr* 'nose-slime,' used here in the original sense of dirty ground . . . . The farm Horr is surrounded by large stretches of moor."

This Orkney farm lies in a wet valley and on the fringe of heathery moorland.

Horsick: Horssack 1567 (R.E.O.). A small farm representing an old 'quoy' attached to Horrie in 1567. It is situated near the shore at the head of a bay now called the Bay of Suckquoy, but it is probable that in former days this branch of the bay was named after the farm of Horrie just above—hence **Horres-vik.*

Kirkgair: 1595. A 3d. land. No longer in existence as a farm, but the name Kirgar is still remembered, and applied to an old track near the mill of Tankerness. O.N. *kirkju-garðr,* 'kirk-garth': no reason for the name can be offered.

Linksness: Linxness 1548 (R.E.O.); Linksness 1595. An old 4d. land, once the property of Sir James Sinclair of Summerdale fame. As Links is a Scots term this name has evidently replaced an older; see further—Essinquoy.

Lidda (lɪd·i): Luddal 1455; Luddale 1495; Luddell 1559 (R.E.O.); Lidday 1665. An old 6 markland. Termination O.N. *dalr,* valley; first element obscure.

Messigate: Meissegere 1509; Messeger 1519; Messager, Mesiger 1583 (R.E.O.). Formerly (1583) one of the 'onsetts' or satellite small farms attached to Sabay; a moderate-sized and good farm today. Messigate is a fairly common place-name in Orkney for an old track leading to a church: O.N. *messu-gata,* 'mass-road' (to where mass was performed). But the earlier form (which is still remembered locally—Messigar—is somewhat puzzling; the termination would normally indicate O.N. *garðr*, a farm or enclosure, but its combination with *messa* would seem strange.

Ness : Nesse 1507; Nes 1570 (R.E.O.). A former 3d. land. O.N. *nes*, headland.

Occlester (okl·stər) : Okkilsetter 1492; Okilsetter 1500, 1595. A former 1½d. land near boundary of Holm parish, which was originally part of St. Andrews but was in some curious way incorporated with Paplay, now conjoined with Holm. The transfer seems to have been made in the 15th century, and probably was connected with Earl William Sinclair's acquisition of Paplay lands in excambion for Sabay lands in St. Andrews.

Second element O.N. *setr*; prefix O.N. *ǫxl*, a shoulder, used also of a shoulder or projecting ridge of hill. The change of xl (ksl)>kl is regular in Orkney dialect; cf. *yackle* (molar tooth)—O.N. *jaxl*.

Quybewmont : 1492; Quoybewmont 1500, 1595; quye callet Bewmonte 1546 (R.E.O.). An old 2 markland in Tankerness. Today there is no such farm, but the name survives in a place-name Point of Beman—east of the Hall of Tankerness.

Second element presumably a personal surname, though Beaumont is a curious name to find in Orkney place-names.

Quoykea (kwike·) : Quoykay 1567 (R.E.O.). A quoy formerly attached to Horrie, and a small farm still today. Second element obscure.

Quytob : 1566 (R.E.O.). A small quoy in Toab, q.v.

Sabay (se·be) : Sabaye 1492; Sabay 1500, 1595. A former 9d. land or half-urisland, and still one of the largest and best farms in the East Mainland. O.N. *sæ-bœr*, ' sea-farm '; a common farm-name in Norway, and in N.G. x it is suggested that in some cases such a farm may have been the nearest portion to the sea of a larger unit. In Orkney, however, -bae, -bay (*bœr*) farms (see Part III) would seem to have been large original units dating from the time of the original settlement. In point of fact, Sabay, which abuts on the coast, is adjacent to another old urisland—Toab, which lies inland from it, between Sabay and the ' hill,' and the two may well have formed one original unit. But the name Toab (see *infra*) would appear to be of later date than Sabay from which it may have been an offshoot.

One curious feature of Sabay is however of special interest. In R.E.O. is to be found a Court decree of 1509 in regard to the marches or boundary-lines between Sabay and Toab, and after defining these the decree proceeds : " be resone that the IX penne land of Saba lyis in ane *inskeyft* within hyttself in lentt and breyd that the neboris of Toop and all wdyr personis quhatt sumewir sall keyp ther guydis [animals] of[f] the grownd of Saba beyth summyr and winthir."

Now Inskift is a name that occurs several times in Orkney applied to a field, but is never on record elsewhere applied to a whole tunship. The word represents O.N. *eignarskipti*, which signifies a division of landed property in such a way that each part becomes the sole property of one person, and is in contrast to *hafn-skipti* a portioning of land among joint-owners in such a way that the owner of a portion has only occupation-right of it; the property as a whole remaining *pro indiviso*, and after a time another of the joint-owners may have occupation-right of this portion. Taranger in *Norske Besiddelseret* expresses it thus : " óðalskipti (som eignaskipti) er udskiftning af sameiet; hafnskipti er undskiftning af sambruget."

It is difficult to know in what sense the term has to be understood in connection with Sabay, for by the 16th century there were several portioners of the tunship who presumably occupied it in accordance with the runrig system. The most probable solution would seem to be that at some earlier date one heir of a third of the larger tunship of Sabay and Toab secured Sabay for himself as a completely separate slice, thus renouncing any interest in the other two-thirds, viz. Toab. In any case the fact that Sabay was an 'inskeyft' implies that a division had been made at some time by which it was separated from something else.

Sanday: Sandaye 1492; Sanday 1500, 1595. A 1½d. land of old. A small district now goes under the name of Upper Sanday. See this name—Deerness.

Stembister: Stanbuster 1465; Stenbister 1576 (R.E.O.); Stambustar 1595. A 3d. land of old, and a considerable farm today. O.N. *stein-bólstaðr*, 'stone-farm'; so named from a standing-stone still to be seen there.

Swartabreck: Swartabrek 1578 (R.E.O.). A 6d. land of old in Toab, and a moderate-sized farm today. O.N. *svarta-brekka*, 'black or dark slope.'

Tankerness: Tannskaranes, Tannskaarunes (O. Saga — different MSS); Tankskerness 1455 (R.E.O.); Tangskeriness 1492; Tanskernes 1495 (R.E.O.); Tankarnes 1500; Tankernes 1595.

A 12d. land of old, and the seat of the 12th century chieftain Erling of T——. The Hall of T—— is a large farm still, and here was the seat of the Baikies of T——for over three centuries. The name Tankerness has now an extended use and embraces the whole northern part of St. Andrews parish.

The above first-noted Saga form may be interpreted as signifying the ness of a man nicknamed Tannskári—a very rare name, but the variant Saga form renders such an origin rather doubtful.

Toab (to·b); Tollop 1439, 1584, 1590; Toop 1509; Toep and Thoep 1519; Thoip 1537; Tohope 1579; Tolop 1589 (all in R.E.O.); Tollope 1492 R.; Tohop 1500 and 1595 R.

A tunship rather than a farm-name, and from the 15th century at least, it consisted of two parts—"T—— above the yard," and "T—— beneath the yard." Though the exact skattable valuation is difficult to ascertain from the Rentals, it would seem that the two parts amounted to 2 urislands.

The 'yard' referred to certainly indicated a dyke of some kind (O.N. *garðr*), and a parallel may be found in Papa Westray where the island was divided into two parts—North Yard and South Yard—the dividing 'yard' being one of the old gairsty dykes. North Ronaldsay was similarly divided into three parts by two such gairsties, one called The Muckle Gairsty; and a part of the parish of Sandwick is still known as North Dyke since it apparently had such a gairsty as its southern boundary. These gairsties or treb-dykes are quite prehistoric: see further *sub* Treb (North Ronaldsay).

The origin of the name Toab (still in common use) is obscure, but the early forms seem to point to an O.N. **toll-hóp*, 'toll-bay or -hope,' i.e. a harbour where foreign ships paid toll on arrival or departure. In that case the 'hope' must have been some part of the present Deer Sound. It is interesting to note that the same name—Tolob—occurs in Shetland, where it is applied to a small tunship above the present Pool of Virkie in Dunrossness.

NAME MATERIAL 87

Toll in form of *landeyrir* appears to have been imposed on imported goods in Norway from the time of King Harald Fairhair, but records in regard to Orkney are completely lacking.

Twinness (twɪn·ɟez) : Twyngnes 1583 (R.E.O.). A former 'onsett' attached to Sabay (1583); now two small farms—Muckle and Little T—— situated near a low headland or 'ness.' Probably O.N. *tungu-nes*, 'tongue-ness': see Twiness (Westray).

Vedder : Weda 1595. A David Wader is on record from 1665, and also a Harry Wadder. An earlier Henry Veidaye *alias* Androwsone appears as a witness in connection with Yinstay lands in 1550 (R.E.O.). A 3d. land of old, and there is a farm Vedder still today. Origin obscure.

Vcy : Woy, Voy 1595. Same name occurs in Sandwick. O.N. *vági*, dat. case of *vágr*, a 'voe' or bay.

Whiteclett : Quhittlet 1492; Quhytclet 1500, 1595. A 2d. land (quoyland) of old, and a small farm of today. See Cleat (Sanday).

Yinstay : Zenstaith 1492; Yinstaith 1500; Yensta, Yinsta, Yinstay 1595. An 8d. land of old, and there is a farm—the Hall of Yinstay—still. One of the few -*staðir*-names in the East Mainland. The first element no doubt represents a personal name which is perhaps much contracted and certainly obscure; it has even been suggested that the name may represent a much-truncated form of the Saga *Skeggbjarnarstaðir*, but that seems improbable.

OTHER NAMES

Aikers (ɛ·kərs) : a farm in old Yinstay tunship. O.N. *akrar*, plur. of *akr* (or *ekrur*, plur. of *ekra*), arable fields.

Biggings : a farm in Upper Sanday. This name which occurs a few times in Orkney is probably of Scots rather than of Norse origin. Cf. Bigging, Westray.

Boondatoon (bønd·ətun) : a farm near Point of Rerwick. See this name Stronsay.

Bossack (bɔs·ək) : a small farm. The name, like the similar Norwegian farm-name Bosvik probably represents O.N. *botns-vik*, the inner end (bottom) of a wick or bay; this farm lies at or close to the inner end of an arm of Deer Sound.

Calset : a small farm in Yinstay. Probably O.N. *kald-setr*, 'cold setter or farm.'

Cockburnsquoy, now usually shortened to **Cobrance** (ko·brans) : a small farm in Yinstay. This is another case in which -*quoy* is found compounded with a Scottish personal surname.

Comely (kʌm·li) : Clumley and Clumlie 1620 (Viking Club Sasines); a small farm near beach of Peter's Pool, the inner end of Deer Sound. O.N. *kumli*, dat. case of *kuml*, a cairn or prehistoric mound, a term frequently occurring in Orkney place-names.

Copenageo (now usually pron. kɔpne·gi) : a small farm near Head of Yinstay. Origin rather uncertain, but may well be O.N. *kobbanna-gjá*, 'geo or inlet of the seals.' O.N. *kobbi* was a popular name for a seal, and here at the

Head of Yinstay is a skerry which is a noted haunt of seals. On the Holm of Scockness (Rousay) another point much frequented by seals is called Cuppataing.

Crofty: a small farm in Toab. The name is apparently of Scots origin.

Gears (ge.ərs): a small farm in Upper Sanday. Origin doubtful.

Grind (grınd): a small farm on northern border of Toab. See this name—Deerness.

Holland: a small farm in what was probably Linksness tunship. See this name—North Ronaldsay.

Langskaill: a farm in Toab. O.N. *langi-skáli*, 'long hall.'

Nearhouse: a farm in Toab. A corruption of Nether-house; cf. Appiehouse (Sanday).

Niggly: a small farm in Foubister. The same name occurs in Evie. Apparently O.N. *knykli*, dat. case of *knykill*, a 'knuckle,' hillock or protuberance. In Rousay Nuggle is a place-name applied to a hillock in Wasbister.

Purtabreck: a small farm in Upper Sanday. Origin of first element obscure. The same name occurs in North Ronaldsay.

Quoyburray: a farm in Toab. The second element of this name is almost certainly the personal surname Burray. That name is rare in Orkney, but a St. Andrews man Barnard Burray is on record in 1559 (R.E.O. 111).

Rerwick: Reirweik 1597. A small farm at Point of same name at north-east corner of Tankerness. At this point a rough stony ridge (so much resembling a breakwater that some have regarded it as at least partly artificial) encloses a sort of natural harbour, and that ridge is probably what the first element of the name refers to: O.N. *reyrr*, a heap of stones.

Scarpigar: a small farm. The second element is no doubt O.N. *garðr*, a farm, field, enclosure, etc., and the first is the O.N. adjective *skarpr*, which was used of dry, shallow, infertile ground. In the Orkney dialect today such poor, shallow, barren ground is still termed 'skarpy land' or 'poor skarps.'

Valdigar (vald·igər): a small farm in the old Essonquoy area. First element obscure; cf. Waldgarth (Sanday).

Veltigar: small farm on outskirts of Toab. A *-garth* name of which the first element must be O.N. *velta* (used of something 'turned over') which in Færoe is applied to land that has been turned over (with spade), and hence means cultivated ground. From its situation one may suggest that this farm was originally a 'cultivated enclosure' outside the tunship dyke.

Vinikelday (vın·i–kɛld·ˑi): a small farm on Campston Ness. An interesting example of an O.N. *vin* name: O.N. *vinjar-kelda*, 'pasture-well'; well on the *vin* or pasture. By an unfortunate blunder the farm appears on the O.S. map as Olnekelday.

Weethick (wið·ək): a farm on north coast of Tankerness adjacent to a salt-water lagoon or 'oyce' (øs). Probably O.N. *við-vík*, 'wood-wick or -bay.' Cf. Woodwick (Evie).

HOLM AND PAPLAY

Home, 1481, 1605 (R.E.O.); Hom, 1482 (R.E.O.); Holme, 1492 R., 1552, 1564, 1587 (R.E.O.), 1500 R., 1595 R., 1614, 1642 (P.R.); Holm, 1492 R., 1552, 1603 (R.E.O.); Ham, 1614, 1642 (P.R.).
Papuli, Papule (O. Saga); Paplaye, Paplay 1492 R., Paplay 1500, 1595 R.

The latter name corresponds to the old Icelandic place-name Papyli, which is generally regarded as a contraction of O.N. *Papa-býli*, 'home or settlement of Papae,' i.e. of early Celtic clergy. In the 12th century this Orkney Paplay was the abode of Sigurd who had married Earl Erlend's widow Thora (mother of St. Magnus). There is another Paplay in South Ronaldsay.

The name Holm is much more obscure. Its spelling is extremely puzzling as the pronunciation is regularly Ham, a form which appears on record sporadically from the early 17th century; yet the normal spelling continues down to the present day as Holm.

As most of the parishes on the Mainland of Orkney bear names of coastal features—an island (e.g. Birsay and Orphir), a bay (e.g. Sandwick and Firth), or a headland (e.g. Deerness and Stromness)—and as in the case of Holm there are two adjacent small islands or holms (Lambholm and Glimsholm) it might be thought that the parish-name had reference to these. That possibility, however, would seem to be ruled out by two facts: (1) *holm* is still in common use in Orkney both as a generic and a place-name, and is always pronounced exactly as the English word *home*; (2) if the parish-name referred to the holms in question it should show some trace of the old plural—O.N. *hólmar* or if Englished—*holms*.

In Orkney and Shetland place-names Ham is the normal spelling and pronunciation of a place-name for a haven or good anchorage—derived from O.N. *hǫmn*, Norwegian *hamn*—and as there is such a haven (adjacent to the present village) in Holm it is pretty certainly that haven from which the parish takes its name, a name which seems to appear again in that applied to the hill from which the parish slopes down—viz. Hamly Hill.

How then is the traditional spelling Holm to be explained? No certain interpretation is of course possible, but the most feasible would seem to be that the spelling is due to some scribal error repeated from a very early date through confusion of the umlauted O.N. form *hǫmn* (haven) with *hólmr* (holm or small island).

Though Holm and Paplay are entered as separate parishes in the early Rentals there would seem to be no justification for such division. The two form one continuous stretch of coastal belt, Paplay being the name applied to the eastern end of the parish. By itself, also, Paplay was much too small to form a parish as it contained only about one urisland of skatted land together with another of quoyland; Holm and Paplay together indeed (Lambholm included) did not amount to quite 6½ urislands. The parish church, however, was situated in Paplay (probably at the site of an earlier Celtic church), and about the middle of the 15th century practically the whole of Paplay had been acquired by Earl Wm. Sinclair from the old native Paplay family (by excambion for Sabay in St. Andrews)—a fact which may perhaps help to explain why it was entered in the Rental as a separate parish.

After acquisition of this property it would appear that considerable rearrangement of farms was carried out by the Earl. Greenwall and some smaller units were combined to form what was termed a 'grange,' and further adjustments are indicated by the fact that almost every farm is entered as composed partly of skatted land and partly of quoyland. That being so, the original size of each farm in Paplay must be regarded as uncertain and the pennylands noted do not therefore necessarily indicate the original size of each farm.

PAPLAY

RENTAL NAMES

Angusquoy: 1492, 1500. A 1d. land; site unknown. The prefix indicates a Scottish incomer.

Annynsdeall: 1739 (not previously on record). 3½ farthing land; site unknown. For second element cf. Brendale, Rousay; first element probably a personal surname: Scots Annand?

Banks: Banks 1492, 1595; Bankis 1500, 1614, 1642. 1d. skat land, but along with it in 1492 went two ½d. lands of quoyland—viz. Carlqwy and Quythom. Banks is still a farm today, near the shore; whence its name O.N. *bakkar*, banks.

Breckquoy: Braqui, Braquy 1492; Braquoy 1500; Braquyis 1595; Brea Quoy 1614; Breackquoy 1739. A 2d. skat land plus 1d. quoyland. Represented today by Upper and Nether Breckquoy. Despite the early spellings this is pretty certainly O.N. *brekku-kvi*, 'slope-quoy.'

The Bu: The Bow of Scale 1492; The Bull of Skaile 1500; The Bull of Skaill 1595. A 5d. skat-land of which in 1492 ½d. land went with the 'grange.' This Bu—represented by Upper and Nether Bu today—almost certainly represents the old head-house of Paplay, and the residence of the old Paplay family. See Bu—Part III.

Carlquoy: Carlqwy 1492; Carlquoy 1500. A ½d. quoyland attached to Banks. O.N. *kvi* (quoy) of a man named or nick-named Karl; though unlikely to have been the same person, Sigurd of Paplay's son (12th century) was named Hakon Karl. He was a half-brother of St. Magnus.

Chenziebrek: Mekill Chenzebrak, Little Chenze-brekin 1492; Meikilchenyebrek 1500. Absent from later records. Small farms of uncertain value; situation uncertain but evidently near some small loch or tarn, probably that in the hill above Greenwall.

O.N. *tjarnar-brekka*, 'tarn-slope'; the form with the *-in* ending represents *brekkan*, 'the slope.'

Cornquoy: Cornequy 1492; Cornequoy 1500, 1642; Cornquoy 1595; Corniquoye 1614. A 1d. skat-land with ⅜d. quoyland (1492). A farm still, near east end of Paplay. Name self-explanatory.

Ducrow (du·kro): Dowcrow 1492, 1500, 1595; Dowkrow 1614; Ducro 1642. A farm adjacent to Cornquoy, and of same valuation in 1492.

The termination suggests O.N. *kró*, a pen or enclosure for animals, but such an origin for an old farm like this would be improbable. Both elements of name obscure.

Fea: Feaw 1492; Feall 1500; divided later into Esterfea and Wosterfea 1595; Easter and Waster Fea 1739. 1d. scat-land and 2d. quoyland 1492. Site now probably known as Braehead. O.N. *fjall*, hill.

Feaquoy: Feaquy 1492; Feaquoy 1500. Not on record later, and site now unknown. O.N. *fjall-kví*, hill-quoy.

Flaws (fla:z) : Flawis 1492, 1500, 1595, 1642; Flaws 1739. 1d. skatland and 1d. quoyland 1492. A fertile farm still today. An interesting name. As a farm name it seems peculiar to Orkney where there are at least five farms of this name : in Birsay, Harray, Evie, South Ronaldsay and here. In the Appendix to my *Orkney Norn* I quote references to 'flaws' used as a generic term in a lawsuit of 1825, from which it would appear that a flaw was some kind of division of arable land that might embrace considerably more than one ' rig.'

The term is most probably an Anglicising of the O.N. plural *flár* (Sing. *flá*) which in Iceland is used of strips of meadowland.

Greenwall: Grenewell 1492; Grenewall 1492, 1500, 1595; Greenwell 1595. 4d. skatland and 3d. quoyland (1492), forming the major part of the lands of 'the grange' in Paplay. A good farm still today. O.N. *grœn-vǫllr*, green-field.

Holmes: The Holmys 1492; The Holmes 1500; Holmeis 1595; Holmes 1614, 1642. 1d. quoyland, 1492. Both name and site now forgotten.

There is no island in Paplay to give rise to such a name, and it would thus appear to be O.N. *hólmar*, plur. of *hólmr*, a small island, but used in the secondary sense of a patch of ground situated like an island in the mid t of ground of a different nature. Cf. Holm (North Ronaldsay).

Orklandquoy: Orklandquy 1492; Orklandisquoy 1500. This together with Suthirquoy formed a ¾d. quoyland which went along with 'the grange,' and does not appear in later rentals. The first element of the name gives it a special interest and may be compared with Ork in Shapansay, q.v.

Quoybernardis: Quybernardis 1492; Quoybarnerdis 1500; Quoybarnettis 1595; Quoybarnets c. 1640; Quoy Barnet 1739. A ¾d. quoyland; name and site forgotten. The quoy of a man Bernard.

Quoyingabister: Quyingebres (*sic*) 1492; Quoyingyebister 1500, 1595; Quoyingabuster 1739. A 1d. quoyland : name and site forgotten. A compound of rather doubtful meaning; it obviously describes a quoy related in some way to a *bólstaðr* (farm), but the exact sense of the middle element is uncertain; it might represent either O.N. *inni* or *inni í*, in or within.

Quoythom: Quythom 1492; Quoythome 1500, 1595; Quoytam 1642. A ½d. quoyland attached to Banks (1492); name and site forgotten. 'Thomas's quoy.'

Roy: a small farm at the extreme east end of Paplay, and adjacent to the headland known as Roseness. That headland almost certainly has its name from a prehistoric cairn situated thereon—O.N. *hrøysi*, a cairn,— and it is difficult not to associate this farm-house therewith. Origin however obscure.

Skaill: The Bow of Scale 1492; The Bull of Skaile 1500; The Bull of Skaill 1595, 1739; Bow of Skaill 1642. A 5d. land, all skatland (1492). O.N. *skáli*, a hall. See also Bu.

Sotland: 1492, 1500, 1595; Soitland 1642. A ¾d. quoyland; name and site forgotten. Apparently same name as the Norwegian Sodeland, from O.N. *Sóta-land*, named after a man Sóti.

Stout Farding : 1500. A ¼d. quoyland; name and site forgotten. A neighbouring ¼d. land was called Wallis Farding, the first element of each being a personal surname.

Suthirquoy : 1492. See Orklandquoy.

Uttesgarth : 1492; Uttisgarth 1500. Not in later rentals. With Meikle Chenziebrek it formed 2d. scat land. Probably signifies ' outermost farm '; with the first element may be compared Fær. *uttastur*, outermost.

Vigga : Vega 1492, 1500; Viga 1595; Vigga 1614, 1642, 1739. A 1d. quoyland near western boundary of Paplay; still a farm today. Origin quite uncertain.

Wallis Farding : 1500. See Stout Farding.

HOLM

Holm proper (i.e. excluding Paplay) embraced four fully skatted urislands together with the 3d. quoyland of Hunclet. Three of these urislands each formed a tunship by itself; the fourth was divided into two separate half-urisland tunships—Easterbister and Westerbister. The boundaries of those five old tunships may be observed on Mackenzie's Charts (1750).

The tunship of Swartaquoy was bounded by Paplay on the east, and on the west apparently by the Loch of Graemeshall and a burn which runs into it. The old rentals show that Swartaquoy included the 3d. land of Hurtiso and the 4½d. udal land of Netherton, together with other unnamed lands. To the west of Swartaquoy was a tunship sometimes referred to as Aikerbister and so marked on Mackenzie's Charts, though Aikerbister proper formed only a quarter of this urisland tunship which included also Meal (now Graemeshall), Graves and a 4½d. land termed How or Valay. West again of that tunship and extending to Scapa Flow, were Easterbister and Westerbister, the former lying to the south rather than east of the latter, and being divided from Aikerbister tunship apparently by a tiny stream entering the sea near the old Storehouse on the shore of Holm Sound. North of Westerbister lay the tunship of Hensbister, and then up in the hill by itself was the 3d. quoyland of Hunclet, very obviously a later settlement.

RENTAL NAMES

Aikerbister : Akyrbustir, -er, 1492; Aikarbustar 1500; Aikerbuster 1595, 1739; Akirbister 1614; Aickerbuster 1642. A 4½d. land or quarter urisland, but the name was sometimes applied to the whole urisland tunship. O.N. *akra-*, or *ekra-bólstaðr*, a farm settlement notable for its tilled fields. Cf. Aikerness (Westray and Evie).

Barnettsdeall : 1739. A small farm apparently in Swartaquoy tunship. First element a personal name; for second cf. Brendale (Rousay).

Blomuire : 1739. Still a farm today. Probably O.N. *blá-mór*, blue moorland or heath.

Cott : 1739. A small holding in Swartaquoy. O.N. *kot*, a cottage or cot.

Easterbister : Estirbuster, Easter- 1492; Eisterbuster 1500; Easterbuster 1595. A half-urisland tunship. O.N. *austr-bólstaðr*, east farm-settlement. The conjunction of the two half-urislands Easter- and Westerbister is fairly clear evidence of an earlier undivided urisland settlement.

Fenzie-land (prob. fɛn·ji-) : 1739. A small unit in Swartaquoy. Cf. Finyo (Rousay).

Graves (grevz) : Gravis 1481 (R.E.O.), 1492, 1500, 1642; Grawin 1595, 1614; Gravin c. 1640; Graves 1739. A 4½d. land in Aikerbister tunship; still a farm today. In various ways a rather obscure name. The root is pretty certainly O.N. *grǫf* (plur. *grafar* or *grafir*) which meant a grave but which in place-names usually indicated a pit or hollow of some kind either natural or artificial. The plural *grafar* would not normally become *graves* in Orkney, but the name has probably been assimilated to the Scots word so spelt. The variants Grawin, Gravin present difficulty, and their origin is uncertain. From Shetland Jakobsen cites place-names ' de Gravens ' and ' de Grevens ' which would appear to derive from an O.N. plural with def. article attached. But in Norway a farm-name Graven has been explained simply as a singular with def. article attached. (G.N. iv. 1. 170.)

Hensbister: Hensbuster, -ir, 1492; Hensbistare 1500; Hensbustar be east the gate and H——be west the gate 1595, etc.
A whole urisland tunship through which ran the ' gate ' or roadway to Kirkwall. First element almost certainly the O.N. personal name Heðinn : hence *Heðins-bólstaðr*, H's farm-settlement.

Horriequoy: 1595; Horraquoy 1614; Horroquoy 1642. A small quoy—site unknown. First element uncertain but most probably the Orkney personal surname Horrie.

How: 1492, 1500. A 4½d. land in Aikerbister tunship. O.N. *haugr*, a mound. See further Valay.

Hunclet: Hundclett 1492; Hunclett 1500, 1642; Hunclet 1500, 1595; Huncleat 1595; Hunklitt 1614; Hunklett 1739.
A 3d. quoyland. Second element O.N. *klettr*, a rock, but the first is quite obscure. Cf. Hunclet (Rousay).

Hurtiso (hʌrt·əzo, -əso; but often hʌrt·sək through the East Mainland partiality for changing a final vowel to -ick (-ək). Thurstainshow 1492; Thurtishow 1500; Hurteso 1595, 1614, 1642, 1739. A 3d. land in Swartaquoy tunship, q.v.
O.N. *Þorsteins-haugr*, Thorstein's cairn or mound. For the change of initial þ (th) to h cf. Haroldsgarth (Shapansay) and Horraldsay (Firth).

Kelday: Easter Kelday 1739. A small unit in Swartaquoy. Probably a mistake for Hestikelday q.v. *infra*.

Meal: Mele 1492; Milne 1500; Male, Mail 1595; Maill, Meall 1614; Maill, Meall 1642. A 4½d. land in tunship of Aikerbister. The name was changed to Graemeshall after coming into possession of the Graeme family in the 17th century. O.N. *melr*, sand; here is (or used to be) a sandy beach.

Netherton: Nethirtown 1492; Nethirtoune 1500; Neather Town 1739. There is a farm so named today. See further Swartaquoy.

Skailtoft: 1595, 1642; Skeall Toft 1614. A ½d. land; name and site now forgotten, but it seems to have been somewhere in west of Holm. O.N. *skála-topt*, ' hall-steethe,' the site of a *skáli* or hall.

Swartaquoy: 1595, 1614, 1642; Swartaquy 1739. Though this name appears on Mackenzie's charts (1750) as a tunship name, and is entered in the

1739 Rental as an 18d. land (or urisland), it must have been a later or secondary name which had displaced the original urisland-, or tunship-name. That indeed is implied by the very name itself, a *kvi-* name being well-nigh incredible as an original-settlement name. Swartaquoy first appears on record in the 1595 Rental where under that name are specified 4½d. lands *pro rege* and 10½d. lands *pro episcopo*. These lands however are entered in the earlier 1492 Rental under the name Nethertown. The other 3d. lands of this urisland formed the lands of Hurtiso, and as that farm is situated on the higher ground of the tunship it would seem that it constituted the 'upper-town,' and it is possible that Hurtiso represents the original name for this whole urisland tunship. Swartaquoy must in the first instance have been the name of one of the later-formed constituent farms in the tunship, and in course of time it has somehow come to be used of the whole tunship just as Aikerbister did, and Beaquoy in Birsay.

Valay: Wala 1595; Wallay 1614; Valey 1642; Vallay c. 1640; Valay 1739. A 4½d. land in Aikerbister tunship which in the 1492 and 1500 Rentals appears as How. Neither name is now in use but the farm must have been in the northern part of the tunship. Origin of name uncertain, but as it replaced How (a mound) the most probable would seem to be O.N. *hváll* (dat. *hváli*) a rounded and somewhat isolated hillock. That word was synonymous with O.N. *hóll*, and is a form used in western Norway, occurring also frequently in Færoe, e.g. *Úti i Váli*, or *Uppi á Váli*.

Valdiger: 1739. A small unit in Hensbister. Cf. this name in St Andrews.

Westerbister: 1492; Westerbuster 1492, 1595; Westerbustar 1500, 1595. A 9d. land or half-urisland tunship. O.N. *vestr-bólstaðr*, 'west bister' (relative to Easterbister).

Withaquoy: Quhyitqwy, Quhitqwy 1553 (R.E.O.); Whataquoy 1739. A farm in Westerbister still in existence. O.N. *hvíta-kví*, 'white quoy.' The reason for the adjective 'white' in several Orkney place-names is doubtful. Cf. Quatquoy, Firth.

Wilderness: 1739. Still a farm today—in Hensbister tunship. Reason for such a name unknown.

OTHER FARM NAMES.

Ayre: represented today only by some grazings near St Mary's village, but in the 16th century the Sinclairs of Ayre (or Air) were one of the notable Orkney families. See Ayre (Deerness).

Backakelday: a farm in Westerbister. O.N. *bakka-kelda*, 'slope spring-well.'

Button: Nether Button is a farm today on the south bank of a deep valley in which is another farm known today as Deepdale. That name is a translation of O.N. *djúpi-dalr*, and as Button represents O.N. *botn*, bottom (of a valley) it would seem that the Deepdale lands were formerly part of the lands of Button.

Creabreck (krɛ-, kre·brɛk): a farm in what was probably the old Westerbister tunship. See next word.

Crearhowe (krɛr.-, kre·ər–hɔu): a farm probably in the old Swartaquoy tunship. This farm is a mile distant from Creabreck, but the first element in each would appear to be from the same source, and may well be the same as appears in the Deerness Creya, q.v. The second element of each is clear—O.N. *brekka*, a slope, and *haugr*, a mound.

Gorn: the Hall of Gorn is a farm today, probably representing the old How and Valla. See Goir (Sanday).

Hestikelday (hɛst·ə-): O.N. *hesta-kelda*, 'horse-well'; *hestr* properly meant a stallion.

Hestimuir: O.N. *hesta-mór*, 'horse moor or heath.'

Hestwall: O.N. *hest[a]-vǫllr*, 'horse-field.'

Lyking: a farm today situated northwest of Hurtiso, and probably outside the limits of the old Swartaquoy tunship. O.N. *leik-vin*, 'sports-field or pasture': see Lyking (Sandwick).

Russamoa: O.N. [*h*]*rossa-mór*, 'horse-heath'? Cf. Mo (Westray).

ST. OLA

Parochia Sancti Olavi 1492; —— Olaui 1500; —— Sanct Olave 1576 (R.E.O.); ——Ollaw 1587 (R.E.O.); St. Olas Parochin 1614; St. Olaw 1642.

In the heart of this parish lies the town of Kirkwall, and its name is derived from that of the old church of St. Olaf in that town, a church supposed to have been built about the middle of the 11th century by Earl Rognvald Brusason in honour of his friend the sainted king of Norway who had been slain in 1030.

The original skatting of this parish is somewhat uncertain as the old Rentals are very confusing thereon; there is also some reason to suspect that several of the quoys attached to Kirkwall may have previously formed parts of surrounding farms. There were however roughly about seven urislands in all.

RENTAL NAMES

Aisedale: Aisdaill 1536 (K. Chart.), 1642; Aisdale 1595. This together with Quoybanks formed a 3d. land in 1595. The lands lie in a valley debouching into the main Scapa valley from the east. First element uncertain but perhaps simply indicating ' east.'

Backbighouse: 1492. A 1d. land. This curious name never appears again on record, but from its position in the Rental the lands would seem to have been somewhere in the Hatston-Saverock area. No reason for the name can be suggested.

Berstane (bɛr·stən, bɪr·stən) : Birstane 1595. A 3d. land of old, and a large farm today.
As this farm-name does not appear in either the 1492 or 1500 Rentals (where names derived from O.N. *staðir* appear as *-stath* or *-staith*) it is uncertain whether this can be regarded as a *-staðir* name though that is most probable. The problem is complicated by the existence of a farm of the same name in South Ronaldsay which in a deed of 1677 appears as a 1d. land only, forming a part of the tunship of Paplay (*Scot. Hist. Review* xvii, 19). Such a unit would appear to be much too small for a normal *-staðir* farm, though originally it may have been larger. If these are genuine *-staðir* names the first element can hardly be other than the O.N. personal name Bersi.

Caldale: 1595. A 3d. land. The present farm occupies a ridge between two side-valleys debouching into the large valley on the south side of Wideford Hill. Such a side-valley was commonly named *kalfa-dalr* (calf-valley) by the early Norsemen, but as O.N. *kalfr* in Orkney has come to be pronounced exactly as the English calf (e.g. Calf of Eday, Calf of Flotta) it is most probable that Caldale represents O.N. *kald-dalr*, ' cold dale.'

Cannigill: Cannigil 1595; Cannigill 1536 (K. ch.), 1614; Cannagail 1642. A small farm today which in 1595 formed with Fea a 3d. land. The second element is O.N. *gil*, a narrow ravine or valley, but the former is of uncertain origin.

Cleat: Clet 1595, 1642; Cleatt 1614; Cleat 1739. An old 9d. land and a large farm today. See Cleat (Sanday).

Clova: 1492, 1614, 1642; Clowa 1595, 1614; Clove 1739. A ¼d. land (1492). Now incorporated with Fea. Evid. O.N. *klofi*, a cleft on a hillside, etc. Cf. Clowally (Orphir).

Corse: Corss 1492, 1614; Corse 1595; Cors 1642. A 3d. land in 1492 and a good farm of today. O.N. *kross*, a cross. Exact reason for that name here is unknown; see Corse (Rousay).

Crantit: Crannystoft 1492; Grantet 1595, 1642; Grantit 1614; Crantit 1739. An assizeman in 1578 signs as 'Robert Grantoft.' (R.E.O.) A ¾d. land in 1492, and a biggish farm today. First element no doubt a personal name, probably O.N. Grani; initial 'g' occasionally appears in Orkney and Shetland as 'k'; e.g. O.N. *gren* (an animal's lair) is found in Orkney as *krane*—(for a lobster's hole). Hence probably—the house-site of a man Grani.

Fea (fiə): 1595, 1614; Fia 1642. Together with Cannigill it formed a 3d. land in 1595, and it survives as a farm today. O.N. *fjall*, hill.

Foreland: Forland 1492, 1642, 1739; Foreland 1595; Foirland 1614. In 1492 it formed along with Nettillhil a 4½d. land. Precise meaning of name uncertain. In Norway *forland* is interpreted in G.N.iv.2 as = *fodr-land* (fother-land), i.e. rich in hay, but in G.N. xiii we read: "It can be assumed that *forland* has originally had the meaning 'flat land in front of hills near the coast,' i.e. = the Icel. *forlendi*, flat land between hills and the sea. With Foreland in Shet. Jakobsen compares No. *forar-lengd*, a field strip, and from Færoe Matras cites *fodlendi* which he thinks can be best regarded as representing an old *forlendi* (with the Icel. sense). This Ork. Foreland is on flattish ground at the foot of hilly ground, but is in the middle of the valley stretching from Kirkwall to Scapa, and thus not 'between hills and the sea.'

Gaitnip (gɛt·nɪp, gjɛt·nɪp): Gaitnepe 1492; Gaitneip 1595, 1642; Gaitnip, Gettnip 1614; Gaitneep 1739. A 3d. land in 1492. O.N. *geit[a]-gnipa*, 'goat-headland.'

This farm is situated on high ground near steep cliffs about a mile south-east of Scapa, and must certainly be the farm referred to in the O. Saga as '*at Geitabergi*' where a man Borgar lived in the mid-12th century, and from which he spied the ship in which Sweyn Asleifson was conveying Earl Paul Hakonsson to captivity. In an earlier chapter of the Saga it is stated that Borgar and his mother Jaddvǫr (a base-born daughter of Earl Erlend and thus a half-sister of St. Magnus) lived '*a Knarrarstauðum*' (Knarston) which is situated on low ground on the west side of Scapa. That, however, would seem to be a mistake from more reasons than one, and in the Danish translation of the Saga (c. 16th century) it is stated that Jaddvǫr and her son Borgar lived '*paa Jadvorstodum*.' This duality of name is referred to in Part III where it is suggested that here we have another example of a *-staðir* name displacing an older name for a time as in the case of Flenstath in Deerness. *Geitaberg* and *Geita-gnipa* are practically synonymous terms.

Garth (gɛrt): Garth 1492, 1576 (R.E.O.); Gairth 1595. With Langarth it formed a 1d. land in 1492. Disappears from Rentals after 1595 but is still a small farm today. O.N. *garðr*, a farm.

Glaitness (glɛt-nez): Gletnes 1536 (K. Ch.), 1642; Glaitness 1595, 1739; Glettnes 1614. A 3d. land in 1595 and a farm still.

As there is no other farm of this name in Orkney this must presumably be that referred to in the O. Saga as—'a Glettunesi,' the home of a man Grimkell in the 12th century.

The name is unusual. In the first place the farm lies on the south-west shore of the tidal lagoon known as the Peerie Sea (adjacent to Kirkwall), and, though the shore-line may have altered somewhat, there can never have been here anything more than a very inconsiderable 'ness.' The first element is still more puzzling. 'Glette' occurs once as a farm-name in Norway (G.N. x), and Prof. M. Olsen could only suggest that formally it might be either *glette*, a clearing in the sky, or *gletta*, smoothness, or a slippery way. The Saga form of this Orkney name indicates O.N. *gletta*, which meant banter, or teasing, something to irritate one, and, literally, the name would thus seem to denote the 'ness of teasing or irritation.' Such a name may have arisen from some notable but unrecorded incident.

Grain (grɛːn): Gyrn 1492 and 1500 (*sub* Evie); Graine 1536 (K. Ch.); Grynd 1595. A 6d. land in 1595. The old spellings are confusing, that of 1595 suggesting that the name was then regarded as *grind*, a gate, and there may indeed have been at that time a gateway through the old hill-dyke somewhere near this farm. The earlier forms, however, as well as the present-day pronunciation, seem to indicate an O.N. *grœn[a]* in the sense of a green patch of ground, a term which appears in Færoe as *grøna*, and according to Jakobsen also in Shetland. Grain occurs again in Rousay as the name of a hill croft.

Grimsetter (grɪmˑsɛtər, grɪmˑstər): Grymsett, Grymesett 1492; Grymesetter 1500; Grymsetter 1595; Grimsetter 1642; Grimsitter 1739. A 1d. land in 1492; today forms part of the aerodrome of this name. O.N. *Grims-setr*, the farm of a man Grímr.

Grimsquoy: Gramesquye 1550 (R.E.O.); Gremisquoy 1576 (R.E.O.); Grymesquoy 1595; Grymsquoy 1595 (*sub* Deerness).

A small quoy-farm adjacent to Grimsetter, and probably the *kví* of the same man Grímr.

Hatston: Haitstach 1492 ('c' is a misreading of 't'); Hatstoun 1536 (K. Ch.); Hatstane 1595. A 6d. land in 1595, most of it now forming part of the aerodrome of this name. An O.N. *staðir* name, and in such the first element is no doubt a personal name: thus perhaps *Hadds-staðir*, the 'stead' of a man Haddr.

Heatherquoy: Hadderquoy 1595, 1642; Hetherquoy 1614. A small unit on the shores of Inganess Bay which survives as a small farm today. Evidently a late formation, as the first element is Scots.

Holland: 1595. A 3d. land on the Head of Holland, now part of the farm of Seatter. See this name North Ronaldsay.

Instabillie: Instable 1492; Enstabillie 1500, 1595; Instabillie 1500. A 1½d. land in 1492 which disappears from the Rentals after 1595. Situation unknown, but from its place in the Rentals between Orquil and Warbuster it is now probably part of the farm of Orquil or of Tofts.

O.N. *Innsta-bœli*, 'innermost b—— or farm? It is impossible to tell the point of reference to which it was 'innermost.' The original sense of *bœli* here is also doubtful as that term was also used of a 'buil' for animals, i.e. a place where animals were gathered together for the night, a cattle-fold, etc. This name occurs again in Sandwick parish.

NAME MATERIAL 99

Jadvarstodum: (O. Saga). See Gaitnip.

Knarston: Knarrarstauþum (twice), a Knarrarstaþi (twice), a Knarrarstodum (O. Saga); Ovir and Nethir Knarstane 1492, 1500; Knarstane 1595. An old 4½d. land now forming part of Lingro farm.
   O.N. *Knarrar-staðir*, the 'stead' or farm-settlement of a man Knǫrr.

Lingro (lɪŋ·ro): Lyngrow 1492; Lingro 1500; Lingrow 1595. A 'skatland' or 4½d. land in 1492 and a large farm today. The first element is certainly O.N. *lyng*, heath or ling; the second is more doubtful. From Shetland Jakobsen cites Lingerø which is probably the same name, and which he regarded as O.N. **lyng-rjóðr*, the second element of which meant an open space in a forest, a clearing, or a grassy patch amid heather, etc. It is difficult, however, even in Norway to distinguish between names derived from *rjóðr* and those from O.N. *ruð* which also signified a clearing, and which forms an element in a vast number of relatively late farm-names in that country. In parts of Norway *ruð* is now pronounced *-ro*, and it may very well be that word which is to be seen in Lingro. In either case the meaning is much the same—that of a clearing among heather.

Muddisdale: Muddisquoy and Mudisquoy 1536 (K. Ch.); Muddisquoy 1595, 1706 (Hossack). The change to -dale seems to have taken place in the 18th century. First element almost certainly a personal name though the precise name is uncertain. O.N. *Móði?*

Natural: see next word.

Nettillhil: 1492; Naturall 1595, 1642, 1739; Netural 1614. Name and situation now forgotten; in 1492 Nettillhil and Forland went together to make a 4½d. land. This name is a complete mystery. Both forms are pretty certainly Scots corruptions or mis-spellings of an earlier Norse name, but what that was is quite obscure.

Newbigging: name forgotten; see Tofts.

Orquil (ɔr·kwəl): Orqwill 1492; Orquile, Orquhil 1500; Orkill 1595. A 1d. quoyland in 1595, and a large farm today. See this name Orphir.

Papdale (pab·del): Papdall, Papdell 1536 (K. Ch.); Papdale 1573 (R.E.O.), 1595; Pabdaill 1614; Papdaill 1692, 1739. An old 9d. land or half-urisland, and a good farm still.
   O.N. *papa-dalr*, 'valley of papas,' i.e. old Celtic clergy.

Quoybanks: Quoybankes 1536 (K. Ch.); Quoybankis 1536, 1595, 1642. With Aisdale it formed a 3d. land in 1595. Almost adjacent to this farm and situated at the foot of the Clay Loan in Kirkwall there used to be an ancient building known as the House of Banks. The early history of that house is unknown but it is probable that there was a farm of that name (O.N. *bakkar*, slopes) here before there was a town of Kirkwall at all, and that its lands and some of its 'quoys' were incorporated in Kirkwall. Quoybanks may have been a quoy attached to that farm, or perhaps been named after its owner—some member of the family of Banks whose surname was adopted from the farm name.

Saverock (sɛv·ərək): Saverok 1492; Savereck 1536 (K. Ch.); Saverock 1595. In 1492 Zardshow and S—— formed a 5d. land; the former is long obsolete, but Saverock is a large farm still. The farm-buildings, however, were rebuilt at a fresh site last century; the old houses were at the shore near the remains of an old mound—probably a broch-site. O.N. *sœvarhaugr*, ' sea -mound.'

Scapa: Skalpeið (O. Saga); Scalpaye 1492; Scalpay 1500; Scapa 1595; Skappa 1614.
There are two farms bearing this name—Upper and Lower Scapa. In the 1492 Rental the former was entered as Ouir Scalpaye, a 6d. land, and the latter as " Are viz. Nethir Scalpaye," a 4d. land. The name Are (ayre—O.N. *eyrr*) never appears again, but it may well have been the name for both farms prior to their separation; there is a distinct ' ayre ' here still.
The name Scapa (*Skalpeið*) has been commonly regarded as signifying ' ship-isthmus,' but the same name occurs as a farm-name in Norway (Skalpe) where Falk (G.N. v. 301) interprets it otherwise. O.N. *skalpr* was a poetic term for a ship, but its more prosaic sense was that of a sword-sheath (etymol. something cleft in two) and it could be used in place-names for ' a long hollow or depression in the terrain.' That is exactly what there is between Scapa and Kirkwall, and the name might be understood as ' long valley isthmus (*eið*) '. On the other hand, if the original idea of ' something cleft in two ' survived in the term *skalpr*, Scapa might be interpreted as the isthmus cleaving the Orkney Mainland in two. While the original form of the name is not in doubt, its precise interpretation is thus still uncertain.

Seatter (set·ər) : Seter 1595; entered merely as a ' quoyland but scat,' and not appearing in any other Rental. A large farm today embracing the old Holland and perhaps part of Westermeall. O.N. *setr*, a ' seat ' or farm-settlement.

Soulisquoy (su·les-kwai) : 1536 (K. Ch.); 1595. A 1d. land pro rege. First element evidently a personal name : O.N. Solli, or Sǫlvi?

Tofts (tafts): Tofts (a 1d. land pro rege) and Tofts viz. Newbigging (a 3d. land pro epo.) 1492; Toftis 1500, 1595. Tofts is still a farm today, but Newbigging has vanished. See Tafts, Sanday.

Warbuster : 1492, 1595; Warbustare 1500. A 1¼d. land in 1492 and 1500; later it appears linked up with Tofts. The site of this old farm-house was remembered until recently. It seems to have been on the north side of the Orquil road, above Tofts. The name clearly indicates an O.N. *varðbólstaðr*, ' ward- or beacon farm,' but there is no memory of a beacon-site near here, though there may have been one on the top of the ridge behind —near the present Braehead.

Westirmele : 1492; Waster Maill 1500; Wostermale, -maill 1595; Meall 1739. A 3¾d. land now part of Cleat (and perhaps Seatter); a field of Cleat near the shore of the sandy Bay of Meal is still known as the Meal field. O.N. *melr*, sand.

Weyland (wi·lən[d]) : Weilland, Weiland 1536 (K. Ch.); Weyland 1595, 1642; Weland 1614. A 6d. land bordering Kirkwall on the east. Almost certainly the oldest parts of Kirkwall must have been built on land previously part of Weyland. Origin of name doubtful; see Weland, Shapinsay.

Wideford : South and North Widefirth 1492; Southweidfuird and Northweidfuird 1500; Whytefurd and Whitfurd 1536 (K. Ch.); S. and N. Weidfurd 1595; S. and N. Witfirth 1614; Wydfoord 1739. Both farms made up a 7d. land in 1492, and this is a large farm today.

An obscure ame in itself and doubly so through its being confused with Wideford Hill which is on the other side of Kirkwall—a different name altogether. This farm lies at the inner end of a bay now known as Ingaess Bay, a branch of the firth or channel leading into Kirkwall from the east. The earliest spelling of this farm-name seems to link the name with a firth, and there is no other firth to come into consideration here. But the first element of the name Widefirth still requires explanation, and there are at least three possibilities : (1) O.N. *viðr*, wide; (2) O.N. *viðr*, a tree or timber; (3) O.N. *víðir*, willow. As for (1) the firth outside Ingaess Bay is from three to five miles wide. As for (2) the present Woodwick in Evie was spelt Withwit (final *t* a mistake for *k*) in the 1492 R., and Weithweik in 1500, but the O.N. word has been Englished to *Wood-* since that time; and as for (3) the O. Saga *Víðivágr* in South Ronaldsay has now become Widewall Bay (cf. Kirkwall for older Kirkwa); and near the mouth of Wideford Burn which runs into this bay is a wide stretch of marshy ground with considerable growth of brushwood including no doubt willow.

On the whole, possibility (3) would seem to be the most likely origin. Incidentally it may be noted that on the south shore of this firth is a lagoon near a farm called Weethick which, there is little doubt, must be regarded as O.N. *við-vík*, timber-bay, the same name as the present Woodwick in Evie.

Work : Wirk 1560 (R.E.O.); Work 1595. A former 9d. land or half-urisland, and a big farm still. The farm-steading here is built on top of the ruins of some prehistoric structure—perhaps a broch. O.N. *virki*, a fortification.

Yairsay : Zardshow 1492; Yarsay and Yairsay 1536 (K. Ch.); Yairsay 1595. This farm, now long out of existence, went along with Saverock in 1492 to make a 5d. land.

The earliest form points to O.N. *garðs-haugr*, ' garth-mound,' in which the first element refers probably to some old dyke or wall, perhaps an old ' gairsty.' The site is quite unknown, but the houses probably were adjacent to that old mound, and there is a bare possibility that the prehistoric earth-house encountered in excavations for the construction of Hatston aerodrome may have indicated the site of that mound.

OTHER FARM-NAMES.

Carness : ' ness illud nuncupatum lie Carness ' (K. Ch. 1536). A relatively modern farm. The name is of uncertain origin, but as this headland is the nearest point of the Mainland to Shapansay—a mile distant across ' The String '—it may perhaps be a truncation of O.N. *kallaðar-nes*, ' calling ness,' i.e. a point at which one might call for a ferry-boat, etc., from the farther shore. Cf. Callarnish, Lewis.

Quanterness : ' ness de Quanterness,' (K. Ch. 1536). This now large farm lies on north side of Wideford Hill and is relatively modern. Origin of first element most obscure.

Quoyberstane : not in old rentals and apparently a farm broken out on what was formerly styled the East Hill of Kirkwall. Here (as in the case of Quoybanks *supra*) it is difficult to determine whether this has been a quoy pertaining to an older farm—Berstane—or whether it was one belonging to a man of that name. In this case the latter alternative would seem preferable as the lands of another farm (Seatter) lie between it and Berstane.

Quoydandy: a relatively late date is apparent from the Scottish personal name—Dandie.

Smerquoy (smer·kwi) : a small farm on west slope of Wideford Hill. Origin of first element rather uncertain. In Shetland are several Smer- names which Jakobsen would refer to O.N. *smjǫr* or *smør*, butter, and the same root is assumed for various Norwegian names, applied as a term of praise in respect of the fertility of each place. In Orkney, however, the word *smero* (smɛro) is used of natural clover, as is No. *smæra* and Mod. Icel. *smári* and *smæra*, and such an origin would seem preferable in this case. Hence probably—' clover quoy.'

ORPHIR

(i) Aurfuru, Aurfioru, Orfioru, Iorfioru (O. Saga); Orphare, Orphir 1492; Orphair 1500, 1595; Orpher, Orphir 1595.

This parish name is that of the old tunship in which the parish church stood. The early Rentals show that tunship as including the 9d. land or half-urisland Bu of Orphir, together with some twelve smaller units, which together made up 3d. lands. Some confusion however had occurred by that time, for while the 1500 R. included two of these—Banks and Grega—among the 3d. lands, the 1492 R. explicitly excludes them, stating "Banks and Gregay, of my Lords will, with the Bull [Bu]," and enters their rents separately. The 1500 R. shows each of these to have been ¼d. land, and in this tunship also were included Quoyclerks, a half-farthing land, and another unit, Myre, of which the value is not noted. Hence we may reasonably conclude that a 1d. land has been somehow lost sight of in the Rentals, and that the Orphir tunship included not 12 but 13d. lands.

Adjacent to this tunship on the west was the 8d. land tunship of Midland, while west of that again was the 6d. land tunship of Howth. The name Midland of itself implies that that tunship was the middle part of a once larger unity which must have included Howth on the west and Orphir on the east. Those three together would thus have included 27d. lands or 1½ urislands, and would to all appearance represent one large original settlement or 'land-take' such as is referred to in Part III.

What name, it may be asked, did that larger unit have? Half-surrounded by the old Midland and Howth tunships is a picturesque harbour known now as the Bay of Howton, but memorable as the Meðallandshǫfn where King Hakon Hakonsson laid up part of his fleet for the winter after his return in 1263 from the disaster at Largs. For an early settler from Norway that cove must have seemed an ideal refuge, for in the mouth lies a small island—the Holm of Howton—which makes it an entirely safe and almost land-locked anchorage. The entrance on the east side of the Holm is open for small vessels at any tide, while that on the west side (now called Holm Sound) ebbs completely dry at spring tides.

Now the name Orphir represents O.N. (Icel.) *ørfiri* or *ørfjara* which meant "an outgoing, ebbing" (C. and V.), and Vigfusson adds "Ørfiris-ey is the pr. name for islands which, at low water, are joined to the mainland by a reef which is covered at high water." That definition applies exactly to the Holm of Howton, and we may thus confidently assume that its old name would have been Ørfirisey, and that the neighbouring part of the mainland was Ørfiri or Orfjara, i.e. Orphir. And the part to which that name was originally applied can have been no other than the unity referred to above which included the three later tunships of Howth, Midland and Orphir.

When that earlier unit was divided into three tunships one cannot tell (though it must have been prior to 1263 when a 'Midland' was already i existence), but it would appear that its name was thereafter restricted to the largest of the three, even though it happened to be farthest from the geographical feature that gave rise to the name. The old earls' seat or 'Bu,' so often referred to in the O. Saga, was also located in that tunship about a mile distant from Meðallandshǫfn, one explanation perhaps being accessibility to a plentiful water supply from the adjacent burn all the year round.

When parish churches were set up, and parishes delimited, the name of the Orphir tunship where the parish church stood was applied as the name for the whole parish. That included rather more than 4½ urislands, and we find the constituent tunships enumerated in a report of 1627 printed in Peterkin's Rentals as follows: Howbuster, Groundwater, Tuskebuster, Kirbuster, Swanbuster, Orphir, Midland, Howtoun, Kowbuster, and the two Claistranes (besouth and benorth the burn). These were all termed 'touns,' but in addition there were three 'roumes'—Naversdaill, Smogro and Orakirk—these probably consisting of single farms only.

RENTAL NAMES

Aikers: Akiris 1492; Akirris 1500; Akarris 1595; Aikiris 1614. A 1½d. land in Swanbister tunship; a farm no longer. O.N. *akrar* or *ekrur*, arable fields (plur. ending Englished).

Banks: 1492; Bankis 1500, 1595, 1614, 1642. A ¼d. land in Orphir tunship near Head of Banks. O.N. *bakkar*, banks, slopes.

Breck: Brekindwaith 1492; brek et Wayth 1500; Breck 1595; Brek 1614, 1642. A ½d. land in Swanbister tunship. The early spelling suggests two small units—Breck and Waith, the first of which is O.N. *brekka*, a slope, and the latter perhaps *vað*, a wading-place or ford.

The Bu of Orphir: Bull of O. 1492, 1500, 1595, 1614; Bowll 1614; Bow 1614, 1642. A 9d. land tunship in which was a main seat of the old Norse earls (often referred to in the Saga) and adjacent to which was the parish church. Some of the foundations of the old hall have been revealed in recent excavations. For the name Bu see Part III.

Clestran (klɛst·rən): Clatestrand, Cletestrand 1492; Claistrand 1500; Clestrain 1595; Clestran 1595. This old tunship was divided into two :—C——benorth the burn, an 8d. land, and C——besouth the burn, a 3d. land probably originally. The same name occurs in Stronsay, q.v.

Coubister (ku·bəstər): Cowbuster 1492; Cowbustar 1595; Cowbister 1614; Kowbustare 1627. A 3d. land. See Coubister, Firth.

Crowall: Corwell 1492; Crowell 1500; Crovale 1595; Crowall 1614, 1642. A ½d. land in Orphir tunship, immediately east of Inkster. Origin doubtful.

Crya: (krai·ə): Gregay 1492; Grega 1500; Crega 1595, 1614, 1642; Craga 1614. An old ¼d. land in Orphir tunship and a small farm today. A very puzzling name which occurs also in Rousay, Evie, Deerness and Stromness and perhaps elsewhere. In my *Place-names of Rousay* I discussed the name at some length, but the derivation I there suggested would seem ruled out by the early spellings of this Orphir case which had for the time escaped my notice. Origin therefore quite uncertain.

Fea: Feaw 1500, 1595; Fea 1614; Fia 1642. An old 1d. land in Midland tunship and a small farm still. O.N. *fjall*, hill.

Gara: Garay 1492, 1595, 1642; Gara 1500; Garray 1614. A 1d. land in Swanbister tunship, and a small farm still. Origin uncertain.

Gossaquoy: Gossaquy 1492; Gossaquoy 1500, 1595, 1614, 1642. With another quoy it made a ¼d. land in Orphir tunship. Probably O.N. *gása-kví*, geese-quoy.

Grindally (grɪnd·əli) : Gryndale 1492; Grindale 1500; Grindela 1595; Grindulla 1614; Grindelay 1642. A 2d. land in Midland tunship. O.N. *grindar-hlið*; see this name—Sanday.

Groundwater: Grimswater (evident misspelling) 1492; Grundwattir 1500; Groundwattir 1595; Groundwater 1642.

An old 9d. land (½ urisland) tunship on north-east side of a loch from which it had its name, though now known as the Loch of Kirbister. O.N. *grunn*[*a*]*-vatn*, shallow loch or lake.

Gyre (gai·ər) : 1492, 1500, 1614, 1642; Gara in 1595, probably through confusion with the Swanbister Gara. In 1500 Gyre with Scalebuster formed a ½d. land in Orphir tunship. Name now changed to Gear. O.N. *geiri*, a 'gore' or wedge-shaped piece of land.

Hangaback : Hangabak 1492, 1500, 1642. In 1500 formed together with Grind a ½d. land in Orphir tunship. Situated near shore banks, whence O.N. *hengi-* or **hangi-bakki*, 'hanging bank or slope,' so named from a steep declivity. Cf. Hangeland, N.G. ix., 281.

Hobbister (hɔb·əstər) : Howbustir 1492; Howbustirland 1500; Houbister 1595; Howbister 1614, 1642; Hobbister 1739. A 6d. land tunship. There is here a very conspicuous mound (a probable burial mound) to which the name must refer—O.N. *haugr*, a mound or cairn—and it is thus of different origin from the Sanday Hobbister, q.v. There is a third Hobbister in Stenness characterised by several mounds.

Howth (hɔuþ) : Howe 1492; Howth 1500, 1614; Howt 1595. A 6d. land tunship fringing the Bay of Howton. O.N. *hǫfuð*, head (or in place-names—headland). That must have been the original name for the adjacent bluff headland now called The Head of Howton. The name Howton itself represents an old *hǫfuð-tún*, i.e. the tunship of Howth. Cf. Howth near Dublin, which has the same origin.

Ingamyre : 1595, 1614, 1642; Ingamyir 1595. An old 1 mark land; site unknown. Probably O.N. *engja*[*r*]*-mýrr*, 'meadow-bog.'

Inkster : Inkesettir 1492; Inksetter, Ingsetter 1595; Ingsiter 1614; Inksetter 1642. A ¼d. land in Orphir tunship, and a small farm still. O.N. *engja*[*r*]*-setr*, meadow-seat or -farm.

Kirbister (kɪr·bəstər) : Kirkbuster 1492; Kirkbustir 1500; Kirkbustare 1574 (R.E.O.); Kirbister 1614, 1642. A 3d. land tunship. O.N. *kirkju-bólstaðr*, 'kirk-bister' or church farm.

Lerquoy : Lerquy 1492; Lerquoy 1500; Larquoy 1595; Lairquoy 1614. A ½d. land in Swanbister tunship, and a small farm today. O.N. *leir-kví*, 'clay-quoy.'

Midland : Midland, Mydland 1492, 1500. An 8d. land tunship. O.N. *meðal-land*, 'mid-land'; see *sub* Parish Name *supra*.

Mussaquoy : Mossaquoy 1500; Mussaquoy 1595, 1642. A 1d. land in Midland tunship, and a small farm today. O.N. *mosa-kví*, "moss-quoy.'

Myre : 1492, 1500, 1595. There were two farms so named, one in Orphir tunship called in 1614 Nethermyre, the other a 1d. land in Midland called Owermyre, i.e. Upper Myre. O.N. *mýrr*, mossy or boggy land.

Naversdale: Naversdaill 1627; Naversdeall 1739. An unskatted farm—probably a late outbreak—up in the hill. Probably O.N. *Nafars-dalr*, Nafarr's valley; *Nafarr* was a fairly common O.N. name; cf. Navershaw (Stromness).

Orakirk: Oikirk 1492; Orakirk 1500, 1595, 1642; Orrakirk 1614. A ½d. quoyland in the Coubister (Petertown) district; still a farm today. Here is the site of an old chapel dedicated it would appear to St. Peter. The site is on the north bank of the Burn of Coubister, almost at its mouth. The first syllable Or- plainly indicates O.N. *ár*, genitive of *á*, a stream, but the next syllable—a—is of uncertain origin. It might perhaps represent O.N. *haugr*, a mound, but that is uncertain.

Orquil: Orquyll 1492; Orquile, Orquill 1500. A vanished ¼d. land in Orphir tunship. The situation, as appears from a map of Orphir published in Saga Book of the Viking Society, Vol. III, p. 184, was near the junction of two burns—The Burn of Linnadale and the Burn of Orquil—where a Brig of Orquil is marked on the map. Though the name here in Orphir is obsolete it occurs also in St. Ola, Evie and Rendall, applied in each case to a farm near the meeting of streams. Hence O.N. *ár-kvísl*, lit. 'stream-fork,' which is found as a farm-name also in Norway. For the dropping of the 's' sound cf. the Norw. dial. variants—*kvihl*, *kvitl*, *kvikl*, etc. (Aasen and Ross), and the dropping of that sound in Ork. and Shet. dialect words and place-names from the combination xl(ksl) in the case of words derived from old Norse: e.g. yackle (a molar tooth) from O.N. *jaxl*; Slap of Aklar (Rousay) from O.N. *axlar*, gen. of *oxl*, a shoulder; cf. also Occlester in St. Andrews *supra*.

Quoyclerks: Quyclerk 1492; Quoyclerk 1500; Quoyclarkis 1614, 1642. A farthing land in Orphir tunship (1500), and a small farm still. Evidently a quoy pertaining to, or assigned for the use of some cleric or churchman. Cf. Canker, Sanday.

Quoys: Quyis 1492; Quoy 1500. A small unit which went with Gossaquoy to make a ¼d. land in Orphir tunship.

Skelbister: Skelbustir 1492; Scalebustar 1500; Skakebustar 1595; Skobister 1614; Skobuster 1642. Together with Gyre it formed a ½d. land (1500). The name is now obsolete, and from the strange variations in spelling since 1500 one suspects it must have been obsolete even then. The earlier spellings, however, seem to point to an O.N. *skála-bólstaðr*, a farm with a *skáli* or hall. There is another Skelbister still in this parish (in Groundwater tunship), and there are others in Sanday and Gairsay (Skelbist). As these are all more or less inland the prefix can have no reference to O.N. *skel*, shell, or a shelly beach.

Skidgibist (skidʒ·i·bəst): Sketybustir 1492; Sketebustar 1500, 1595; Skettebuster 1614. A ½d. land in Swanbister tunship (1500), now merely a field-name on Swanbister farm. The same name is to be found in Sanday. Origin uncertain; cf. Skidge (Birsay).

Smoogro (smu·gro): Smogrow 1595; Smougrow 1614; Smugro 1642. A 3d. land (1595). The name is now applied to the mansion-house of this estate, the lands forming the farm of Yarpha. Origin of name obscure.

Sowlie: Sowly 1492; Soulie 1500; Sowlie 1595, 1642. A 1d. land in Swanbister tunship, and a small croft today. Origin obscure.

NAME MATERIAL

Sorpool: Sowrpow 1500; Sorpow 1595; Soir Pow 1614; Soirpow 1642. A 2d. land in Midland tunship. The farm no longer exists but it was on low ground near the Bay of Howton : hence probably a combination of O.N. *saurr,* mud, and *pollr,* a pool (Scots pow).

Swanbister: Swanbustir 1492; Swanbuster 1492, 1500, 1595; Swanbister 1492; Swambuster 1642.

A ½ urisland or 9d. land tunship of old, and a large farm today.

The prefix here is puzzling. In Orkney there are two farm-names besides this having Swan- as first element—Swandale (Rousay) and Swanney (Birsay). The former appears in the 1595 R. as Swindale, which suggests an old ' swine-dale,' but no early form of the latter is known. In Norw. place-names Svan- seems capable of various interpretations. In the case of this Orphir name the first element (associated with a *bólstaðr*) is most probably the personal O.N. name *Sveinn* which was well known in Orkney, and which seems to survive in the surname Swanson (common in Caithness): in Orkney we find a record of 1488 having reference to a "Johne Swansonne . . . sone to Swannie of Gruthill " in Deerness (R.E.O. 197) which would seem to suggest that the vowel change had taken place before then.

Swartabreck: Swarthbak 1500; Swartabreck 1595; Swartabrek 1642. A 1d. land in Midland tunship. O.N. *svarta-brekka,* black or dark slope.

Tuskerbister: Custarbuster 1492 (an obvious misreading of 'c' for 't'); Tuskarbustar, -er 1500; Tuskibuster 1614; Tuskebuster 1642. Skatted as a 9d. land tunship in 1492 though entered in rental as an 8d. land.

A curiously interesting name, as the first element can hardly be other than O.N. *torf-skeri,* ' peat-cutter,' a kind of special peat-cutting implement; a term which survives in Orkney today in the form ' tusker.' In combination with *bólstaðr* such a prefix has presumably been a personal name, but there seems to be no record of its being so used in Norway or elsewhere. It may be noted that this tunship is the farthest inland in Orphir, and lies in the midst of peat-country.

Windbreck: Wyndbrek 1492, 1500, 1642; Windbreck 1595; Wind Braik 1614. An old ¼d. land in Orphir tunship. Probably O.N. *vin-brekka,* pasture-slope.

OTHER FARM-NAMES

Clowally (klou·əli) : a small farm at the summit of the public road between Howton and Clestran. Here there is a slight hollow or saddle in the hill ridge with a somewhat steep slope up on the north-east side. The second element is pretty certainly O.N. *hlið,* a slope, and the first is probably O.N. *klofi* (orig. ' something cleft '), a term which in Fær. place-names has acquired (*inter alia*) the meaning ' track,' espec. a cattle-track. Dr. Matras informs me that *Klovalið* (klo·vali·) is a place-name actually occurring in Færoe, apparently the same name. Here in Orphir a track or road has certainly run over the hill-ridge past this farm from ancient times.

Gerwin (gɛr·wən) : a small farm in the Orphir tunship. In the absence of early forms of the name its origin must be regarded as uncertain, but the second element may conceivably be O.N. *vin,* pasture. It is adjacent to Windbreck. q.v. Cf. also Geroyne (gjərɔ·in) the name of a now vanished cot in Rousay.

Greeniegeo (grin'i-gjo): a farm at the head of a ravine running down to Scapa Flow. O.N. *grœna-gjá*, green ravine.

Kebro (ki·bro): a farm on the brow of a hill in Tuskerbister tunship. Origin obscure.

Oback (o:bak): a farm by the side of a burn in Tuskerbister. See this name—Deerness.

Scows (sku:z): in Groundwater tunship; most probably O.N. *skógar*, pl. of *skógr*, woodland, forest. Cf. Skogar, Birsay. The occurrence of such names in Orkney can denote only scrub or brushwood, as large trees have not flourished naturally in Orkney since the Boreal period some 6000 years ago or more.

Skaill (skel): a farm in Tuskerbister. O.N. *skáli*, a hall.

Yarpha (jar·fə): the home-farm of the Smoogro estate. See this name—Deerness.

CAVA

Calfay (Fordun) c. 1375. O.N.*kálf-ey*, 'calf-isle'; probably metaphorically as a calf alongside a cow, which would thus be Hoy.

This small isle does not figure in the early rentals prior to 1595, but in the Report of Orphir (1627) printed in Peterkin it is stated: " The ile of Cava, ane pendicle of the parochin of Orphar friers-land perteining to the blak friers of Inverness, and be the act of annexation was annexit to the Crown. . . . ."

The isle had only about three small farms, and is now uninhabited.

STENNESS

a Stæinsnese (O. Saga); Stanes, Staness, Stenes 1492; Stanehous 1500, 1595; Stennis 1546 (R.E.O.).

O.N. *Stein-nes*, 'stone-ness,' so named from the famous Standing Stones near the Brig of Brogar: the medial 's' in the Saga form is probably a mistake (perhaps for a genitive plur. 'a').

The early Rentals are confusing in regard to Stenness, but by correlating the data from all three (1492, 1500 and 1595) it becomes apparent that the parish embraced 3½ urislands.

As the Saga dates from pre-parochial days in Orkney the name must have applied then to the tunship of Stenness, a 6a. land in which the Stones referred to are situated, and wh'ch as in the case of Orphir has given its name to the whole parish.

RENTAL NAMES

Bigswell (bɪgz·wəl): Bigiswell 1492; Bigiswall 1500; Bixwell 1551 (R.E.O.); Bigswall 1595. A 3d. land in a valley up in the hill. The second element is no doubt O.N. *vǫllr*, a field, while the first is probably a personal name.

Clouston (klu·stən): Clowstath 1492; Clonstaith ('n' by mistake for 'u') and Johne Cloustane 1500; Clustay 1522; Clouchstay 1527 (R.E.O.); Cloustane 1595; Cloustoun 1607 (R.E.O.). A 6d. land.

This is one of a group of no fewer than five *staðir* names in this parish. In the vast majority of such names in Iceland and elsewhere the first element is a personal name, and such is almost certainly the case here also. The late Mr. Storer Clouston argued with much cogency that the actual individual whose name appears was the 12th century Hakon Kló (son of Havard Gunnason), a man who is referred to repeatedly in the O. Saga. A 12th century date is admittedly late for the application of a *staðir* name, but another Orkney example appears in the O. Saga—Jadvorstodum in St. Ola, q.v.

Colston (kʌl·stən): Cottistaith (the 'tt' a palpable mistake for 'll') 1492; Culstane 1565 (R.M.S.). From some obscure reason this tunship does not appear in any later rental, but it figures in old legal documents, etc., as Culston, Culstane, etc., and the name is still remembered today. It was a 9d. land or half-urisland. Probably O.N. *Kollz-staðir*, or *Kolla-staðir*, the settlement of a man Kollr or Kolli.

Dowscarth (dɔuz·gər): Dowaisgarth 1492; Dowskarth 1500; Dowascarth 1595. A 3d. land. As the 1492 form suggests this is pretty clearly a 'garth,' not a 'skarth' name; the site is a valley up in the hill, but not in a skarth (O.N. *skarð*, a col or saddle in a hill ridge).

The first element is probably the O.N. personal name Tófi. Shortly before his death the late Mr. Storer Clouston had collected a deal of evidence which seemed to indicate that the patronymic Toveson or Tuison, etc., was an old alternative name for people of Clouston ancestry like himself—the surname Clouston of course being derived from the name of the tunship Clouston which is a neighbouring tunship to Dowskarth. One striking piece of evidence he had noted was that a Sir Audulf Toveson whose seal was attached to a document of 1287 in Norway used the same coat of arms as that of the Cloustons, and a John Dwison in the 16th century owned land in Ness in Tankerness which Mr. Clouston believed to have been old Clouston property. Oddly enough, I do not recall his ever having thought of Dowskarth in connection with the Tovesons, but its proximity to Clouston tunship is certainly suggestive.

Germiston: Grimistith 1492; Garmistane, and as a surname Garmiscath (where ' c ' is an obvious misereading of ' t ') 1500; Garmistoun, Garmiston 1595. A 6d. land tunship. Today the name is used of the district, not of a single farm. O.N. *Geirmundar-staðir*, the ' stead ' or settlement of a man Geirmundr. With the compression of that name cf. Hermisgarth, Sanday.

Ireland: Areland 1492; Irland, Airland 1500; Irland 1595; Iyrland 1627. A whole urisland tunship. Today the name is used of the district, and the principal old farm in it is called the Hall of Ireland. O.N. *eyrland* or *eyrarland*: O.N. *eyrr* was used in a general sense of a gravelly beach, but in Orkney in the form ' ayre ' it is applied nearly always to a narrow spit of land (generally a pebbly ridge) partially shutting off a salt-water lagoon from the outer sea with which it has a connection through a narrow gap at one end of the ' ayre.' On the shore of this Ireland tunship near the Bridge of Waithe, there is a curving spit of land partially enclosing a small bight, and it was probably from that coastal formation that the tunship had its name.

Unston (ʌn·stən): Onsta 1546, 1576 (R.E.O.); Onestone, and surname Onsta 1576 (R.E.O.); Unstane 1595; Vnstoun 1627.

A 1½d. land adjacent to Clouston; still a farm today. This also is an old *staðir*-name, but the first element is obscure, though almost certainly a personal name.

Ottergill: 1595. A 1½d. land between Ireland and Clestran (in Orphir). Second element O.N. *gil*, a ravine, but it is uncertain whether the first represents O.N. *otr*, an otter, or the personal O.N. name *Óttarr*: a small burn entering the sea here might lend greater probability to the animal source.

Tormiston: Tormystaith 1492; Turmiston 1627. A 9d. land tunship; still a good farm today. O.N. *Þormóðs-staðir*, the ' stead ' or settlement of a man Thormothr.

OTHER FARM-NAMES

Aglath (ag·ləþ): id. 1735. Small farm in Ireland tunship, lying towards the lower end of a ridge on west side of the Burn of Ireland. Origin doubtful. The same name occurs on the border between the parishes of Harray and Firth where the north-western slopes of the hill known as The Ward of Redland are called The Braes of Aglath. The first element—*Agl-*, which occurs in several Norwegian names, has not been satisfactorily explained even in that country.

Anderswick: Anderswik 1563 (O.S.R.). A small farm near Dowskarth. The second element is apparently O.N. *vík* which Rygh notes as being used of a bight or indentation in a hill-ridge, as well as in the more usual sense of a bay. Here the hill side is indented by such a valley. The first element is probably a personal name which might well be O.N. *Andrés* (Andrew), a name which was borrowed in Norway as early as the first half of the 12th century, and which was the name of a son of Sweyn Asleifsson in Orkney (12th century also).

Appiehouse (ap·i-hus) : there are two houses so named in this parish, one in Clouston, the other in Germiston tunship. This type dates back to a time when the tunship came to be divided, and new farm-houses were named from their relative geographical position to others. In Clouston e.g. there is an Appiehouse and a Nether Bigging (the old head-house), while in Germiston there are an Appiehouse, a Nisthouse, an Eastabin and a Nistaben (East and Lowermost Bigging). From Færoese and Shetlandic parallels the original form of Appiehouse was O.N. '*uppi í húsi*,' lit. 'up in house.' The change of *uppi* to Api- is in line with the old Orkney pronunciation of English *up* as *ap*.

Barnhouse: the main farm now in what was the tunship of Stenness. Its early history is unknown.

Bea (now pron. bi·ə through influence of spelling, but the old Orkney pron. was simply be :) : a farm in Ireland tunship. O.N. *bœr*, a farm-settlement; see Part III.

Brogar (bro·gər, bro·djər, brəd·jər): Broager, -ar, 1563 (O.S.R.). A farm in Stenness tunship situated on the point at the north-west end of the Brig of Brogar. O.N. *brúar-garðr*, 'bridge-farm,' a name indicating the existence of a bridge here dividing the Lochs of Harray and Stenness from a very early period; perhaps only stepping stones at first.

Button: a small farm in Ireland tunship. See Nether Button, Holm.

Coldamo (kɔld·əmo) : farm in Ireland tunship. O.N. *kaldi-mór*, cold moor.

Cummaness (kʌm·ənəz) : Cumanes (1618 Sas.). A farm on a small headland in Ireland tunship. Here stands a prehistoric mound (a probable broch-site) known as Cummy Howe or Cummany Howe. Origin obscure.

Cuppin (kʌp·in) : a small farm up in hill behind Germiston. O.N. *kopp[r]-inn*, 'the cup' (or hollow).

Eastabin (est·əbɪn) : small farm in Germiston. 'East Bigging'; see Appiehouse.

Hobbister: a small district where two farms bear this name—Upper and Nether H——. O.N. *hauga-bólstaðr*, 'mounds-bister' or farm; so named for mounds in the area.

Housequoy (hus·kwai) : a farm in Stenness tunship. Reason for prefix uncertain.

Kethisgeo (keð·ez-gjo) : a small holding above Clouston. The last element is O.N. *gjá* (Ork. geo) a ravine or cleft in the landscape, especially at shore, but the first element is quite obscure.

Moa (mo) : an old farm in Culston tunship. O.N. *mór*, moor or heathland.

Nether-bigging: the old head-house of Clouston tunship. See Appiehouse.

Nistaben: see next name.

Nisthouse: both this farm and the last-named (Nistaben) are in Germiston tunship and about half a mile apart. Both names mean practically the same thing—the lowermost house or 'bigging' (O.N. *neðsta hús*, or *bygging*), and the only apparent explanation of such a duplication would seem to be that the tunship had been divided and that each of these was the lowermost in its own part. Cf. also Appiehouse.

Overbigging: a farm in Culston. O.N. *øfra-bygging* upper-b.

Queena (kwin·ə): a farm in Germiston. O.N. *kvi* (quoy, cattle-fold, etc.) with def. art. either in accus. sing. *kvina* or nomin. plur.—*kviarnar*.

Queena-marion: the old name for the present Lochside in Culston. For the first element see last name; the second is obscure but strongly suggests a woman owner—Marín, Marína, Marion, etc.

Queenamoan (kwin·ə-mɔ:n): a farm in Tormiston tunship. A compound of O.N. *kvi* and *mór*, moor or heathland: *kvin á móinn*, ' the quoy on the heath,' or perhaps in dative—*mó(i)num*.

Quoyer (kwai·ər): a farm in Culston. O.N. *kviar*, nomin. pl. of *kvi*, cattle-fold, etc.

Ramsquoy: Rammisquoy (1619 Sas.). A farm in Ireland tunship. First element almost certainly the personal O.N. name *Hramn (Hrafn)*.

Savedale: a small holding above Clouston tunship. First element uncertain, but probably O.N. *sef-dalr*, ' sedge or rush-valley.'

Scuan (sku·ən): a farm in Germiston. O.N. *skóg-inn* (accus.), the shaw or woodland. Cf. Scows, Orphir. From Shetland Jakobsen cites a place-name Skuen brenda (O.N. accus. *Skóginn brenda*) ' the burnt wood.'

Skethaquoy (skeð·ə-kwai): a small farm in Ireland tunship now joined to Settersquoy. Cf. Skae, Deerness.

Twatt: farm in what was probably the old Tormiston tunship. An interesting occurrence in Orkney of a name found again only once (in Birsay). It must represent O.N. *þveit* (Eng. *thwaite*) a word difficult to interpret exactly in Norwegian place-names, but which Dr. Kristian Hald has recently shown (*Vore Stednavne*, København 1950) may be interpreted in the sense of ' a clearing ' (*rydning*); here probably of heather or brushwood.

FIRTH

i Fiaurþ (O. Saga); Firth 1500, 1595. O.N. *fjǫrðr*, fjord.

Here again, as in the case of Stenness and that of Orphir, the parish took its name from that of a tunship. The present village of Finstown occupies part of that old tunship of Firth, at the inner end of a bay which penetrates into the heart of the West Mainland, and is now called the Bay of Firth. In early Norse times it was known as *Aurriðafjǫrðr* (O. Saga), i.e. Trout Firth or Fjord. The later parish of Firth stretched out along each side of that firth, and embraced rather more than three urislands of skatted land. The parish church is in the Firth tunship also.

RENTAL NAMES

Benzieroth (bɪn·jəro): Binyereth 1500; Benyereth 1595; Binzæreth 1601; Benzieroth 1794. A 3d. land on northern shore of bay. A difficult name: see Binscarth.

Binnaquoy (bɪn·jə-kwai,-kwi): Binzequoy 1595; Binzæquoy 1601; Binaquy 1627; Bingaquoy 1794. A ½d. quoyland adjacent to Binscarth, q.v.

Binscarth: Benyeskarth 1500; Bynzescarth and Binzescarth 1595; Binzæskarthe 1601; Benascart 1627; Bingascarth 1794. A 4d. land west of Firth tunship, and situated in a deep gap or 'skarth' in the hill-ridge there. The first element of this name is common to a group of Orkney place-names, and in several cases difficult to explain. Here it is relatively clear, the name pretty certainly representing O.N. *bœ[r]inn i skarði*, 'the' bae or farm in the skarth (gap in the hill-ridge). Adjacent to this farm on the same hill-slope is a smaller old farm Binnaquoy (bɪn·jə-kwai, -kwi), q.v., which may perhaps have been established as a 'quoy' of Binscarth, the name thus signifying the 'quoy of the bae.' At the shore, however, about half a mile farther from Binscarth is the old farm of Benzieroth (bɪn·jəro) q.v. As the lands of Horraldsay intervene, this can scarcely be regarded as a further outgrowth of Binscarth. Yet the first element of the name seems the same, while the second is doubtful, perhaps representing O.N. *rjóðr*, a clearing (as in Lingro).

In Sandwick parish are two farms each called Bain, but that name is pronounced as a disyllable—bɛ·ɒn, a form which undoubtedly represents O.N. *bœr* (a farm) with the def. article attached—*bœ[r]inn*, 'the' bae or farm. In Sandwick also, in the tunship of Skaebrae is a farm Bea which represents O.N. *bœ[r]* without the article. But in the same tunship is another farm which (probably to differentiate) is called Benziecleft (bɛn·ji-klɛt), which is fairly clearly O.N. *bœ[r]inn á kletti*, 'the' bae (farm) at a klett or rock.

In the North Isles Benyie- is found as first element in several names. At the beach in Stronsay on the farm of Houseby (an old *bœr*-name) is a prehistoric cairn called Benyiekuml (bɛn·ji-kʌml) where there seems little doubt that the name indicates the *kuml* or burial mound of this *bae* or farm. In Sanday, in the Evirbist district, is a small farm Benyiecot (bɛn·ji-kɒt), which name again suggests a cot of a *bae*, and though there is now no *bœr* farm in that vicinity there is a farm with the relatively modern name—Beafield, which suggests that a *bœr*-named farm once existed here. The same name Benyiecot occurs in Westray applied to a small farm in Rapness area, and in Rousay a field of Saviskaill goes by the name of Benyiecot. Then again in Westray a part of the beach near Noltland Castle is called Benyieber, while another beach in Sanday is known as The Ebb of Benyasty (ben·jəsti).

Some of those Benyie- names thus appear in association with farms whose names incorporate the element *bœr* as part of their *nomen proprium* (e.g. Houseby), which others do not. It is certain, however, that *bœr* was used also as a *generic* term for a farm (as we see it used e.g. in the *O. Saga*), and it would thus seem reasonable to assume that the prefix Benyie- does have reference to some farm even though that farm does not always have a *bœr* proper name. In fact the most feasible explanation of the three farm-names *supra* would seem to be as follows. There can be little doubt that the original farm-settlement in this area would have been that at the head of the bay, known as Firth tunship later. Though *bœr* did not form part of its *nomen proprium* yet it was a *bae* or farm-settlement, and when another farm was established in the 'skarth' above it might very probably have been differentiated by the name of the 'bae in the skarth.' Binnaquoy would have been a quoy attached to one of these two *baes* or farms, and Binyero might then very well be regarded as a 'clearing' or extension made to the original *bae*, viz. Firth. In some of the examples of the name cited above, however, there is real difficulty in divining the O.N. grammatical form from which Benyie could develop, and in the absence of intermediate forms no suggestion is at present hazarded.

Finally, to add to the difficulty it must be mentioned that a hill in Walls on the north side of Longhope is called Binyafea!

Burness: Burnes 1500, 1595; Burness 1794. A 6d. land tunship at a headland on the north shore of the Bay of Firth where there is still visible a prominent broch-mound. Hence O.N. *borgar-nes*, 'broch-ness.'

Coubister (ku·bəstər): Cowbustir 1500; Cowbustar 1595; Coubister 1794. A 1d. land. First element doubtful; it might phonetically represent O.N. *kúa-* genit. plural of *kýr*, cow, but curiously enough, on the south side of the Bay of Firth immediately opposite is a farm Cursetter, a name which almost certainly represents O.N. *kýr-setr*, cow farm or settlement, and a different 'cow-prefix' in such close proximity would be quite unlikely. Hence the first element here is probably a personal name— O.N. *Kúga-bólstaðr*, Kugi's farm. A man of that name is recorded as living in Westray in the 12th century (O. Saga).

There is another Coubister in Orphir parish.

Cursetter (kʌr·sətər): Cursetter 1500, 1595, 1627; Cursater 1794. A 3d. land on south side of Bay of Firth. Probably O.N. *kýr-setr*, 'cow farm.'

Firth: an old skattland, i.e. a 4½d. land. See under parish name.

Grimbister: Grymbustar 1500, 1595; Grimbuster 1601,1794; Grembister and Grimbister 1627. A 4½d. land or old skattland between Firth tunship and Cursetter. Though this name might represent O.N. *grœni-bólstaðr* (greenbister) with dropping of inflexional 'i' and labialising of 'n' to 'm' before 'b,' it is more probably O.N. *Grims-bólstaðr*, the bister or farm-settlement of a man Grímr: cf. Grimsetter and Grimsquoy, St. Ola; Grimeston, Harray, and Grimness, South Ronaldsay.

Heddle (hɛdl): Hedal 1425 (R.E.O.); Hadale 1500; Haddale 1595; Heddell 1601; Heddle 1794. A 3d. land up in the hill. O.N. *hey-dalr*, hay-dale.

Holland: id. 1500, 1595, 1601, 1794. A 6d. land. See this name, N. Ron.

Horraldsay (hor·əldze): Thorwaldishow 1500; Horraldshay and Horraldsay 1595; Horralescho 1627; Horraldshay 1794. A 3d. land. An interesting name as illustrating the change of initial 'th' to 'h' as in Hurtiso (Holm). O.N. *Þorvalds-haugr*, the mound of a man Thorvaldr.

Quatquoy (kwat·kwi): Quhytquoy 1500, 1595, 1601; Quytquy 1627; Whitquoy 1794. A 1d. land near north shore of Bay of Firth. There are now three farms—North, South and East Q——. Same name as Withaquoy, Holm, q.v.

Redland: in the *Complaint* of 1425 against Menzies of Weem a "Magnus in Renaland" is referred to, almost certainly this farm. In R.E.O. also, p. 76, an entry appears from 1504 of "Jhon Flet, Bere of Randell and Will Flet, etc.," In P.O.A.S. X, however, Mr. Clouston states in a footnote on p. 37 that subsequent examination of the original showed that this should read: "Jhon Flett sere of Randelland. etc.," the reference being again to this farm. Earlier forms therefore are: Renaland 1425; Raynland 1500; Randelland 1504; Redland 1595, 1601, 1627, 1794.

A 6d. land tunship on east side of Redland Hill.

As is obvious from Mr. Clouston's difficulty in deciphering the document of 1504 the writing must be far from clear and the form Randelland can therefore be accepted only with hesitation, though it might help to explain the otherwise inexplicable appearance of the medial 'd' in Redland. A burn runs down here, and one might thus compare the name with Rendall, q.v., but for the present the origin must be deemed quite obscure.

Rennibister: Rannebustar 1500; Ranybustar 1595; Ranibister 1627; Rennibister 1794. A 6d. land adjoining St. Ola parish. The second element is plain, but the first uncertain. A burn enters the sea here, and the present pronunciation of the name might suggest an O.N. *Rennu-bólstaðr* (cf. Rendall), which would imply that the name of this burn was Renna in Norse times just as the burn of Rendall was so named. Such a stream-name however is rare in Norway, and it is more likely that a personal male or female name appears here as first element. In the absence of sufficiently early forms of the name its origin must be deemed quite uncertain.

Saville (sa·vəl): not in earlier Rentals; Savale and Benyereth 1595; Savell 1601; Sauiel 1627.

With Benyereth it formed a 3d. land. See Savil, Sanday.

Seater (set·ər): id. 1500, 1595, 1794; Sæter 1601. Along with Wasdale this formed a 2d. land. O.N. *setr*, farm.

Settiscarth: Setscarth 1500, 1595; Settiscarth 1595; Setskairth 1601; Settascart 1627; Settiscarth 1794. A 3d. land tunship situated in a valley running up to a prominent 'skarth' or gap in a hill-ridge. O.N. *setr-i-skarði*, farm at a 'skarth.'

Wald: Wathill 1500; Wale 1595; Vaill 1627; Waal 1794. A ½d. land. Today there are two small farms—North Wald and South Wald—situated near the Oyce [øs] (O.N. *óss*, river-mouth) of the Burn of Isbister. From the opposite side of this estuary a point of land projects, leaving a rather narrow channel across which one can wade at certain times of tide. Hence O.N. *vaðill*, a wading place or ford.

Wasdale: id. 1500, 1595, 1794; Wesdale 1601; Waisdell 1627. This together with Seater formed a 2d. land. Though this name is commonly regarded as 'West-dale' the first element refers to the adjacent loch now known as the Loch of Wasdale: hence O.N. *vaz-dalr*, 'lake-dale'; (from *vaz* = *vatns*, genit. case of *vatn*, a lake).

OTHER FARM-NAMES

Appiehouse (ap i–hus): a small farm in Settiscarth tunship. See this name—Stenness.

Breckan: a farm in old Firth tunship. O.N. *brekkan*, 'the slope.'

Cruan: in Grimbister tunship. O.N. (Icel.) *kró*, Norw. *krú* (with def. article suffixed), 'the krue or crue (krø).' The Norse word, which is used of a pen for animals, was a borrowing for Early Celtic *cró*. Today in Orkney the term krue (krø) is used still in a generic sense for a small enclosure used rather for growing cabbage plants than as a pen for animals; but it is becoming obsolete.

Damsay: appears repeatedly in the O. Saga where the MSS spellings vary between Damisey and Daminsey. It is a small island in the Bay of Firth, and though not entered in the early rentals it appears to have been skatted as a 3d. land. There was a castle of sorts here in the 12th century, but the isle is uninhabited today and no traces of the castle are now visible.

Origin of name obscure, but as there are two small islands near together in this bay (the other being called the Holm of Grimbister) the name may be compared with the Celtic loan-word *Dímun* which occurs in Norway and in Iceland for twin heights, and in Færoe for twin islands —Stóra and Lítla Dímun.

Grandon (grand·ən): a small farm in the old Firth tunship on the shore of the 'Oyce,' the name here applied to a lagoon at the mouth of a burn running through the Binscarth valley. Origin not quite certain, but a low point of land projecting into the Oyce near this farm may have given rise to the name: O.N. *grandi*, a sandbank or gravelly bank in or by water. In that case, here with def. art. *grandinn*, 'the grand."

Ingshowe (ɪŋz·ou): a small farm across a burn-mouth from Rennibister. The name is taken from that of an old broch-mound at the shore here. Origin of first element uncertain—probably a personal name.

Lettaly (lɛt·əli): a small farm high up in the Settiscarth valley. The termination is no doubt O.N. *hlíð*, a slope, and the first element is probably O.N. *leiti*, a place of far outlook. In his *Indleðning* Rygh states that this term was used of 'a place where one can see far, as in a valley where a stream runs for a considerable distance in a straight line and no projecting elevations block the view.' Here there is only a tiny winding stream but otherwise the definition exactly fits the situation. Hence probably—'a slope with a far view.'

Lyde (laid): a small farm at head of Settiscarth valley. As this place is at the top of a steep slope one might suppose that the name is O.N. *hlíð*, a steep slope, a term which appears in South Ronaldsay as well as in North Yorkshire and Lancashire as Lythe. In Orkney, on the other hand, *hlíð* normally becomes *lee* (li). Here, however, is a 'skarth' (saddle or col) in the hill-ridge which separates Firth and Rendall from Harray, and over that skarth a public road now follows the course of an ancient track which formed a connecting link between these parishes and was (and is) known as The Lyde Road. Hence we can here confidently regard Lyde as representing O.N. *leið*, a road or track.

Midbigging: a small farm in Grimbuster tunship. See Netherbigging, Stenness.

Midhouse: a small farm in Settiscarth tunship.

Moan (mɔ:n): a farm near hill west of Holland. O.N. *mó-inn*, (accus.) 'the moor or heath.'

Netherbigging: in Grimbister tunship. Cf. Midbigging.

Quear (kwi·ər): a very small farm. O.N. *kvíar*, plur. of *kví*. a quoy.

Queena (kwin·ə): see this name—Stenness.

Rossmire: 'meadows of Rosmyre' 1587 (R.E.O.). Evidently not a really old farm. O.N. *hrossa-mýrr*, horse-bog.

Smogarth (smo·gər): a farm between Horraldsay and Savil. Probably O.N. *smá-garðr*, small garth or farm.

Thickbigging: the chief farm in old tunship of Firth. O.N. *þykkva-bygging*, 'thick-bigging': probably as distinct from a long-house—the common Orkney type of farm-house. Perhaps two houses were built back to back.

Upperbigging: in Grimbister tunship. Cf. Midbigging.

Vindon (vɪn·dən): a small farm in face of hill above Holland. Origin obscure. It is possibly an O.N. *vin* (pasture) name, but in that case the termination would present difficulty. As an alternative one might suggest that it represents what Rygh termed an Imperative name such as the Norw. *Vendom*, 'turn round.' An ancient road northwards from Firth tunship to Redland and beyond runs round the shoulder of a hill, passing close to this farm near which a bend shuts off a view of the tunship behind. In Orkney, however, I know of only one other possible example of such a type of name—Falldown (fa-dun) in Rousay at the foot of a steep declivity, and in both cases the evidence is slender. Origin thus quite uncertain.

RENDALL

til Rennadals, and *til Rennudals* (O. Saga MSS.); *Kolbeinn i Rennadali, and Reinadali* (Hakon Saga MSS); Randale 1492, 1500, 1595, 1601; Rendale 1601. As a surname—Rendell 1516, 1542; Rendale 1539, etc. (R.E.O.).

The first element represents what must have been the name given to the small stream which runs down through the old Rendall tunship—*Renna*, a stream-name attested by several Norwegian place-names: hence *Rennudalr*, valley of the Renna.

Here once more we find the parish name to be that of an earlier tunship. In this case the old parish church was not located in that tunship, but in the adjacent tunship of Gorsness. There is reason however to suspect that originally Gorsness formed part of a larger settlement-unit which embraced Rendall tunship and others as well. Rendall itself was a 9d. land or half-urisland, immediately south of which was the large 24d. land of Gorsness, while adjacent on the north was the 3d. land of Queenamuckle—the three together forming exactly 36d. lands or two urislands. The farms again (of which Tingwall was one) between Queenamuckle and the Evie boundary together made up another exact urisland, and several facts suggest that this whole long seaboard of three urislands was originally one large settlement-unit.

Firstly we read in the Hakon Saga of a man Kolbeinn of Rendall who c. 1230 A.D. was a leading kinsman of the great Wyre family descended from Kolbeinn Hruga. Mr Clouston, in *The Orkney Parishes*, suggested with high probability that this Kolbeinn of Rendall was a son of Sweyn Asleifsson's son Andres, who, it is recorded, married a daughter of Kolbeinn Hruga.

Then in the Ork. Saga it is stated that in the previous century Helgi Rolfsson, an uncle of Sweyn Asleifsson, lived at Tingwall.

Next, in the 1492 Rental it is stated with regard to Gorsness that "of the said land there was 3d. terre be David Rendale given to Saint Ninian" (as church land to a chapel or 'stouk'). Mr. Clouston has pertinently pointed out in a footnote to the Rendall genealogy (R.E.O. p. 463) that "this may be taken as not more than one-sixth of his estate—the tent penny and ferd—which [as an odal landowner] he had the right to give away."

From these records we have evidence of men of the same family (or closely related families) owning lands in the whole area referred to above, and from the Uthell Book of 1601 we find that two men of Rendall surname owned two-thirds of the Rendall tunship still at that late date.

Putting all those facts together we have the strongest presumption that the whole area in question had formed one original estate, which may also have included the island of Gairsay opposite, the home of Helgi's brother Olaf and his redoubtable son Sweyn in the 12th century. The original settlement of course would have been made several centuries earlier, but bearing in mind the tenacity of odal succession there can be little doubt that the various men referred to above were the heritable successors of the original 'land-taker' or settler. And it may not be irrelevant in that connection to state that as evidence of still earlier occupation and cultivation by Picts no fewer than four brochs (or probable brochs) may still be traced on this stretch of coast: one at Tingwall, a second at Waswick, a third near the Hall of Rendall, and a fourth at Dishero near the old parish church in Gorsness.

RENTAL NAMES

Banks: Bankis 1595; Banks 1727, 1794. In 1595 Cruik and Banks were entered together as 1½d. land. This farm exists no longer, but the approximate site of the farmstead was probably near the point below Crook appearing still on the 6" O.S. map as—Banks. O.N. *bakkar*, pl. of *bakki*, a bank or slope.

Cottascarth (kot·ə–skarþ) : Cotscarth 1595; Cottascarth 1727; Cottiscarth 1794. A 3d. land (quoyland) in Settiscarth valley. O.N. *kot í skarði*, 'cot in skarth': see Settiscarth (Firth).

Crook (krøk) : Cruik 1595; Crook 1727, 1794. A farm south of Tingwall; see Banks. O.N. *krókr*, a crook or bend; here applied with reference to the coastline.

Ellibister: Ellibustar 1500; Alibustar 1595; Elebuster 1727; Elibister 1794. A 5d. land in a valley between hills. Second element O.N. *bólstaðr*, farm-settlement; the first almost certainly a personal name, but it is difficult to say which.

Garson: Garsent 1492, 1500; Garsetter (an obvious mistake) 1595; Garson 1727, 1794. With Midland this formed a 6d. land next to the Evie border. O.N. *garðs-endi*, garth- or dyke-end. The reference may have been to an old 'gairsty' dyke; see Treb, North Ronaldsay.

Gorsness: Goryssness 1492; Goirisness 1500; Gorseness 1595; Gorsness 1727. A 24d. land tunship next south from Rendall tunship. See further under parish name. Though curious the name perhaps represents O.N. *garðs-nes*, though it is difficult to say in what sense *garð* should be interpreted here. The vowel change may be compared with the farm or field-names Goar (Sanday) and Gorn (Holm); and the *garð* may possibly have had reference to the head-house farm in Rendall tunship with which (as suggested under the parish name) this tunship was once probably conjoined. Even so, the name would appear unusual, and its origin must therefore be regarded as uncertain.

Hackland: Halkland 1500, 1595; Halckland 1794. A 6d. land tunship on the slope of a spur or 'shoulder' of Enzie Hill. First element O.N. *ǫxl*, a shoulder; here probably in genitive sing. *axlar-land*; for change of xl (ksl) to kl in Orkney cf. Occlester (St. Andrews).

How: 1492, 1500, 1595. Wasweik and How together formed a 3d. land. Farm no longer existing, and site even forgotten. It may have been near the large mound (a presumed broch-site) close to the beach at the bay of Waswick. O.N. *haugr*, mound.

Howaquoy: 1595. Entered along with Tingwall and Midgarth. Almost certainly to be identified with the present Hacco or Huiko (høk·o) a small farm south of Tingwall. May have been a quoy pertaining to the old farm of How, or a quoy at a mound itself: O.N. *haugr*, a mound.

Isbister (aiz·bəstər): Ossbustir 1492; Osbustare 1500; Osbuster 1595; Isbuster 1794. A 12d. land tunship near mouth of the Burn of Isbister or Sweenalay. O.N. *óss-bólstaðr*, 'oyce-bister,' i.e. farm-settlement at the mouth of a stream which widens out to form an 'oyce.'

In Orkney there are three other Isbisters, but not all from the same origin: one in Birsay (Estbuster 1492; i.e. East B.), a second in South Ronaldsay (Ystabustare 1500), and a third in Grimeston, Harray.

Midgarth (mɪdˑjər) : id. 1492, 1595; Mydgarth 1500. A 3d. land; so named apparently from its position between Midland and Tingwall. O.N. *miðgarðr*, 'mid-farm.'

Midland : Mydland 1492, 1500; Midland 1595. With Garson this formed a 6d. land adjacent to Evie parish border. The occurrence of such a name in that situation is puzzling, and would seem to suggest that in pre-parish days the adjacent tunship of Woodwick in Evie formed part of a unity that embraced much of Rendall parish also.

Mosseter : Morsettir 1492; Morsettar 1500; Marsatter 1595; Mossater 1727. A 1½d. land (quoyland). O.N. *mýrar-setr*, 'marsh-, or bog-setter.' In Norway *mýrr* sometimes appears with vowel sound changed—*mør-*, *maar-*. (*Indled.*)

Queenamuckle (kwinˑə-mʌkl) : Quyna-mekle 1492; Quoyname:kill 1500; Quoynamekle 1595; Quinamekill 1601. A 3d. land immediately north of Rendall tunship. O.N. *kvín-mikla* (or in accus. sing. *kvina-miklu*), 'the big quoy.' This was a very early 'quoy' as it was fully skatted.

Rendall (rɛnˑdəl) : a 9d. land or half-urisland tunship. See under parish name.

Tingwall : á þingavǫll (O. Saga); Tyngwell 1492; Tyngwale 1500; Tingwall 1595. A 6d. land tunship. A stream runs down here on the bank of which near the shore is a prehistoric mound (a broch-site) which was probably the meeting place of the old 'thing.' O.N. *þing-vǫllr*, 'thing-field.'

There is no evidence that this was the principal Orkney thing-stead or moot-site; from Saga references that seems to have been near, or at, Kirkwall. This was probably the site of a district thing similar to what was apparently held at Dingishowe, the conspicuous mound on the isthmus between Deerness and the rest of the Mainland.

Waswick (wazˑwik) : Waswit 1492; Wasweik 1500, 1601; Waswick 1595. This together with How formed a 3d. land (1595). There is no farm of this name today, but a bay north of Queenamuckle is so named. On the shore of this bay is a probable broch-mound, immediately inland from which is a wet boggy expanse which in former times must have been a loch at certain times of the year. Hence O.N. *vaz-vik*, 'loch-bay'; cf. Wasdale (Firth).

OTHER FARM-NAMES

Appietoon (apˑi-tun) : Uppetoon (1629 Witch Trial, Hossack); farm in Gorsness tunship. Cf. Appiehoose (Stenness).

Avidale : farm in Gorsness. Origin of first element uncertain.

Blubbersdale : small farm in a valley of same name up in the hills. O.N. *blábers-dalr*, blae-berry dale.

Bught (bʌut) : a farm near a small bend or bight in the Gorsness coast-line. The same name is applied to a small farm in Eday. Icel. *bugt*, a bight or bay, which Vigfusson regarded as a foreign and rare word in Iceland. The same word occurs in Norwegian, and is stated by Falk and Torp to be a borrowing from L. German.

Breck: farm in the Rendall tunship valley above the Hall of Rendall. O.N. *brekka*, a slope.

Brettobreck: in an older recorded form—Brattobreck. A farm south of Tingwall. O.N. *bratta-brekka*, steep slope.

Crowrar (krəu·rər): farm at roadside above Tingwall and near the old tunship dyke. Origin uncertain.

Bu: referred to as The Bu; an alternative name is Upper Inkster. This was the old Bu of Gorsness. See Bu (Part III).

Curcabreck (kʌrk·ə–brɛk): farm near Tingwall. O.N. *kirkju-brekka*, church slope: there has evidently been an old church or chapel near here.

Ettit (ɛt·et): Aetoft, 1620 Sas.; Aithtoft, 1629 Witch-trial (Hossack). There are two farms now—North and South Ettit, the latter being down below the former close to the beach adjacent to the old parish kirk of Rendall. Locally, South Ettit is also known as Ayre.

Origin of first element obscure; at North Ettit is a large prehistoric mound which may indicate the site of old buildings from which the second element—toft—arose. O.N. *topt* (*tupt*), id.

Corn: farm in Isbister tunship. See Goir (Sanday).

Grind (grɪnd): farm at roadside east of Isbister farm. Near here there must have been a 'grind' or gateway (O.N. *grind*) in the Isbister tunship dyke.

Gutter-pitten (gʌt·ər–pɪtn) and also formerly Gitterpitten (gɪt·ər–): a small farm in Hackland tunship in a wet boggy situation, caused by numerous springs

A difficult name. The 'n' termination of the second element shows that term to be the O.N. *pyttr* with def. art. attached—*pyttinn* (accus.) which meant a 'pit,' cesspool, wet-hole, etc. The first element is more doubtful. There is a farm in Holm named Gutterpool, or rather Gutterpow, which would seem to be a Scots form = 'muddy pool.' Here, however, in Rendall the formation of a hybrid name (*Gutter* (Scots) and *pitten* (Norse)) would be very unusual, and that fact together with the old alternative pronunciation makes one suspect that the first element may be Norse also. *Gitter-* is found as an element in Norse place-names though very rare, and its significance is uncertain. The origin of the first element here must therefore be left unexplained.

Hacco: spelt also Huckow (høk·o): Howakow (1629 Witch trial, Hossack). See Howaquoy.

Hayon (hai·ən): small holding up on hill. Probably O.N. *heiðr*, heath; with def. art., or perhaps in dat. pl.

Hinderayre (hɪnd·ər–ɛːr): a farm at beach in Gorsness. This farmstead is situated near the southern end of a beach or 'ayre,' while close to its northern end are the farm-houses of South Ettit which is locally known also as—Ayre. Hinderayre thus pretty clearly indicates an O.N. *hindra-eyrr*, 'farther ayre,' i.e. from the point of reference which is unknown but may well have been the parish kirk a short distance north of South Ettit.

Hogarth: a farm on elevated ground in Hackland tunship. A short distance down below is another farm Nearhouse which probably here represents

Netherhouse: hence Hogarth probably is O.N. *há(hó)garðr*, 'high garth or farm.'

Inkster: a farm in Gorsness. See this name—Orphir.

Kewing (kʌu·in): a farm between Tingwall and Hackland. Origin obscure.

Lyking (laik·in): a small farm south of Tingwall. See this name—Holm.

Moa: see this name—Stenness.

Nistoo (nɪst·u: nɛs·tu): Nestahow 1594 (Sas.; Nestow 1652 (Kirkwall Sess. Reg.). Farm now united with Appietun, but formerly belonging to St. Peter's prebend in the Cathedral.

The old form as recorded would seem to point to an O.N. *neðsti-haugr*, 'lowermost mound,' but it is difficult to avoid suspecting that the recorded form is corrupt. In the first place there does not appear to be any mound to give rise to the name, and the relative position of the farm suggests a different origin. It lies immediately below Appietun, while a very short distance above the latter is the farmstead of the old Bu of Gorsness. Appietun obviously indicates the upper part of the old farm-tun (cultivated land) of the Bu, and Nistoo is in the position to be called the 'lowermost tun'—O.N. *neðsta-tún*. It may be suggested that the record form was due to an original misreading of Nestu where the final 'n' was indicated by a contraction mark over the 'u.' The exact origin, however, must be regarded as uncertain.

Orquil (or-kwəl): a farm in valley above Waswick Bay. Two small streams united here, hence O.N. *ár-kvisl*, 'stream-fork.' See further under this name—Orphir.

Puldrit (pʌldrit·): a farm in Gorsness at a beach from which the sea does not ebb out properly and seaweed lies rotting there emitting an offensive smell. O.N. *pollr*, a pool; *drit*, 'dirt' or filth.

Queena: a farm in Gorsness. See this name—Stenness.

Quoyfree: a farm in Gorsness. Origin of second element uncertain.

Riff (rɪf): a farm in Gorsness at a beach where there is a projecting 'reef' of rocks running out into the sea. O.N. *rif*, id. The Orkney pronunciation of this word (which occurs in several places) has probably been influenced by the local pronunciation of the word 'roof.'

Skaill: in Isbister tunship. O.N. *skáli*, a hall.

Skiddy: a small farm near the public road between Hackland and Ellibister. A 'brae' here is known as The Brae of Skiddy.

Probably O.N. *skeið* (in dative—*skeiði*) which was used of a course for horses to race on, but is found used in place-names sometimes simply of a track between fields, etc. A track from Firth to Evie must have existed here from earliest times.

Sweenalay (swin·əli): a farm near public road which gives its name also to a stretch of the burn called The Burn of Isbister lower down. First element probably O.N. *svin*, swine, but as there is no very pronounced slope (O.N. *hlið*) here the second element is of uncertain origin.

Velzian (vɛl·jən): a farm in Gorsness. O.N. *vǫllr*, a field; here probably in def. plur. *vellirnir*, 'the fields.'

EVIE

- - - *til Efiusunds* (O. Saga): Evy 1492; Evie 1500, 1595.

O.N. *efja*. According to Rygh's *Indledning* that term was used in different senses: 1. muddy ground; 2. a currentless bight in a river. From later Norse usage it appears to have been applied also to a bay having a swampy margin, or to a back-current in a river, etc. Here the name has no doubt been applied with reference to the bight inside the strong tidal current which flows through Evie Sound (the Saga *Efjusund*) where as in all such cases a 'back-tide' or current in the opposite direction makes itself apparent.

In this case the parish name is that of an earlier unit which was so large that it can scarcely ever have been a single tunship. The whole later parish embraced eight urislands as follows: Outer and Inner Costa forming one urisland, Outer Evie two, Stanesakir or Stenso one, Garth one, Redland one, Aikerness one, and Woodwick one. The early Rentals unfortunately do not specify exactly how many of these were included in Inner Evie, but from their detached situation it is practically certain that Aikerness and Woodwick were not. With these omitted Inner Evie would have embraced the three urislands of Stenso, Garth and Redland; and Inner and Outer Evie together would thus have been an extensive sweep of country, skatted as five urislands—an area much larger than could ever have been a single tunship. The individual urislands enumerated above were all presumably tunships by themselves, and the early pre-parochial Evie must therefore have been a district name for that whole five-urisland area abutting on the *efja* or back-water whence its name was derived. The old parish church was situated in that area, near the beach in what was probably the tunship of Garth.

Many of the Evie farms were in possession of the Bishopric and do not figure in the early Rentals. The absence of early forms of such names is particularly regrettable.

RENTAL NAMES

Aikerness (ɛk·ərnez): Akirness 1492; Aikarnes 1500. A whole urisland tunship, all of which is probably included in the present farm. See this name—Westray.

Arsdale: id. 1492; Arisdale 1500; Arsdale and Airsdaile 1595. A 1d. land in Costa. First element obscure.

Clouk (kløk): id. 1500. A 3d. land in Costa. Origin obscure. There is a farm of the same name in Outer Stromness, and there was formerly another in Rousay.

Costa: Costa and Costay 1492; Costa, Costaith and Costeth 1500; Costa and Costay 1595. An old urisland tunship (or double tunship—Outer and Inner), but no longer a farm-name. A difficult but interesting name. This is the Evie tunship which borders with Birsay, and between these two parishes rises to a height of nearly 500 feet the rounded mass of Costa Hill or Costa Head. Jakobsen cites O.N. *kǫstr*, a heap or pile, which he considers is represented by names in Shetland, e.g. Kostifel, and such an origin might suit here also. But the 1500 variants Costaith, Costeth would seem to exclude such an origin and to indicate that here is a *-staðir*-name, though it is strange that the ending has not become

-*ston* as in most other cases in Orkney. Another exception, however, is to be found in Yinstay (St. Andrews). The first element is thus doubtless a personal name, probably *Kollr* or *Kolli*: hence we might assume an original O.N. *Kollz-*, or *Kolla-staðir*, the settlement of a man Kollr or Kolli. In that case the name would be of the same origin as Culston in Stenness.

Curquoy (kʌrk·wi): Curquy 1492; Curquoy 1500, 1595. A 4½d. land in Outer Evie. The second element is plain—O.N. *kví*, but the first is of quite uncertain origin. In Rousay there is another Curquoy, but it is a small croft far up 'in the hill,' and obviously a relatively late outbreak on the common; it appears in none of the early Rentals, and may indeed have been named for some reason after the Curquoy here in Evie.

The first element may represent a personal name, e.g. O.N. *Kári* or *Kóri*, though that is hardly likely. It is much more probable that the 'q' represents a double 'k-' sound, and one at once thinks of a possible Kirk-quoy (*Kirkju-kví*). That possibility, however, would seem to be ruled out by the fact that in 1492 4d. of the 4½d. lands in this farm were udal-owned and the remaining ½d. land was earldom property.

The name must also be considered in relation to several other names in Orkney beginning with Curk- (kʌrk-) or Cork- (kɔrk-), e.g. Curkland or Curkoland in Egilsay, Curkabreck in Evie, Harray and Birsay, Corkaquina in Birsay, Corkatae in Sanday, etc. These may not all contain the same initial element, but some at least may derive from O.N. *korki*, oats (a loan-word from Celtic *coirce*, id.) which is used as a tabu-name in Unst in Shetland, and in Foula survived in the compound *korkakost*, oatbread. Cf. the Rousay Bigland which probably represents O.N. *bygg-land*, 'bere (barley)-land.'

In the absence of earlier forms of this name its origin is thus obscure.

Fealquoy (fjal·kwi): Feaquy 1492; Feawquoy 1500; Fealtquoy 1595. A 3d. land in Outer Evie. O.N. *fjall-kví*, hill-quoy.

Georth (gjɔrt, djɔrt): Garith 1492; Garth 1500. An urisland tunship in Inner Evie: no longer a farm today, but the tunship included the farms of Dale, Orquil, Hestival, etc. In Outer Evie, however, there is still today a farm Georth which does not appear in the old Rentals. O.N. *garðr*, a farm-settlement. The phonetic change is remarkable.

Grudgar (gru·djər): Growagarth 1492, 1500; Growgarth 1500; Growagair 1595. A 3d. land in Outer Evie. There is a small farm of the same name in Egilsay. O.N. *Gróu-garðr*, the farm of *Gróa*, a female name which occurs very frequently in Norwegian place-names. Cf. Grobister (Stronsay).

Hammiger (ham·igər): Hamagarth 1492; Hammagarth 1500; Hamigair 1595. A 3d. land in Garth tunship. Origin of first element doubtful.

Howa: 1595. A 2½d. land in Outer Costa. Evidently some form of O.N. *haugr*, a mound, but the terminal 'a' is puzzling.

Mydhouse: 1500. A 3d. land in Garth tunship which formed part of the old Prebend of St. Peter in the Cathedral. Name and site forgotten.

Neigarth (ni·gjər): Negarth 1492, 1500; Neagair 1595. A 3d. land in Outer Evie. O.N. *ný-garðr*, new garth or farm.

Newhouse: Newhouss, Newhouse 1492; Newhous 1500 Newarhouse 1595. A 3d. land in Outer Evie.

Quoys (kwaiz) : Quyis 1492, 1500; Quoyis 1595. A 4½d. land in Redland tunship. The plural of O.N. *kvi—(kviar)* anglicised.

Redland : Roithland, Rothland 1492; Rochland ('c' by mistake for 't') and Rothland 1500. (This last is the MS. reading which Peterkin has wrongly transcribed 'wthland'). An urisland tunship which embraced Flaws, Nistigar, Nigley and Quoys, etc. The name is still remembered locally, but seldom used now.

An almost certain example of faulty translation through confusion of two similar O.N. words—*rauðr*, red, and *rjóðr*, a clearing. Rygh states that the latter was also used of an open space in a forest, of green pasture or of a grassy plot, and he adds that it is sometimes difficult in place-names to distinguish it from O.N. *ruð*. The corresponding verb, however, was *ryðja* (cf. Eng. *rid*) which meant to clear ground.

Now as there is nothing in this Orkney Redland soil or landscape to account for such a term as 'red' it would appear that at the time when Norse speech in Orkney was giving place to Scots the Roth- in Rothland was mistakenly regarded as representing O.N. *rauðr*, red, instead of *rjóðr*, a clearing, and there can be little doubt that the original name was *Rjóð-land*, 'cleared land' (whether of heather or rocks or scrub it is impossible to tell).

There is another Redland in Stromness parish.

Spithasquoy : Spittalisquy 1492; Spittasquoy 1500, 1595; Spithisquoy 1595 (S.C.B.). A 4d. land in Outer Evie which the 1500 R. states had been 'conqueist' (acquired) by Earl Wm. Sinclair, the last Norse earl; hence sometime about the middle of the 15th century.

An interesting but all too obscure name. The origin is not in doubt—O.N. *spitals-kvi*, the quoy of a *spital*, i.e. a hospital, or a hospice for travellers. Such institutions were usually set up by some monastic house, and the existence of one in such an out of the way area as Outer Evie in Orkney would be incredible were it not for the clue given by C. and V. under the word *spitalskr* (leprous) that "the old hospitals were established for incurable lepers."

Leprosy was common in northern Europe in the Middle Ages (and much later), and though I cannot recall any literary reference to the disease in Orkney there can be little doubt that it occurred here also. In a paper entitled "Antiquarian Notes on Sanday" (P.O.A.S. I, p. 28) I referred to a rather pathetic leper legend from that island. Whether there ever was an actual leper hospital at this farm, or whether it was merely an isolation-refuge for such sufferers, or whether its lands had been devoted to the support and maintenance of such a hospital elsewhere, it is now impossible to tell. Cf. Spittal—a name in Caithness.

Stenso (stɛns'o) : Stanesakir 1492; Stanissakir 1500. An urisland tunship in Inner Evie. The modern name cannot represent the old Rental name, but the first element of each has had a common origin, which judging from other Orkney place-names must have been a Standing Stone. For the Rental name we can assume an original *Steins-akr* (or in plural—*akrar*), 'standing-stone cultivated field(s).' The modern name has reference probably to the burn which flows down through the tunship and is now called The Burn of Stenso. O.N. *Steins-á*, 'standing-stone burn.' A house at the roadside on the bank of this burn bears the uncompounded name O.N. *á*, burn or river, but as elsewhere in Orkney it is pronounced oo[u:] and spelt Woo. That spelling is on the analogy of the Scots wool for wool, which is also pronounced 'oo' in Scotland.

Turriedale: Thurisdale 1492; Thorodale 1500; Turkadaill 1595 ('k' by mistake for 'r'); Thurridaill 1595 (S.C.B.). A 1d. land in Outer Evie. Origin of first element uncertain.

Woodwick: Withwit 1492 ('t' by mistake for 'k'?); Weithweik 1500; Widweik 1618 Sas.; Widwicke 1627. An urisland tunship of old, and now a district name for the part of Evie adjacent to Rendall. So named from a bay here between the two parishes. The modern form shows a part-translation of O.N. *við-vík*, ' wood or timber bay.' The ' wood ' here is probably to be understood as driftwood.

OTHER FARM-NAMES

Arwick: Nethir Arweik (1618 Sas.); Nether Erweik (1619 Sas.). A small farm in a valley between Vishall Hill and the Hill of Dwarmo on the north side of the Bay of Woodwick. The second element must be O.N. *vík*, used here of an indentation in hills as in the case of Anderswick in Stenness, but the first element is puzzling. A small stream runs down here, but O.N. *ár* (genit. of *á*, a stream) becomes Or- in Orkney, and cannot be considered here. Ar- as a first element in Norwegian names also presents difficulty, and though there it is sometimes referred to O.N. *ari*, an eagle, such an origin in a situation like Arwick is plainly untenable. Cf. however Arsdale in Costa *supra*.

Though irrelevant here it may be mentioned that from some crags on the side of the adjacent Hill of Dwarmo there is a notable echo which explains the otherwise puzzling name—Dwarmo—O.N. *dverg-mál*, an echo; lit. ' dwarf-talk ' (from the old belief that dwarfs (*dvergar*) lived in rocks.

Bisgarth (bɪz·gər): a farm adjacent to Arwick. Bis, the first element, occurs in a few Norwegian farm-names also (including Bisgaarden, this same name) but the editors of N.G. cannot offer any convincing explanation of its origin, and in the absence of early forms its occurrence here must also be left unexplained.

Breckan: a small farm in Costa. O.N. *brekk-an*, the slope.

Bruar: a small farm on the bank of the Burn of Woodwick. O.N. *brúar*, plural of *brú*, a bridge. There must have been a bridge here in older days, but the reason for the plural form is obscure. As the same form occurs in the genit. sing. the name may have once had another word suffixed which has fallen away; e.g. *Brúar-hús*, etc.

Burgar (bʌr·gər): Burgare (1619 Sas.). An old 8d. land in Outer Evie, and a big farm yet. O.N. *borgar*, plural of *borg*, a fortification, the usual name given of old to a broch-site. Here the plural form arises from the presence of two mounds near or at the beach, one of which is a certain broch-site, and the other may have been mistaken for such. As it chances, however. a second almost certain broch-site known as Ryo is at the beach about ¾ mile to the east, and a third at· Peterkirk rather over a mile to the west.

Creya (kraɪ·ə): a small farm in Woodwick. See this name—Orphir.

Crismo (krɪz·mo): a farm at roadside in Costa near the Birsay border. The second element is no doubt O.N. *mór*, moor or heath, but while it is tempting to regard the whole name as O.N. *Krist-mór*, ' Christ's moor,' the significance of such a name would be inexplicable here.

Cufter (kʌft·ər, køft·ər) : a small farm now going along with Bisgarth. Origin quite obscure.

Cupper : there are two small farms so named in this parish—one in Woodwick, the other in Outer Evie. O.N. *koppar*, plural of *koppr*, a cup; in place-names used of a small hollow.

Dale (del) : ' 3d. land in Deal ' (R.M.S. 1600); a farm in old Garth tunship. Origin uncertain. The farm is on a slope but not in a real ' dale.' One therefore (especially in view of the 1600 spelling) is inclined to compare it with the termination -dale seen in Brendale and Ervadale (Rousay) q.v.

Durrisdale : a small farm in Woodwick situated by a tiny stream flowing into the Burn of Woodwick. Origin uncertain.

Dyke (daik) : a farm in old Stenso tunship. Origin uncertain; it is possibly a translation = gairsty, though I am not aware of the presence of any such old structure about here.

Flaws (fla:z) : a farm in old Redland tunship. See this name—Holm.

Fursan (fʌrs·ən) : a farm near Woodwick border. O.N. *fors-inn*, ' the foss ' or waterfall.

Gallow-ha' : a small farm in Woodwick on S.W. slope of Vishall Hill No tradition survives as to whether there ever was a gallows here or not. The same name occurs in Kirkwall at the head of the Clay Loan where criminals were formerly put to death. In a witch-trial case of 1633 it was referred to as Gallows-hill.

Georth (gjɔrt, djɔrt) : a farm in Outer Evie. See Garth *supra*.

Hellicliff (hel·ji-klɪf, -klɪv) : a small farm in Woodwick on slope of Vishall Hill. The same name occurs in Rousay applied to a now-deserted cottage high up on a slope of the Ward Hill. It also occurs in Foula in Shetland where it was explained by Jakobsen as O.N. *hellu-klif*. O.N. *klif* or *kleif* was used of a steep rocky slope, especially of a slope up which ran a track or pathway. In Orkney dialect the form *klivvy* is used for such a pathway itself, a track leading up over a cliff or steep hillside. The first element Jakobsen regarded as O.N. *hellu*, genit. of *hella*, a flat rock or stone, so that the compound would mean ' steep stone track.'

Hestival : a former small farm in tunship of Garth. See this name—Shapansay.

Howe (hou) : a farm in Stenso tunship—so named from a large mound. O.N. *haugr*, a mound.

Jubidee (tʒub·ədi) : a small holding probably of more recent origin near a dam of same name. O.N. *djúpa-dý*, deep bog or marsh.

Midhouse : large farm in Costa.

Mistra or Mistrae : a former farm in Garth tunship. Origin obscure.

Niggly : a farm in old Redland tunship. See this name—St Andrews.

Nisthouse : a farm in Woodwick. See this name—Stenness.

Orquil (ɔr·kwəl) : a farm in old Garth tunship. See this name—Orphir.

Pow (pou) : a farm in Costa. See this name—Rousay.

Pulkitto (pʌlkɪt·o) : a farm in Woodwick. First element O.N. *pollr*, a pool, deepish hole in a stream, etc. The second is probably the woman's name —Kitto, a somewhat contemptuous form of Kate. The name of a pool in the adjacent burn may have arisen from some accident there; a parallel example is Kirsty Paavie's Pow—a small pool in a burn at Nethermill, Rousay.

Stymbro (staim·bro) : a small farm in Redland tunship. O.N. *stein-brú*, stone-bridge. A tiny stream passes nearby.

Urigar (ør·igər) : a farm in Costa. A -garth name, but first element uncertain. Probably O.N. *urðar-garðr*, from *urð*, a heap of stones on a slope where there has been a land-slide.

Vinquin (vɪn·kwin) : a farm in Costa at a small hill of same name. Apparently O.N. *vin[jar]-kvín*, 'the pasture quoy.'

Walkerhouse: now the chief farm in Woodwick, but the name is Scots. There used to be here a waulkmill (mill for waulking or fulling cloth) which was referred to in a letter of 1727 (v. my *Merchant Lairds of Orkney*, I, p. 117). The same name occurs in Birsay, and at Kirbister, Orphir there is a Waulkmill Bay. The mill in each case has had power from a neighbouring burn.

BIRSAY

Birsay is mentioned nine times in the O. Saga, the commonest form being *i Byrgisherapi* (dative of *Byrgisherap*). One of the best MSS. however has on two occasions the variant *Byrgissey*.

Birsay 1455 (R.E.O.); 1595 R.

There can be no doubt that the above MS. variant represents the origin of the modern Birsay—O.N. *Byrgisey*, the island of a *byrgi*. That word (an i-umlaut of *borg*, a fortress) is explained by C. and V. as an enclosure, fence; by Fritzner as *forskandsning, vold* (intrenchment, rampart); and by Hægstad and Torp as *skans, festning* (intrenchment, fort)*. As to the *byrgi*-island, there is only one to be found—the tidal isle now known as the Brough of Birsay. Recent excavation has proved that prior to the arrival of the Norse there was a Celtic church settlement on this isle, and running along the top of the cliff facing one when approaching from the mainland can still be seen the remains of what must have been a Celtic cashel-wall or rampart. From that rampart the isle has obviously had its Norse designation, and it is noteworthy that the term applied was *byrgi* and not *borg* which, in Orkney, was generally applied by the Norse invaders to that type of stronghold we now call a broch. Sometimes, however, that name was also applied to an islet or rock-formation specially difficult of access, and such must have been the origin of this island's present name—the Brough of Birsay (as of the Brough of Deerness). It may be noted incidentally that the spelling *broch* is (or should be) now restricted to actual broch (Pictish tower) sites.

The second element of the Saga *Byrgisherað* is variously interpreted in Scandinavian lands, and, while it is impossible to tell exactly how it should be understood in this case, the most probable sense would seem to be the Icelandic, which (according to Vigfusson) was "a district, valley, fjord, country as bordered by mountains or within the same river-basin." Here in Orkney Byrgisherað must have included not only the present parish of Birsay but that of Harray as well, for the latter name is simply O.N. *herað*. In the 1492 R. Harray is entered as Burgh St. Michaels, and in the 1500 R. as Herray and Harray Brugh (by a misprint Hurray), while in a document of 1490 (R.E.O. 199) it appears as "the parochin of Burch in the Herray." From early Norse times therefore Byrgisherað must have been a geographical term for a large area of the West Mainland, and it is curious how in course of time the second part of the name became by itself a parish name.

A further very curious feature in regard to Birsay remains to be mentioned. The name does not figure in either the 1492 or the 1500 Rentals, but, instead, a large part of the present parish is entered under the name Marwick as if that were a separate parish. Under that name were included not only the present district or old tunship so called, but in addition the tunships of Isbister, Folsetter, Greeny, Sabiston, Beaquoy and what is now commonly termed the Hillside (Evribist, Skelday, Howally, etc.), in fact—roughly the whole parish except what lies north of Ravie and Greeny Hills.

---

* Both in Færoe and Iceland *byrgi* is a term applied to an enclosure, especially a sheepfold. Matras (p. 84) states that in Fær. place-names the term may also be found used of a place so shaped by nature that it can easily be shut off, and he compares Kålund's description of a place in Iceland—*Asbyrgi*: "an angular valley, or, more accurately, a flat-bottomed ravine bounded by steep walls of cliff."

We are thus reminded of the Rental separation of Holm and Paplay, for here again there is no evidence at all that Marwick ever formed an ecclesiastical parish; nor was there any parish church in the area. In these two cases we seem to have a hint of large pre-parish districts, as to the origin or significance of which we have as yet no inkling. But a third example may be noted hereafter in the Rental separation of North and South Sandwick.

The omission of such a large part of Birsay from the two earliest Rentals was probably due to the fact that it was all (or practically all) Bishopric estate, and, as the Bishop's right to skats therefrom was evidently not challenged, there was no occasion for including that area in these Rentals. By 1595, however, the old Bishopric estate had been acquired by the Stewart earls, and the whole parish was included in the Rental of that year.

This Rental reveals further curious parallels to Paplay. After the acquisition of that estate by Earl Wm. Sinclair in the 15th century we noted the establishment there of a 'grange,' the exact nature of which is obscure; and it appeared from the early Rentals that some rearrangement of lands took place by which practically every farm in Paplay was assigned a certain amount of 'quoyland' along with its skat land. The reason is uncertain, but as the usual skats were not levied on quoyland the intention may have been to equalise the skat burdens for the various tenants.

In Birsay the 1595 Rental shows an exactly similar state of things, practically every farm in the old Bishopric area except "Above the Hill" being shown to have a certain amount of quoyland along with what was here termed its 'townland,' i.e. land subject to skat. And this northern part of Birsay now came to be known as the Barony of Birsay, but details as to its constitution or administration are unfortunately lacking. It may be noted, however, that 'Jo. Ben.,' that mysterious person, writing c. 1590 states: "Birsa baronia dicitur."

RENTAL NAMES

Banks: id. 1492. There were two farms so named—one beside the Loch of Banks west of Greeny tunship, the other near the shore of the Barony. The name is an Anglicising of O.N. *bakkar*, banks or slopes; in Orkney usually of the banks of a stream, loch or beach.

Bea (now with the spelling pronun. —bi·ə): id. 1595. A 3d. land in a valley 'over a ridge' from Birsay proper and known as "Abune (above) the Hill." This farm lies in the upper part of the valley, and was no doubt the original settlement there. Together with Swanney in an adjacent valley this district of Abune the Hill appears to have been skatted as a whole urisland unit.

Beaquoy (be·kwi): Beaqui 1492; Beaquoy 1500, 1595. Entered in the 1595 R. as a 4½d. land and a 10/- land (i.e. ¾ markland = about ¾d. land for skat).

An interesting name illustrating how in the course of time a quoy-name has usurped the place of an earlier tunship-name. The district of that old tunship is now sometimes termed Beaquoy-side (corrupted in pronunciation to beak·wi-said), and in the 1595 R. it was entered as 'Beaquoy, Clok and Housbie.' The tunship lies on the southern slope of Greeny Hill on the east side of the Loch of Sabiston. Near the edge of this loch are the farm-houses of Housebay, below which on a small peninsula jutting out into the loch are the remains of an old chapel. The existence here of an O.N. *bœr* name—*Hús[a]bœr*—and the attendant chapel leaves no doubt as to this having been the original settlement. Beaquoy on the other hand—the quoy of the *bœr* or bae-farm—lies away on the outskirts of the tunship, and its displacement of Housebay as the tunship name is indeed surprising.

Boardhouse: id. 1727. This large farm, one of the best in Orkney, does not appear in any of the three earlier Rentals, and must be one of relatively late origin which embraces several earlier farms—one of which bore the honourable name of Langskaill. That name is now applied as a field name only.

Boardhouse is an interesting though rather obscure name. In the early Ork. Rentals *bordland* (lit. 'table'-land) is used as a generic term for lands which were private estates of the earls and thus exempt from skat. "Na scat quia bordland" is the 1595 R. refrain. The term, however, does not seem to be of Norse origin, but it was comparatively common in Scotland and England where it is on record from c. 1250. In his *Dictionary of the Older Scottish Tongue* Craigie (prefacing it with a question-mark) gives the following definition—" Land providing supplies for the lord's table."

One may thus suspect that the name Boardhouse dates only from the time of the Stewart earls, one of whom it is assumed established a Barony here in Birsay. Earl Robert Stewart also, it will be remembered, built a palace here about 1574, and it is probable that a large mensal farm then became necessary to provide food for the occupants of the palace.

Something similar, however, would seem to have been in existence in the early Norse period when the great Earl Thoffinn had his chief seat in Birsay. Almost adjoining the farm of Boardhouse is another called Wattle. That name derives from O.N. *veizla*, which was a technical term for the guesting to which a ruler was entitled when on tour through his domains, and the same name—Wattle—was used as a name for one of the skats on Orkney lands. O.N. *veizlu-jǫrð*, 'wattle-land,' was also a technical term for land of which a ruler granted the use to one of his officials or courtiers for services rendered: the term might thus perhaps be simply translated as 'food and maintenance land.'

Bow: this farm does not appear in any extant Rental prior to that of 1794, but as it is placed therein next to Wattle it is presumably now represented by a field of Walkerhouse—The Bu field. See Part III for Bu.

Breck: 1595. Two farms of this name are entered in Birsay-besouth (i.e. on the south side of the Birsay Links), one of them as "Brek and Stoif." A present-day farm there is called Breck-besouth. O.N. *brekka, a* slope.

Brettobreck: Bretabreck 1595. An old 1½d. land in Marwick. O.N. *brattabrekka*, steep slope.

Brockan (brok·ən): Broken 1794. A farm in Marwick. An interesting example of *brokka* which occurs in Norse as a side-form to *brekka*; hence O.N. *brokkan*, the slope. This name occurs also in Sandwick and Rendall.

Broomquoy: 1794. Name (in Barony) unknown to me today.

Buckquoy (bʌk·wi): Bikquoy and Biggaquoy 1595; Bigquoy 1627; Biggaquoy 1727, 1794. A farm no longer, but the name is applied to land adjacent to the Point of Buckquoy (facing the Brough of Birsay). Origin doubtful; perhaps O.N. *bygg-kvi*, bere(barley)-quoy, but the Bigga- forms make that uncertain.

Butterland: Buttirland 1595; Butterland 1727; absent from 1794 R. A 3d. land in Birsay benorth. Origin doubtful.

Cleatfurrows (klet–fʌr·oz) : 1794. A farm adjacent to Palace and still existing. Not recorded prior to 1794. Origin quite obscure.

Cloke (klɔk) : Clok 1595. A small farm on southern slope of Greeny Hill. The name Klokk occurs in Norway both as a farm- and as a hill-name, and it has been suggested that the word represents an ablaut form of *klakk*, Icel. *klakkr*, which in Iceland means a lump, knob, etc.; hence its probable use as a hill-name. In this Orkney case it may have been an early name for Greeny Hill—a name applied from the tunship at its western base.

Cot : 1595. A vanished 3d. land farm entered next to Langskaill in Rental.

Cumlaquoy (kʌm·lə–kwi, –kwai) : Cumlaquoy 1595. A 2d. land in Marwick. O.N. *kuml*, a burial mound or prehistoric cairn; hence ' quoy at a *kuml* ').

Curcum (kʌrk·əm) : Curcum 1575, as a personal surname (R.E.O.); Kircum 1595 R.; Curcum 1794; absent from 1727 R. A 3d. land in Abune the Hill district. A somewhat puzzling name. As an old chapel-site is marked on the O.S. map only some 300 or 400 yards from the houses of this farm there seems little doubt that it is a derivative from O.N. *kirkja*, church. The vowel change in the 1575 form is remarkable at that early date, and would seem to suggest that the 'kirk' association was already forgotten. Yet there is no other place-name Curcum from which the surname could have originated. A *kirkja* origin presents the further difficulty that the only case of that word having an *-um* termination is the dative plural, a form which would imply *kirks* not *kirk*. It is possible that the termination represents another word as second element of a compound, but no satisfactory word suggests itself.

Durkadale (dɪrk·ədel and dʌrk·ədel) : Dircadale 1595; Durkadale 1727, 1794. A small farm in a valley of same name through which flows the longest of the Orkney burns—now usually called the Hillside Burn. It may be taken as certain that the first element here represents the O.N. name of this burn, which, from its more recent pronunciation might well be associated with the Norwegian river-name *Dorga* as listed by Rygh. The 1595 spelling, however, renders such identification doubtful, as the vowel change is exactly the same as was seen in Curcum.

Everbist : Evirbuster, Ever Bust 1492; Evirbustar 1500; Overbustar 1595; Overbister 1727, 1794. A 2d. land lying far up in hill in the Hillside district. O.N. *efri* (*øfri-*) *bólstaðr*, upper bister or farm.

Fald : 1595. A 3d. land entered in rental beteween Banks and Nearhouse, but now vanished.

Fea (fi·ə) : id. 1595. A 2d. land in Greeny tunship; now two farms—Upper and Nether Fea. O.N. *fjall*, hill.

Feaval (fe·vəl) : Fewal 1595; Feval 1727; absent from 1794 R. but still a farm today. The second element is probably O.N. *vǫllr*, a field, but the first is uncertain; in Norway apparently the same name—Fevold, Fevolden, etc., occurs sometimes as a farm name, but there is doubt as to the origin of Fe- there also. O. Rygh himself suggested O.N. *fé*, cattle, but others have suggested *fífa*, cotton grass. Origin thus doubtful; but cf. Veval (Graemsay).

Flaws (fla·z) : Flawis 1595. A ½d. quoyland in Northside of Birsay. There is another Flaws in Marwick—not mentioned in rentals. See this name —Holm.

Folsetter (fɔl·stər) : Fulsetter, Fulsatir 1492; Fulsettir 1500; Folsetter 1595. Fulsiter 1727; Folster 1794. A 3d. land north of Isbister tunship of which it may have formed part. The second element is O.N. *setr*, a farm-dwelling, but the first is uncertain. The same farm-name occurs in Norway where the Ful- is conjectured to be derived from a river-name Fulla. Here in Birsay a stream runs near by into the Loch of Isbister, and that may once have had a similar name. But the whole area here is flat, wet and boggy, and the element may thus represent simply O.N. *fúll*, foul, dirty, stinking, etc.

Garsetter (gɛst·ər) : Garsetter 1595; Gersetter 1727; Gerthsetter 1794. A 6d. land in northside of Birsay. Probably O.N. *garðs-setr*, a farm-dwelling at a garth or dyke—no doubt the tunship dyke. 3d. land of this farm was 'townland,' i.e. skat-land, while the other 3d. land was quoyland—a fact that might suggest expansion by inclusion of land from the 'hill' outside the tunship dyke. In view, however, of what was pointed out under the parish name above as to an apparent systematic allocation of some quoyland to each farm in this area, that suggestion should not be unduly stressed.

Garson : Over, and Nether Garson 1727. Two former farms near shore on Southside of the Barony. O.N. *garðs-endi*, dyke-end; at the end of the old 'garth' or dyke.

Gerraquoy (gɛrəkwi) : Garaquoy 1590; Gerraquoy 1794. A farm in Marwick. First element of uncertain origin.

Gersty (gɛrst·i) : Gerste 1794. A farm in South side of Marwick. The 1727 R. records another farm Gersteith, a 3d. land apparently in the Southside of the Barony. The name refers to the neighbourhood of an old prehistoric 'gairsty'—O.N. *garð-stœði*, 'dyke-steethe'; see further sub Treb, North Ronaldsay. It may have been this gairsty at the end of which Garson (*supra*) lay, but on the other hand that farm may have been near where the old tunship dyke reached the shore.

Gloup (glup) : 1595. A 3d. land in Northside of Barony. This was entered in the 1727 R. as "Minister's Glebe," and is now entered on the O.S. map as Glebe (pronounced glaip). It lies about ½ mile inland from a long cavernous indentation in the cliff-edge called Longagleep, the latter element of which is a local variant of gloup (glup); Norse *gluppa*, a variant of *gloppa*, a precipitous chasm or pit, etc. There are many of those gloups in Orkney (formed chiefly through the collapse of the 'roof of the inner end of a cave') the most famous and spectacular being the Gloup of Deerness.

Greeny (grin·i) : Grenying, Grenyg, Grenyng, Grynning, Greming (where 'm' is an obvious mistake for 'n') 1492; Grenie 1500, 1595.

An old tunship at the western base of a hill now called after the tunship—Greeny Hill. This along with the neighbouring tunship of Sabiston composed an urisland.

This is a name of peculiar interest, as, with the exception of Lyking, it is the only old farm name in Orkney that might conceivably have had O.N. *vin* as second element. It is doubtful, however, whether *vin* is present here. The final 'g' in all the diverse 1492 spellings suggests rather that the original name was O.N. **grœningr* (a green place), a form which does not occur in O.N. literary records but is well attested from Norwegian place-names.

Grew (gru:) : Grow 1595; Grew 1727; Grew of Yeldabreck 1794. A 3d. land in Southside of Barony. Origin obscure; there is no stream here which might have given rise to the name.

Gridgar (now by metath.—grɪg·dər): Grutgair 1595. A 2¼d. land on north side of Marwick valley and next farm to shore. O.N. *grjót-garðr*, 'stone-garth.'

Grindally (grɪnd·əli): Grindela 1595; Grindillay 1794. A 1½d. land in north side of Barony. See this name—Sanday.

Green (locally—gaiərn): today there are two small farms—East and West Green—in south side of Marwick valley, near How. These must certainly represent the Gyren of the 1794 R. and Gyir (a 2d. land) of the 1595 R. O.N. *geiri*, a 'gore' or wedge-shaped piece of land: cf. Gyre, Orphir. The final 'n' of Gyren probably represents the def. art.—*Geirinn*, 'the' gore, and the modern spelling Green has evidently been due to a misapprehension.

Harpsquoy: (a corrupt modern form: local pronun.—hɛrp·ə∫ə): Harpschaw 1595; Harpshay 1727, 1794. A 6d. land in Birsay-besouth.
The second element here is probably O.N. *haugr*, a mound (Cf. Horraldsay, Firth) but the first is quite obscure. In Norwegian farm-names Harp- is usually explained as referred to a river-name, but here there is no stream.

Hawin (ha:n): id. 1727, 1794; not in earlier rentals. Skatted in 1727 as a 3d. land in Birsay-benorth. Evidently skatted under a different name formerly. Origin of name uncertain.

Housebay (hus·be,-bi): Housbie 1595. See Beaquoy. O.N. *hús[a]-bœr*; see this name—Stronsay.

How: Howth (evident misreading in Peterkin), How (S.C.H. MS.) 1595; How 1794. A 6d. land in Marwick. O.N. *haugr*, mound.

Howally (hau·əli): Hawlye, Howlye, Howle 1492; Howlicht 1500; Howallie 1595; Hawillie 1727, 1794. With Evirbuster this formed 2½d. lands. Adjacent to the farmhouse, which stands at the summit of a rising on the Hillside Road, is a prehistoric mound, and there may have been others near to give rise to the name—O.N. *hauga-hlið*, mounds—slope.

Howquoy: Howaquoy and Quoyhokka 1595; Quoyhoka 1627; Howaquoy 1794. A 3d. land townland + ¼d. quoyland farm (1595) situated in Birsay besouth. Cf. Howaquoy, Rendall.
An interesting name illustrating the alternation in the position of the quoy-element. Today it appears on the O.S. map as Howquoy, but the name is pronounced locally Huiko (høk·o), the 'quoy' being dropped altogether. O.N. *hauga-kvi*, mound-quoy.

Hunchaquoy: 1794. On O.S. map—Hinchaquoy. A small farm in Above-the-Hill district. This is almost certainly the same farm as appears in the 1595 R. as Hussaquoy (house-quoy), and the change to its present form is remarkable. The present prefix may be due perhaps to the substitution of a man's nickname.

Hundland (hʌn[d]·lən[d]): Houndland 1595; Hundland 1727, 1794. An old 3d. land on east side of loch of same name. First element obscure; see Hunclett (Rousay and Holm). Here it may perhaps be the personal O.N. name *Hundi*. Cf. Hunskarth (Harray).

Hunto (hʌnt·o) : id. 1595, 1794; Huntoe 1727. As a surname, Hunto, 1574 (R.E.O.). A farm situated adjacent to where a burn issues from the Loch of Boardhouse; in 1595 consisted of 3d. land townland and 1d. quoyland. In the absence of older forms of this name the first element must be left unexplained, as in the other Orkney names beginning with Hun-. Cf. however Hunton (Stronsay). The second element here is probably O.N. *á*, a stream or river.

Hussaquoy : see Hunchaquoy.

Hyvel (haí·vəl) : id. 1794. A farm on low-lying flat ground in Isbister tunship. Probably O.N. *hey-vǫllr*, hay-field.

Ingsay (ıŋ·ze) : id. 1595, 1794. As a surname—Ingissay 1574 (R.E.O.). An old 6d. land in Above-the-Hill district. Strangely enough the name occurs frequently in 17th and 18th cent. Birth Regs. as Inzia. Origin obscure. The farm is on high ground and thus O.N. *eng*, (meadow) can hardly enter into the name. The first element may well be a personal name; cf. Innister (Rousay) and Ingashowe (Firth).

Isbister (aiz·bəstər) : Estbuster and Esterbustir 1492; E:sterbuster 1500; Isbester 1534, Ysbuster 1566 (R.E.O.); Ysbustar and Yesbustar 1595; Isbuster 1727.
A half-urisland tunship (1500). O.N. *eystri-bólstaðr*, more easterly-bister. The point of reference from which it lay to the east must have been the tunship of Marwick. See also Isbuster (Rendall).

Kirbuster (kır·bəstər) : Kirbuster 1595, 1727. An old 3d. quoyland. This small tunship lies at the south-east corner of the Loch of Boardhouse, and was evidently (as quoyland) a relatively late settlement. But, as the name (and the O.S. map) shows, it had an old chapel in its bounds. O.N. *kirkju-bólstaðr*, ' kirk-bister.

Langadie : 1595; Langada 1727. Absent from 1794 R., and no longer extant. Second element doubtful. The farm was in Birsay besouth.

Langskaill : Langskail 1595. A former 3d. land townland with ½d. quoyland, now part of Boardhouse farm. See this name—Gairsay. There is another farm of this name in Marwick (Langskaill 1595) which was a 3d. land also.

Lie : 1595. A 3d. townland with ½d. quoyland in Birsay besouth. Name now spelt Lea. In the 1794 R. a Lian appears in Birsay benorth (and also a Liensquoy)—now also vanished. Lie represents O.N. *hlið*, a slope, and Lian the same word with def. art. attached—*hliðin*, the slope.

Leaquoy (lɛk·wi) : Liaquoy 1595. A farm in Marwick. Origin doubtful; it may represent O.N. *leik-kvi*, sports quoy, but more probably (in view of its situation) *hliðar-kvi*, quoy of the slope.

Marwick : Marweke, Marwek 1492; Marvik, Marwik 1500; Mervick, **Merwick**, Marvick 1595. A 36d. land or 2-urisland tunship in a valley running inland from the bay of Marwick In the early rentals this name was entered as a parish name; see under the parish name Birsay *supra*.

Marvik is a not uncommon name in Norway, where the first element has sometimes been regarded as representing O.N. *marr*, a horse. In this case, however, the name should rather be grouped with such Norse names as *marbakke*, sea-slope (beach), *marhalm*, etc., where the *mar*- denotes O.N. *marr* in the sense of sea (= Latin *mare*). In a note on *mar* in N.N.O. Torp states that it would appear the original indo-germanic meaning was 'mire or swamp,' thence 'lake'; and he compares the Swed. dial. *mar* which is used of 'a shallow bay, a bog or a tarn.'

The bay here at Marwick would seem to be a striking confirmation of that view. Across its mouth runs a reef of rough rocks, mostly submerged at high water. Inside is a large, semi-circular, practically landlocked, sea-water lagoon about half a mile wide at the reef, and known as The Choin (tʒɔin, tʒən,), i.e. O.N. *tjǫrn*, tarn. The bay might thus have well been called the 'tarn-wick,' and there is hardly room for doubt that the name Marwick is to be so interpreted here.

Meiklequoy: 1727, 1794; Meikilquoy 1595. In Birsay besouth; with Kingsdale (1595) it formed a 9d. land. O.N. *mikla-kví*, big quoy.

Messigate: 1794. Not recorded in any earlier rental, and there is no such farm today, but the road leading down to the Palace and Church is known as the Messigate road. O.N. *messu-gata*, 'mass-road,' i.e. road leading to church.

Midhouse: 1595. A 3d. land in Marwick. Almost certainly the farm spelt Mauce in 1794 R. and now known as Muce (møs), i.e. *mið-hús*, midhouse. There is another Muce in Sabiston near Dounby.

Netherskaill: Netherskaile 1500 ;Neather Skeall 1601 (U.B.). A farm in Marwick adjacent to Langskaill, and clearly indicating an early undivided Skaill farm-unit. See further—sub Skaill, Part III.

Norton: Nortoun 1595. A 1½d. land entered in Rental in Greeny tunship but almost certainly by mistake. O.N. *norð-tún*, i.e. the northern part of tunship of Sabiston. It is adjacent to Greeny.

Quackquoy (kwaˑkwi): Quhahoy 1595 (Quhaquoy S.C.H.); Quhaquoy 1601 (U.B.); Quitquoy (1631 Sas.). A farm in Marwick. Evidently the same name as Quatquoy (Firth): thus O.N. *hvíta-kví*, white quoy.

Quoygelding: 1595, 1727. A 3d. land townland plus 1½d. quoyland in Birsay benorth. Absent from 1794 R. There is no such farm today, but in the near vicinity is one called Gelderhouse or Meikleplank which may be the same. There is, however, another Gelderhouse rather over a mile to the south-east (probably a comparatively recent farm), and there is a third in Sabiston tunship near Dounby.

A puzzling name; it suggests an O.N. *gildi-hús*, the house of a 'guild' such as existed in Norway and elsewhere in the Middle Ages, but as we have no record of any such social organisation here these names must be regarded as obscure.

Quoyhokka: 1595. See Howquoy *supra*.

Sabiston: Sebuster, Sebustir, Sabuster 1492; Sabustar, Sabistar 1500; Sabister and Sabuster 1564 (O. & S.R.); Sabistane, Sabistone 1595; Sabistoun 1727; Sabiston 1794. A 6d. land tunship (1595) which together with the neighbouring Greeny made up an urisland (1500).

The altered termination of this name is very remarkable and well-nigh unique. The present-day Sabiston (which dates from the late 16th century at least) might imply a *-staðir* origin, but the earlier rental forms leave no doubt that it is a *-bólstaðr* name. No explanation for the change can be offered, and the first element of the name is also obscure; as the tunship is inland, O.N. *sær* in the sense of 'sea' is out of the question, and in the sense of 'lake' unlikely. There is a small adjacent Loch of Sabiston but such a usage of *sær* would be unique in Orkney.

Savaquoy: 1595. A 3½d. land in Birsay besouth which disappears thereafter. In the Rental it follows Nether Garsand, and was thus evidently near the beach where there is a famous prehistoric mound (in which a Celtic church-bell was discovered) known as Saverough (sɛv·əro), i.e. O.N. *sævar-haugr*, sea-mound. Hence probably O.N. *sævar-kvi*, sea-quoy; cf. Savaskaill (Rousay) and Savaquoy (Shapansay).

Seatter (set·ər): Seter 1595. A 1d. land in Above-the-Hill area. O.N. *setr*, a dwelling.

Skanaquoy: 1595; Scannaquoy 1727 (then uncultivated). 1½d. land in Marwick. First element obscure.

Skelday (skɛld·e, skɛld·o): Skaldale 1595; Skalden 1564 and Skelden 1568 O.S.R.); Skelday 1727; Skeldie 1794. A 3d. land in Hillside district. An obscure name. Jakobsen assigns several Shetland 'Skeld'-names to O.N. *skjǫldr*, a shield (in the sense of something shield-shaped), but whether such an origin can be assumed here is extremely doubtful.

Skidge (skɪdʒ): Skedge 1727. A 3d. land in Birsay-besouth, and quite near the shore. Origin obscure, but cf. Skedgibist (Sanday and Orphir).

Skorn: Skorne 1595; Skoren 1794. A 3d. land in Marwick (1595). The farm lies at the foot of a rather steep hill-slope, and thus may be regarded as O.N. *Skorin* (with def. art. attached)—'the' skor. Rygh in his *Indledning* states that *skor* is applied to a terrace on a hillside or to a level piece of ground at the foot of a hill. Cf. Skorwell, Sandwick.

Stagaquoy: 1595. A 1¾d. land in Marwick. This is almost certainly the farm now known as Stedaquoy (stɛd·ja-kwi)—a name which elsewhere I have interpreted as O.N. *steðja-kvi*, stithy- or anvil quoy, taking it to indicate the site of a smithy. In view of the 1595 spelling, however, that derivation must be regarded as doubtful, and the origin is thus uncertain.

Stanger (sten·ʒər): Nether Stansgar and Over Stansgar 1595; Stensger and Stinsger 1621 (Tests.); Stainsgar 1667 (Tests.); Stainsgar 1727; Stannisgair 1794.
The two farms formed a whole urisland in Birsay-besouth. The name refers to a famous Standing Stone, the Stone of Quoyboon (hwibøn·), which may still be seen in a field here. O.N. *steins-garðr* (standing-)stone farm.

Stove: Stoif 1595; in the combination Brek and Stoif. Situated in Birsay-besouth. See this name—Sanday.

Swanney (swan·i): Swonney (in S.C.H. Ms.—Suannay) 1595; as a surname—Magnus Svinnay 1574 (R.E.O.); Swannay 1727, 1794. A 6d. land. This farm occupies a rather deep valley through which a burn runs down to the sea from the Swanney Loch. The 1574 surname-form would suggest that (as in Swandale, Rousay) the first part of the name represented O.N. *svin*, a swine, but the termination is puzzling. It is just possible that the original name was *Svina-dalr*, swine-valley, and that the final element has somehow been dropped; but that would be very unusual, and in consequence the origin must be regarded as uncertain.

Twatt (twat): Tuait 1564 (R.E.O.); Twat 1595, 1727; Twatt 1794. A 3d. land. See this name—Stenness.

NAME MATERIAL 139

Twattland : Twaitland 1595; Twatland 1727; disappears thereafter. A 3¾d. land in Birsay-besouth. See last word.

Vigga : Viga 1595; Vaigga 1595 MS.; Vigga 1727. A 9d. land in Birsay besouth —now vanished. Origin obscure; the same name occurs in Paplay (Holm).

Wattle : Wattill 1595; Wattle 1727, 1794. A 3d. townland with 1½d. quoyland in Birsay-benorth. See under Boardhouse.

Yeldabreck : id. 1595, 1727, 1794. A 3d. land townland with 1½d. quoyland in Birsay-besouth. Second element O.N. *brekka*, a slope; the first rather uncertain, but cf. Yeldavill—Harray.

OTHER FARM-NAMES

Bigbreck : small farm in Twatt tunship. Origin rather uncertain; the second element is O.N. *brekka*, a slope, and the first might possibly be O.N. *bygg*, bere (barley)—cf. Bigland, Rousay,—but as the main farm in Twatt is nearby and called Bigging this element more probably represents a contracted form of that name.

Bigging : 1. a farm in Twatt; 2. another in Greeny tunship. Probably O.N. *bygging*, a building, rather than the Scots *bigging*.

Cavan (kɛ:vən) : small farm in Twatt tunship. Origin obscure.

Deesbreck : a farm near south end of Hundland Loch in a very mossy boggy area. O.N. *dýs-brekka*, bog-slope.

Eastabist : a farm in Hillside region which does not appear in rentals. Of the older farms in this Durkadale valley this lies farthest up the valley which itself runs in a south-easterly direction. It is thus doubtful whether this name is an O.N. *aust-bólstaðr*, 'east-bister,' or *yzti-bólstaðr*, 'outermost bister' : most probably the latter. Cf. Isbister (Birsay and South Ronaldsay).

Farafield (far·əfil, fɛr·əfil) : a farm in Durkadale valley at base of a hill named from a neighbouring farm—Skelday Hill. See this name—*sub* Faraclett, Rousay.

Fidgarth (fø·djəþ) : small farm in Birsay benorth on flattish ground near east shore of Boardhouse Loch. Perhaps O.N. *fitjar-garðr*, from *fit*, meadow—hence ' meadow-garth.'

Hammar : a farm in Greeny tunship. See this name—Westray.

Hass : a small farm near the narrowest part of the Durkadale valley. O.N. *hals*, neck, throat. This word occurs frequently in Orkney, applied to a *col* over a hill-ridge (e.g. The Hass of Watnaskar, Rousay) or to a narrow channel (e.g. The Hassie between Thieves Holm and the Point of Carness in St. Ola) or as here to the narrowest part of a valley. The term is also Scots in the form halse or hause.

Hayan (hai·ən) : a farm in Birsay-benorth. Origin uncertain; as the farm is in the midst of some very fertile ground the name can hardly derive from O.N. *heiðr*, heath.

Howan (hou·ən) : Chief farm in Sabiston tunship. O.N. *haug[r]-inn*, ' the mound.'

Hozen or **Hosen**: a farm adjacent to Dounby. Apparently the same name as Hosen—a Norwegian farm-name recorded once in N.G., and of obscure origin there also. The possible derivation from O.N. *Hóstrinn* suggested by Prof. Sophus Bugge would seem farfetched and inapplicable to the situation of this Orkney farm.

Linnabreck: a farm on southern slope of Greeny Hill. Cf. this name—Stronsay.

Lobady (lo·bədi): a farm in Durkadale valley in Hillside. The first element can hardly be other than the local dialect-word *loba* or *lubba*, a general term applied to hill grass of the coarser sorts. I have not found that word in use in any other part of Orkney, but it is known also in Shetland. In the first *Statist. Account of Scotland* the writer on Birsay (Rev. Geo. Low) stated: "They (i.e. the hills) are covered with heath and what we call *lubba*, a sort of grass . . . . it generally consists of different species of carices, plain bent and other moor grasses." Jakobsen compared this word with the Icel. *lubbi*, a shaggy-haired dog, etc., and No. *lubben*, adj. leafy. The second element is no doubt O.N. *dý*, a bog, marshy ground.

Nisthouse: a farm in Hundland. See this name—Stenness.

Quoylanga: small farm in Marwick. O.N. *kví-langa*, long quoy.

Quoyscottie: small farm on southern slope of Greeny Hill. Second element uncertain; perhaps a nickname.

Ritquoy: small farm in Marwick. First element uncertain, but possibly referring to a 'sheepright,' i.e. O.N. *rett*, an enclosure in which sheep were driven for 'rooing' (wool-plucking) and marking. The day fixed for this was known as Sheepright Day.

Skrutabreck (skrut·ə-brɛk): small farm in Birsay-benorth. First element obscure.

Spurdagrove (spʌrd·əgrov, -of): small farm on south side of Ravie Hill. Origin obscure.

Stara: a farm at roadside in Marwick. Origin uncertain.

Tufter (tʌft·ər): small farm in Twatt. O.N. *tuptir*, plur. of *tupt*, an old house- or building-site.

Velzian (vɛl·jən): a farm in Sabiston. See this name—Rendall.

Vinbreck: a small farm in Birsay-benorth. O.N. *vin-brekka*, pasture-slope.

Walkerhouse: a farm at burn-side between Boardhouse and Palace. See this name—Evie.

Windbreck (wɪnd·brɛk): there are at least two small farms so called, and the name is rather common in Orkney. Evidently same name as Vinbreck *supra*.

HARRAY

For derivation of the parish name see under Birsay (parish name). From the data there adduced and other records there would seem to have been considerable vacillation before Harray emerged as the final parish name. In the earliest extant record (1490) we read of "the parochin of Burch in the Herray"; in the 1492 R. we find it as "Burgh Sancti Michaelis" and as "Parochia de Burgh"; in a record of 1561 (R.E.O.) as "Sanct Michaelis parochin"; in the 1500 R. as "Herray" and as "Hurray Brugh"; while in a sasine of 1617 (O. & S. Sas.) we find "St. Mitchellis kirk in Harrow in the parish of Harray."

The 'Burch' or 'Brugh' referred to was the tunship of Over (Upper) Brough which lies in the heart of the parish, and in which the parish church is situated. Here then, once more, we find that a tunship name became (for a time at least) the parish name. The church, dedicated to St. Michael, occupies a commanding position on the top of an extensive hillock, and is built on the site of an ancient broch from which the tunship had no doubt derived its name, though there are other broch-sites in the vicinity.

Harray embraced 4½ urislands of skat land together with a few pennylands of quoyland. It is the only parish in Orkney with no sea-coast and udal ownership of lands persisted here longer than anywhere else in Orkney.

RENTAL NAMES

Bimbister (bɪm·bəstər): Bimbustir 1492; Binbustare 1500; Binbustar 1595; Bimbister 1727. A 6d. land tunship. A *bólstaðr* name. In the first element, despite the 1492 'm,' it is fairly certain that Bin- represents the original form, in which case one might perhaps associate the name with Binscarth in Firth, but in the absence of older forms of the name its origin must be regarded as obscure.

Conyer (kɔn·jər): Quoyconze 1595; Conzier—a quoy 1727. A small farm on border of Sandwick parish. Origin uncertain; cf. Conziebreck, Sandwick.

Corrigill (kɔr·ɪgəl, kɔr·əgəl): Corgill 1492, 1500, 1595; Corrigill 1727. This and the following name Corston must be considered together. The two tunships are contiguous—Corrigill lying somewhat higher up in the basin of a stream now known as the Burn of Corrigill. There is some reason to suspect that originally both went together as a single tunship or early settlement, because in the 1492 and 1500 Rentals Corgill appears as a 2d. land and Corstath or Corston as a 4d. land, while in the later Rentals of 1595 and 1727 these values are reversed—Corston being the 2d. land and Corrigill the 4d. land. Such exchange of values in the rentals is unique.

As for the names, the following facts may be first noted: (1) Each has the same first element—Cor-; (2) the same stream runs through each tunship; (3) in Corrigill the qualifying element Corri-, defining the 'gill,' might well be assumed to be the name of that gill-forming stream; in G.N. xiv, 122 (under the name Kaarli) it is stated: "The rather frequent occurrence of names such as Kaardalen and Kaarevik seems however to suggest that there has been a river-name which might be the origin...."

From these facts there might well appear to be a presumption that the first element in both names is the old stream name. On the other hand Corston (Corstath) is an undoubtedly O.N. -*staðir* name, a type in which in Iceland and Norway about four out of every five have personal names as first element. It is thus uncertain whether we have here a personal name e.g. O.N. Kárr or Kári, or an old stream name.

A further complication arises from the occurrence of the name Corrigill in Graemsay, q.v.

Corston (kɔr·stən) : Corstath 1492, 1500; Corsta 1595. See last name—Corrigill.

Garth : id. 1492; Garth and Mydgarth 1500; Garth and Kingishouse 1595; Gairth and Kingshouse 1727. A 4½d. land adjoining Corston. O.N. *garðr*, farm. There is also a farm Garth in Nether Brough tunship.

Grimeston (graim·stən) : pronunciation probably affected by spelling; Grymestath 1492, 1500; Grymestoun 1595; Graemiston 1727. An urisland tunship adjacent to Stenness parish. O.N. *Grims-staðir*, stead or settlement of a man Grímr. Cf. Grimsetter, St. Ola.

How (hou) : How and Ranyisgarth 1492; How and Rannisgarth 1500; How and Ramsgair (Rainsgar) 1595; How and Rainsgairth 1727. A 4½d. tunship. O.N. *haugr*, mound.

Hunscarth (hʌn·skarþ) : Hundisgarth 1492; Hundskarth 1500; Hundscarth 1595; Huntscarth 1727. A 3d. land far up in a hill valley.

Another difficult name owing to absence of earlier forms. The valley in question runs up eastwards through the range of hills dividing Harray from Firth and Rendall. At the head of this valley is a col or ' skarth ' in the ridge, and the name (O.N. *skarð*) has been applied to it as we know from farm-names in the valley on the other side—Settiscarth and Cottascarth. That being so, it is possible that this name Hunscarth in the western valley is also a skarth-name, and the first element in that case might most probably be O.N. *hundr*, a dog, applied from some obscure reason as happens in Norway and elsewhere.

On the other hand, the earliest recorded spelling implies that it is a garth-name, in which case the first element would almost certainly be a personal name. Here again, however, further difficulty arises. The O.N. personal name *Hundi*—a name recorded mainly from ' west over sea '—is a translation of Celtic *cuilen* or *madadh* (dog), and the genitive case is not *Hundis* but *Hunda*. It is conceivable of course that (as in Hermisgarth, Sanday) the old genitive ending might have been superseded by -s. But in Norwegian names beginning *Hunds*- there is extreme doubt also as to the actual source in each case, whether the personal name *Hundingr*, or *Húnn*, etc.

In these circumstances the origin of Hunscarth is quite uncertain.

Kingshouse : see Garth *supra*. A farm still today, but origin of name uncertain. It would seem to have taken the place of the 1500 R. Mydgarth. In that rental it is stated that 1d. land (of this Garth tunship) which had been ' ald erledome ' land had been given in exchange by Earl Wm. Sinclair to Elezabeth Urving (Irving) for some land in Clestran (Orphir), and it is possible that this 1d. land parcel of *pro rege* land had thereafter been given the name Kingshouse. (' Ald erledome ' land here meant land which had been acquired by one of the early earls, but the skats therefrom would still have been regarded as king's scats, though collected of course by the earl.)

Knarston (nar·stən) : Narstain 1492; Nerstaith 1500; Knarstane 1595; Knarston 1727; as a surname—Knarstane 1571 (R.E.O.). A 4½d. land tunship—the most northerly in Harray. A -*staðir* name, but it is uncertain whether the first element indicates the O.N. personal name Narfi or Knǫrr: hence either *Narfa-staðir* or *Knarrar-staðir*, the stead or settlement of a man Narfi or Knǫrr.

Mirbister: Mirkbuster 1492; Myrkbuttir (an obvious error of 'tt' for 'st') 1500; Myrkumbuster 1504 (R.E.O.); Mirkbustar 1595; Mirbister 1727. A 3d. land tunship between Knarston and Garth.

An O.N. *bólstaðr* name, but if the 1504 form can be trusted the first element cannot be O.N. *myrkr*, dark, but rather a personal name. There is no name cited by Lind which this element could represent, and the possibility of a Celtic name cannot be excluded: cf. the early Irish king's name—Myrkjartan (*Lax. Saga*).

Midgarth: See Garth.

Netherbrough (-broχ): Nethirburgh 1492; Netherbrugh 1500; Netherbrught 1595; Neither Brugh 1727. This and the adjacent tunship of Rusland went together in the Rentals to make up an urisland.

See under parish name as to Overbrough. Possibly Nether- and Over-Brough (which are adjacent) may have at one time formed a single settlement, but as there are two broch-sites in Netherbrough itself it had no need to borrow a name from the broch in Overbrough. O.N. *borg*, a fortress, here used of the old brochs.

Netlater (nɛt·lətər): Noutland 1492; Noltland 1500; Noltclet 1595; Notclett 1601 (U.B.); Netclet 1630 Sas.; Nettleter 1727. A 4½d. land between Bimbister and Overbrough. The change in this name is astonishing and inexplicable. The earlier forms leave no doubt as to an original O.N. *naut[a]-land*, cattle-land, a very common name, and the first element of the present name must still represent the old *naut-*. The second element of this later name is as clearly O.N. *klettr*, a rock (or the plural *klettar*), but no suggestion can be offered for the substitution any more than for the appearance of *klett* in so many other farm-names in Orkney. The absence today of any noticeable 'rock' at any of those farms is itself noteworthy.

Overbrough or **Upperbrough**: Ovirburgh 1492; Ovirbrugh 1500; Ouerburght 1595; Overbrugh 1727. A 9d. land tunship. O.N. *øfra-borg*, 'upper-broch.' See further under parish name.

Quoys: Quoyis 1595. A small unit grouped in rental with Noltclet. O.N. *kvíar*, quoys.

Ransgarth?: name now obsolete. In the Rentals it was grouped with How, q.v. Obviously an old *garðr* name, of which the first element must be a personal name, but exactly which is uncertain.

Rusland (rʌs·lən[d]): id. 1492, 1500, 1595. A 6d. land which together with Netherbrough formed a whole urisland. An O.N. *land* name; first element evidently O.N. *hross*, a horse. The same name occurs in Westray.

Winksetter: Wekinsetter 1492; Winksetter 1500, 1595; Winksater 1727. A small 2d. land unit far up in the hill. An O.N. *setr* name of which the first element can hardly be other than the O.N. personal name *Víkingr*.

OTHER FARM-NAMES

Appiehouse: there are two farms so named; see this name—Stenness.

Appietun: there are also two of this name, q.v. Rendall.

Bea (bi·ə): id. 1617 Sas. In Nether Brough. O.N. *bœr*, a farm.

Beboran (bebo·rən, bebɔ·rən): a farm in Rusland tunship. The first element would appear to be O.N. *bœr*, a farm, and the accented element *-boran* might perhaps represent O.N. *borgin*, 'the broch.' There is no broch site now quite adjacent to the present farm-house, but about half a mile to the south is a broch-site known as Burrian, while about a mile to the north is another of the same name. The latter is in this Rusland tunship, but the former (and nearer) is in the tunship of Nether-Brough. The origin of the name is however quite obscure.

Biest (bi·əst): id. 1787 planking of tunship of Nether Brough. In the absence of older forms of the name the origin is quite uncertain.

Bigging: in Rusland tunship. A farm Biggings is in Grimestoun tunship. See this name—Westray.

Boardhouse: a small farm in Bimbister tunship. See this name—Birsay. No explanation can be offered for its occurrence here, but it is evidently a recent outbreak as it does not appear on a tunship map of 1831.

Breckan: in Bimbister tunship. O.N. *brekkan*, the slope.

Brettoval (brɛt·əvəl): a farm in Knarston tunship. The first element is O.N. *brattr*, steep, but it is uncertain whether the second is O.N. *fjall*, hill, or *vǫllr*, field: most probably the former.

Buckquoy: id. 1630 Sas. A farm in Grimeston tunship. Origin uncertain; see this name—Birsay.

Caperhouse (kep·ər—): id. 1787 planking of Nether Brough. Origin quite obscure.

Cuppin: farm in Over Brough. O.N. *kopp[r]-inn*, the hollow or cup.

Curcabreck: id. 1622 Sas.; Kirkabreck in Nether Brugh (1681 record); Tumult [i.e. O.N. *tún-vǫllr*, tun-field] of Kirkabreck (1787 planking of N. Brough). These forms all refer presumably to the same farm-name, but no memory survives of any old kirk or chapel here. The adjacent farm, however, is called Maesquoy, and the first syllable in that name may be that which appears in Messigate (Birsay and St. Andrews) and represent O.N. *messa*, 'mass' or church service.

Eastaquoy: self-explanatory.

Fealquoy: see this name—Rousay.

Flaws (fla:z): see this name—Birsay and Holm.

Fursebreck: Forsabreck (1617 Sas.). In Grimeston tunship. O.N. *fors[ár]-brekka*, waterfall-slope.

Furso (fʌrs·o): Forsa (1618 Sas.). A farm in Nether Brough. O.N. *fors-á*, waterfall-burn.

Geoth (gjøð): in Knarston tunship. Origin obscure; cf. Guith, Eday.

Geroin (geroin·, djɛroin·): a small farm in How tunship. Origin obscure; cf. this name—Orphir.

NAME MATERIAL

Gorn: a farm in Grimeston. See Goir—Sanday.

Handest: a farm in Knarston tunship. Origin obscure.

Hindatun (hɪnd·ətun): a farm in Grimeston tunship. Origin somewhat uncertain. The tunship abuts on the shore of the Loch of Harray, and this farm lies back a considerable distance inland. In the absence of old forms of the name it is uncertain whether it represents an O.N. *hindra-tún* in the sense of the farther-away or more distant part of the tunship (cf. Hinnarholmen, N.G. x, 242), or whether it is a contraction of Scots ' ahint the tun ' = behind the township. In either case the meaning is much the same. On a tunship map of Grimeston in S.C.H. Kirkwall this farm is entered "Behind the Town," but that may be merely the surveyor's interpretation of the name. Cf. Hinderayre, Rendall.

Horraquoy: in Grimeston tunship. See this name—Holm; but as it was formerly spelt Harroquoy the origin may not be the same.

Hybreck (hɛi·brɛk): Haybrek and Heybrek 1627. In Grimeston tunship. Probably O.N. *hey-brekka*, ' hay-slope.'

Jubidee (tʒub·əde): in Bimbister tunship. See this name—Evie.

Langskaill: in Grimeston tunship. O.N. *langi-skáli*, long hall.

Linnieth (lɪn·jəþ): in Garth tunship. Origin obscure.

Maesquoy (mez·kwi): Measquoy 1649. In Nether Brough tunship. See further *sub* Curkabreck.

Mithist (mið·ist, mið·əst): in Mirbister tunship. A strange but interesting duplication occurs here. A few yards away from this farmhouse is another known as Midhouse, and on a tunship map of 1844 (in S.C.H., Kirkwall) each is entered as Midhouse, deriving of course from O.N. *mið-hús*, mid-house. Probably both were formerly united, but it is odd how one name has now been modernised, while the other retains more of its original sound. The final -t must be regarded as intrusive, perhaps on the analogy of a farm-name in the adjacent Knarston tunship—Handest. It may be noted that in two cases in Birsay parish an old *mið-hús* is now spelt Muce and pronounced møs.

Moan (mɔːn): three farms so named, one in Rusland tunship, one in Over Brough and the third in Bimbister. O.N. *mó[r]-inn*, the moor or heath.

Ness: in Grimeston tunship on a 'ness' jutting out into Harray Loch.

Netherhouse: Nearhouse 1620. In Grimeston. O.N. *neðra-hús*, lower-house.

Nistaben (nɪst·əbɪn): in Rusland tunship. O.N. *neðsta-bygging*, lowermost building.

Nisthouse (nɪst·hus): in Mirbister tunship and others also. O.N. *neðsta-hús*, lowermost house.

Northbigging: in Mirbister also. Self-explanatory.

Oback: in How tunship. O.N. *ár-bakki*, burn-bank.

Pow (pɔu): in Nether Brough tunship. See this name—Westray.

Queenafinyeth (kwin·ə-fɪn··jəþ): Quinafinia 1749 Sas. In Germiston tunship on boggy land near shore of Harray Loch. This compound apparently signifies 'the quoy on or at the fen (bog),' but the aspirated final consonant is puzzling.

Quoykea: in Bimbister tunship. See this name—St. Andrews.

Rickla: in Grimeston tunship beside a stream called the Burn of Rickla. Origin obscure.

Runa (run·ə): on tunship map of 1835—Runa; in Over Brough tunship. Origin uncertain, but probably derived from O.N. *hraun*, heap of stones, or stony ground, or some cognate form. In Birsay a small cot on hilly ground is called Runa.

Skuan: Skowan 1678. In Over Brough tunship. See Scuan—Stenness.

Skethouse (skɛð-us): in Mirbister tunship. Origin uncertain.

Tufta (tʌf·tə): in Bimbister tunship. See Tufter—Birsay.

Upperbigging: in Corrigill tunship. Self-explanatory.

Velzian (vɛl·jən): Vailyie 1617 Sas. In Grimeston tunship. See this name—Rendall.

Vola: in Grimeston tunship also. Origin uncertain; perhaps a singular form of last name, from O.N. *vǫllr*, a field.

Windywalls: in Grimeston tunship. See this name—Eday.

Yeldavill (jɛld·əvəl): in Nether Brough tunship. Origin doubtful; the second element is most probably O.N. *vǫllr*, a field, and the whole name would thus seem to be an exact parallel to the Norwegian farm-name Gjeldaaker, in which Falk (N.G. v, 150) was disposed to regard the first element as related to No. *gald*, hard (stamped-down) ground. Cf. Yeldabreck (Birsay) and Yeldadee (Sandwick).

SANDWICK

Sandwik 1492; Sandweik, Sandwick 1500.

This large parish, which embraces the mid-portion of the west coast of the West Mainland of Orkney, derives its name—(O.N. *sand-vik*, sand-bay)—ultimately from the sandy bay now known as the Bay of Skaill to which it forms a roughly semi-circular hinterland. The parish church (dedicated to St. Peter) stands at the north-eastern corner of that bay, but as to whether there ever was an old tunship here called Sandwick which (from the presence of the parish church) became as in some other cases the parish name, we have no actual record, but it is more than likely. The parish is very fertile, and like Sanday in the North Isles was relatively highly skatted, a fact implying no doubt that, like Sanday, it was more extensively cultivated than most other areas in early Norse times when skats were first imposed. And the reason would have been the same—the amount of easily cultivable sandy soil in the neighbourhood of the Bay of Skaill.

In the old rentals the parish is entered as two—North Sandwick and South Sandwick, but, as in the cases already noted of Holm and Paplay, and of Marwick and Birsay, there was no ecclesiastical justification for such division: the only parish church of which anything is known is that of St. Peter, just mentioned. The division is all the harder to explain as there are no geographical features separating the two, and the dividing line was apparently drawn somewhat arbitrarily—simply following the boundary lines between old tunships. Nor is the line of division traceable exactly today, now that the old tunship boundaries are forgotten, but it ran roughly eastwards from the Bay of Skaill, the most southerly tunships in North Sandwick being Housegarth, Scaebrae and Hourston. The only explanation that can be suggested for the division would seem to be that it originated from some old system of local government or administration of which nothing is now known. South Sandwick was skatted as $9\frac{1}{2}$ urislands of skat-land, while North Sandwick included 8 urislands of skat-land with some 6d. lands of unskatted quoyland.

In N. Sandwick, however, there are indications of other divisions which may well date back to pre-Norse times altogether. From the present Quoyloo an old road that ran roughly in a direct line to the parish church was known as the Messigate. It ran through a tunship called in the earlier rentals Skorwell (in various spellings)—a 9d. land (now represented mainly by the several Stove farms), but in later rentals the name Skorwell was used to cover other 9d. land areas as well—all now commonly regarded as parts of that tunship.

The present public road running upwards in a north-easterly direction from the parish church follows roughly the line of an old track known as the Deesgate which separated the lands just mentioned (on the south-east) from those to the north-west which still are referred to under the inclusive name of North Dyke. Now in the 1500 Rental Skorwell is entered as " Skrowell . . . alias fermandikis." The latter name is never found again, and it may be suggested that it represents a misreading of a MS. *sunnan-dikis, *sunn-* (with a long s) having been read as *ferm-*. Mr. McInnes of the General Register House in Edinburgh has been good enough to check the reading for me in the MS. Rental, preserved there, and confirms that *fermandikis* is the correct reading, but that Rental itself has been copied from an earlier, and it is impossible to tell if or when the suggested error was made. If my suspicions are just, we should see in *Sunnan dikis* (O.N. *sunnangarðs*, 'south of the garth') the corresponding term to North Dyke—north of the dyke or garth. That a garth did at one time run along that line is obvious from the name North Dyke itself, and as to the nature of that garth we may recall the old 'gairsties' in North Ronaldsay, and the gairsty in Papa Westray which separates North Yard from South Yard, and the garth dividing Tolhop in St. Andrews. And as suggested earlier, those old gairsties in all probability date from pre-Norse times.

RENTAL FARM-NAMES

Aith (e:ð) : Ayth 1492; Aith 1492, 1500; Aithstoun 1595. An urisland tunship in South S. The present farm-houses of Aith stand on the east margin of the Loch of Skaill, but the area included in the old tunship is unknown. It is also uncertain what physical feature gave rise to the name—O.N. *eið*, isthmus, but it was probably either the mile-wide neck of land between the Loch of Skaill and the Loch of Harray, or the rather narrower neck between the Loch of Skaill and the Loch of Clumlie.

Bain (bɛ·ən): Bean 1739. Farm in Housegarth tunship. O.N. *bœ[r]-inn*, 'the farm-settlement.' There is another farm of this name in Hestwall tunship.

Bea (now pronounced as spelt—bi·ə) : id. 1739. Farm in Skaebrae tunship. O.N. *bœr*, farm-settlement.

Benzieclett (bɛn·ji-klɛt): Banziecleat 1739. In a MS. record of 1657 the name appears twice, spelt Banyeclave [sic], and I myself have heard it pronounced locally Banyieclaith (ban·ji-kleþ). Cf. Hammerclet *infra*. Farm in Skaebrae tunship. First element rather doubtful.

Borwick : Burwyk 1573 (R.E.O.); Burvik 1579 (R.E.O.); Burwik 1595, 1614; Burwick 1642. On west coast; together with Forswall formed a 9d. land. O.N. *borg[ar]-vík*, 'broch-bay'—so named from the broch ruins there.

Botulfyord : Bowtulffiszord 1492; Bowarisyord alias Botulfyord 1500; Boysland 1595, 1642, 1739; Boisland 1614. A 9d. land entered in 1739 R. as part of the 45d. land of Skorwell. Exact location unknown today.

Here we meet for the first time a type of name confined in the Rentals to North Sandwick only, where five of the six examples were adjacent, forming one continuous block of territory, while the sixth was a smaller unit about a mile distant. These names in the earlier rentals are characterised by the presence of the term *-yord* as second element, a term which obviously represents O.N. *jǫrð*, land or earth, while with one possible exception the first element in each is a personal name. These names are as follows (1500 R.) : Mobisyord in North Dyke (but separated only by the dyke or garth from the others); Brekisyord, Erikisyord, Sowlisyord, Bowarisyord alias Botulfyord—all in a group forming in the 1739 R. four-fifths of the tunship of Skorwell; Stoddisyord alias Gryndlenth in Housgarth tunship. This last was only a 2½d. land, but each of the other 'yords' was a 9d. land or half-urisland. It is curious to note that in the later rentals (1595 and after) the second element is rendered *-land* (Mobisland, etc.), but not one of all those names is in use today.

Those names thus present a nice problem for which no solution seems at present possible. The lands in question may probably be regarded as among the very earliest to be settled over here by Norsemen—very fertile lands as they are, sloping down to the Bay of Skaill. It may be suggested that the five 9d. lands formed parts of a single chieftain's 'land-take' which he portioned out among his followers, but that cannot be confirmed at all, and it must suffice here merely to draw attention to the curious group.

The alias names for this unit are also curious. One is plain—O.N. *Bótolfs-jǫrð*, land of a man Bótolfr, but the other (from which the later Boysland is probably derived) is obscure.

Brekisyord : 1500; Brekks 1492; Brekisland 1595; Braikisland 1614. A 9d. land in Skorwell. See Botulfyord. First element obscure, but presumably a personal name also : ?O.N. Bragi.

Brockan (brɔkˑən) : Brocken 1739. A farm in Wasbister tunship. Today there are two adjacent farms here—one on the shore of the Loch of Harray called Bockan (bɔkˑən), the other half a mile up the slope to the west and now known as Bookan (buˑkən). This last is adjacent and gives its name to the ancient monument—The Ring of Bookan. There seems little doubt that both names are the same, and corruptions of the older Brocken, i.e. O.N. *brokkan*, the slope; a variant West Mainland form of the usual Breckan. Cf. Brockan (Rendall).

There is another farm Brockan in Southerquoy, south of Skaill in this parish, which may probably represent a Brek in the old Rentals.

Clumlie (klʌmˑli) : Cluñy 1492; Clumlie 1500, 1595, 1614; Cumly 1739. A 6d. land in South S. sloping down northwards to a loch known as the Loch of Clumlie.

This name occurs also in Shetland and has given rise to a quite unjustifiable association with St. Columba. In his *Place-Names of Shetland* Jakobsen wrote as follows : "According to G. Goudie's researches . . . . it can doubtless be regarded as settled that this name originates from a 'Coluimchille.' A Clumlie is found in the Orkneys . . . and on an old map of the Orkneys (in a Dutch edition of Camden's Britannia, 1617) is found 'St. Columban' in the place where the present Clumlie is situated. Thus there was a church here dedicated to the well-known priest and saint, and the same must have been the case in Dunrossness in Shetland."

I have not been able to consult the map referred to, but on the first printed map of Scotland (by Ortelius, Antwerp 1570) a church marked 'S. Columban' is entered on the west shore of the only loch appearing in the West Mainland of Orkney. And the entry must have apparently been copied by others for it appears again on Speed's map of 1610. It may be noted, however, that the scale of those maps is so small, and the survey so inaccurate, that it is quite impossible to tell whether the site was anywhere near Clumlie at all. And it should be added that there is n o trace or record or memory of any church or chapel having ever existed in the neighbourhood of Clumlie. Nor is there any other record or memory of a church dedication in that part of Orkney either to St. Columba or 'S. Columban.'

A further objection to the suggested origin is to be found in the name itself which makes Jakobsen's adoption of it all the more surprising. Goudie's theoretical development (*Antiq. of Shet.* p. 22) was Coluimcillie —Columlie—Clumbly—Clumlie or Columbalie. It is hardly necessary to point out that such an origin for a place-name would be—to say the least —unusual, unless the word were attached as an adjunct to some other noun, e.g. kirk, etc.

Near the Shetland Clumlie Goudie records some traces of an old chapel, but, as just noted, nothing of the kind can be adduced in the case of this Orkney Clumlie. One form of ancient monument is common however to both—in the shape of a broch mound, and it is not impossible that the name may have reference to that in each case, though on the whole unlikely.

A few hundred yards from the south-east end of the Loch of Clumlie, however, are the remains of a denuded cairn known as the Stones of Via, an interesting and rather famous ancient monument. These are nearly adjacent to the farm-house of Via (q.v. *infra*), while at the margin of the loch below is a house named Lee. Here in all probability is to be sought the origin of the name Clumly. O.N. *kuml* was the name applied to an old cairn or burial mound, and an O.N. *kuml-hlið*, 'cairn-slope,' which by simple metathesis could become Clumlie, would be an entirely satisfactory explanation of the name.

Confirmation of such origin may be found in the case of a farm-name Cumley in St. Andrews parish. In a sasine of 1620 (O. and S. Sasines, Viking Soc., p. 111) that farm-name is on record as Clumley and Clumlie. And as for the name in Shetland it is significant to note Goudie's own words in his *Antiquities of Shetland*, p. 6: " I have expended some labour and outlay on one mound—the 'Bokie (i.e. Bogie) Brae' or 'Knowe of Willol' near Clumlie in the parish of Dunrossness." In view of that fact it is pretty clear that both names are from the same O.N. source indicated, and that the Columban association is a mere red herring.

Consgar: Coginsgair 1568 (R.E.O.); Consgair 1595; Congsgair (MS. copy of 1595). A farm in either Aith or Hurkisgarth tunship. For some time back and until recently this was the residence of the parish minister of Sandwick, but some years ago it was sold and re-named Flotterston by the new owner—a retired sea-captain who gave it the name of his old ship.

Origin of earlier name uncertain; it suggests an O.N. *konungs-garðr*, 'king's farm,' but its early history is unknown. Cf. Kingshouse, Harray.

Croval (kro·vəl): Crowale 1631; Crovell 1739; Crovall in Aithstoun (Sas. of 1807). Origin uncertain.

Cumley: Kimly 1739. Now spelt Cumbla. Farm in Newgarth tunship. See this name—St. Andrews. A.M. Inventory records several ancient mounds here.

Curkabreck: id. 1739. Farm in Hourston tunship. See this name—Harray.

Doehouse: Duthoishouse 1595; Duthois hows 1614; Doeahouse 1739. A 4½d. land in Tenston tunship which had been assigned to St. Duthus Stouk or Prebend—a foundation of one of the Sinclair earls.

Eriksyord: Crykiszard 1492 (initial 'c' in error for 'e'); Erikisyord 1500; Erikisland 1595, 1614; Eriksland 1595. A 9d. land in Skorwell; see Botulfsyord. O.N. *Eiríks-jǫrð*, land of a man Erik.

Fea: Feall 1492, 1500; Fea and Feaw 1595. A small unit in Newgarth tunship. There was another Fea in North Dyke. O.N. *fjall*, hill.

Feaval (fi·vəl): Feavell 1739. Farm in Tenston tunship. See this name—Birsay.

Fiddlerhouse: 1739. In Easter Voy; a Scots name applied for reason unknown.

Flaws: 1739. In Hurkisgarth tunship. See this name—Holm.

Forswell: Forsewall 1595, 1614, 1642; Forsewell 1739. A 3d. land which went along with Borwick to compose a half-urisland.
First element O.N. *fors*, a waterfall; the second probably O.N. *vǫllr*, a field (or *vellir*, pl.).

Garson: Ovir and Nethir-Garsend (Garsent) 1492, 1500; -Garsend 1500; -Garsand 1595. Each of these formed a 9d. land or half-urisland in North Dyke. They lie near the west shore and evidently near the end of an old garth or dyke—a probable ' gairsty ' : O.N. *garðs-endi*, garth- or dyke-end. There is another Garson in Skaebrae tunship.

Goldiger (gold·igər): id. 1739. Farm in Hurkisgarth tunship. The second element is no doubt O.N. *garðr* or some derivative—signifying a field, farm, enclosure, etc., but the first is obscure; in view of the number of ' garth ' names in this parish with a personal name as first element one is disposed to assume the same here.

Gorn: Gorne 1568 (R.E.O.), 1642; Gorn 1595. An 8d. land in Southerquoy. See Goir, Sanday.

Grind (grɪnd): id. 1739. A small farm in Easter Voy. See this name—Rendall.

Grindilo: 1739; in the 1500 R. it appears in the corrupt form Gryndlenth as an *alias* for Stoddisyord q.v. The name is obviously another example of the common Orkney name Grindally (see this name Sanday).

Halkland (hak·lən[d]): Halkland 1581 (R.E.O.); Haulkland 1619 Sas. A farm south of Skaebrae tunship and probably in what was formerly Lerely tunship. See this name—Rendall. There is no hill here in this part of Sandwick, but the farm is on ground rising somewhat above the surrounding levels.

Hallbreck (ha·brɛk): id. 1739. Farm in Hourston tunship. See Habreck, Wyre.

Hammer: 1642. Vanished farm in Aith tunship. See this name—Westray.

Hammercleat: Hammerclet 1500; Hammaraclaith (1612 deed); Hammercleat 1739. Farm in Skaebrae tunship. With the 1612 spelling cf. the pronunciation noted above of Benzieclett, another farm in this tunship. The name is a puzzling combination of two almost synonymous words—O.N. *hamarr*, a rock projecting from a hillside or slope, and *klettr*, a rock. A similar duplication is found in Rousay where a well-known fishing rock is called Klettber (O.N. -*berg*, rock). See also note on Cleat, Sanday.

Hestwall (hɛst·wəl): Hastwill 1492; Haistwale 1500; Hestwall 1595, 1614, 1739. A 9d. land tunship. See this name—Holm.

Heyland: 1739. A vanished farm—apparently in Tronston tunship. Evidently O.N. *hey-land*, ' hay land.'

Hourston (hur·stən): Thurstath 1492; Thurstacht ('c' in error for 't') 1500; Hurstane 1595; Hourstaine 1642; Hourston 1739. A 6d. land tunship adjacent to Harray border. An old -*staðir* name illustrating the initial phonetic change of ' th-' to ' h-,' cf. Hurtiso (Holm) and Horraldsay (Firth). The first element is undoubtedly a personal name, but exactly which is doubtful : O.N. *þórir*? *þórðr*? *etc.* The meaning of course is the ' stead ' or settlement of a man Th——.

How : 1492, 1500. A vanished 1d. land in Housgarth tunship. O.N. *haugr*, mound.

Housgarth (hus·gər) : Houssgarth 1492; Housgarth 1492, 1500, 1530 (R.E.O.); Howsgarth 1500; Howsgair 1595; Housegair 1739. An urisland tunship adjacent to Skorwell. The first element of this name is rather doubtful, but as in the Sanday case of the same name it is probable that it represents O.N. *haugr*, a mound, rather than *hús*, house. Adjacent to the farm-house of Bain in this tunship are conspicuous mounds known as the Knowes of Bain.

Housenia (husni·ə) : id. 1739. A farm in Newgarth tunship. O.N. *hús-nýja*, new house. There used to be a farm of the same name in Skorwell—Housnia, 1809 Sasine), but it is now spelt simply Snea, a curious simplification in which the old adjectival part has clung to the ' s ' of the noun.

Huan (hu·ən) : Howon 1739. A farm in North Dike. The houses stand on an elevation overlooking the main road : hence probably O.N. *hauginn* (accus.), the mound.

Hurkisgarth (hʌrk·esgər) : Thurkingisgarth, Thurkynsgarth 1492; Thurkingsgarth 1500; Hurgisgair 1568 (R.E.O.); Hurkisgair 1595, 1642; Hurtisgairth 1739. A 9d. land tunship in South S.

An O.N. *garðr* name of which the first element, though a personal name, is somewhat uncertain; it might represent the female name—*Þorgunnr*, or perhaps even the rare male name *Þorgnýr*, but the very common name *Þorkell* with change of 'l' to 'n,' is perhaps the most likely. N.B. here once more the initial change of ' th ' to ' h.'

Hyval (hai·vəl) : Hyvell 1739. Farm in Hourston tunship. Probably O.N. *hey-vǫllr*, hay-field. The same name occurs in North Dyke.

Instabillie : Enstabilly 1492; Enstabillie 1500; Instabillie 1595, 1614. A 9d. land tunship in North Dyke. See this name St. Ola.

Kierfold (kir·fə) : a relatively modern farm on slope of a hill which is mentioned in the 1642 R. as " Hill of Keirfia." The name is thus apparently identical with that of the Rousay hill—Kearfea, and represents probably O.N. *kýr-fjall*, ' cow-hill.'

Kirkness : id. (U.B. 1601). Oddly enough this old farm is never mentioned in any of the old Rentals, though one of the oldest and most important of early Orkney families derived their surname therefrom. A Sir Thomas of Kirkness—apparently a knight—appears as a witness in 1391 (R.E.O., p. 27), while a John of Kirkness was Lawman of Orkney c. 1430. In the Uthall Book of 1601 Kirkness is entered as " Quoyland butt scatt," and appears to have been a 6d. land. It was probably included in the tunship of Lerely q.v. The name springs from an old chapel which stood here.

Lerely : 1492; Leirly 1500. The name of this old urisland tunship is quite obsolete today and its situation can only be deduced as having probably been the peninsula jutting out into the north end of the Loch of Harray, including the farms of Kirkness, Halkland, Bankhead and probably Vetquoy. In the 1500 R. it is entered, but with a note—" the haile scattis doune with half the land male (rent)," while in the 1595 R. the only entry is " Quoy-Lelie. Ley." There does not seem to be any later record of the name.

NAME MATERIAL 153

The facts here are very puzzling, because the lands of that area today are very fertile, and how they can have been ' ley ' (uncultivated) in 1595 is incomprehensible. Origin of name also obscure.

Linahow (lɪn·əhou): Lynehow 1492; Lynhow 1500; Lenehow 1595; Linehow 1614; Lina How 1739. A 9d. land in North Dyke. The second element (O.N. *haugr*, a mound) refers no doubt to a mound known as The Castle adjacent to the farm-buildings, while the first is perhaps O.N. *hlein*, a lesser slope.

Linda: Lyndale 1492; Lyndie 1500; Linday 1614, 1642, 1739. A 1½d. land in Newgarth tunship. Origin of name uncertain.

Linklater (lɪŋ·kletər): Lynclett 1492; Lynkclet 1500; Linkleter and Linkletter 1595; Linklatter 1739.
A 3d. land tunship north of Skaebrae. See Linclet—N. Ronaldsay.

Looath (loð): Looth 1739. Farm in North Dyke. Origin obscure.

Lyking (laik·ɪn): Lyking, Nethir and Ovir 1492, 1500; Lykin 1579 (R.E.O.); 1595; Lyking 1614; Lycking 1739. A 12d. land tunship in South S.
A particularly interesting name as it would appear to be the only really old settlement in Orkney having O.N. *vin* (pasture) as second element: O.N. *leikvin*, ' sports-field.' There are two other Lykings—one in Holm, the other in Rendall, but these are small outlying units that cannot be classed as really old original settlements.

Moan (mɔ:n): 1739. Farm in Skaebrae tunship. See this name—Firth.

Mobisyord: 1500; Mobisland 1492, 1595; Moibissland, Moibseland 1614; Moabisland 1642; Moabs Land 1739.
A 9d. land in North Dyke, separated only by an ancient roadway—The Deasgate—from the lands of Skorwell. See M—— land, Westray; also under parish name *supra*. This name is probably obsolete now locally.

Ness: id. 1739. Farm in Wasbister tunship on a ness projecting into Loch of Harray.

Newbigging: Newbiggin 1595; Newbigging 1614, 1642. A 5d. land in Southerquoy.

Newgarth (nø·gər): id. 1492; Negarth 1500; Neugar 1536 (R.E.O.); Newgair 1595, 1642; Newgger 1614. An urisland tunship—the most northerly in South S. and adjacent to Housgarth tunship in North S. Self-explanatory.

North Dyke: Northdik 1565 (R.E.O.); Northdyck 1566 (R.E.O.); North Dyk 1595. A district, not a farm-name. See under parish name *supra*. The name is still in general use today.

Overhouse: Or. house 1739 — a contraction for Over-. Farm in Hestwall tunship. See this name—Harray.

Pow (pou): Poll 1492, 1500. Chief farm in Housegarth tunship today. O.N. *pollr*, a pool.

Quoys: two farms of this name are recorded in Skorwell in the 1739 R.—Quoys in Sowieland and Over Quoys in Boysland. Another Quoys is on record in Newgarth. O.N. *kviar*, pl. of *kvi*, enclosure for animals.

Quoyloo (kwailu:) : Quoylow 1739. Now a small district in Skorwell. Origin of second element obscure; *-lo* in Norw. names can rarely be satisfactorily interpreted.

Rango (raŋ·o) : (Mill of) Vrango 1595; –Wrango 1614, 1642.
This farm and mill are on the banks of a small stream flowing into the Loch of Harray. This is the same name as one of the most famous Saga rivers in Iceland—O.N. *Rangá*, twisting stream. In Southern Norway acc. to Aasen *rang* is pron. *vrang*, but it is doubtful whether the early spellings recorded above indicate a similar pronunciation, or whether the initial letter is due to the spelling *wrong*, *wrang* in Eng./Scots.

Skaill : Scalle 1492; Skail 1595; Skeall 1614; Skaill 1642. An old 8d. land included in what the 1500 R. designated " the 24d. land of Sutherquoy." O.N. *skáli* a hall.

This farm and house (which for over three centuries has been the chief manor-house in the parish) raise interesting problems. The lands lie at the south-east corner of the Bay of Skaill (the original *Sand-vik*, sandy bay) at the foot of a slope rising up to the south, on which are several small farms, which as a group compose the present small district of Sutherquoy. Now as indicated in Part III quoy-names are relatively later settlements, and it is thus plainly incredible that this whole area, including Skaill itself, would have been called Sutherquoy when first settled by Norsemen. We have seen how in the case of Beaquoy in Birsay a quoy-name has superseded an earlier *Hús-bœr* as a tunship name, and something similar must have happened here. It is further unlikely that Skaill itself—primarily a house name—would have been the original *settlement*-name, and we have thus to look elsewhere.

As hinted under the parish name *supra*, though we have no record of any old tunship called Sandwick which, as in many other cases (from the presence of the parish church), gave its name later to the whole parish, yet the existence of such a tunship seemed very probable. It may now be suggested that there was such a tunship or early settlement called Sandwick, which embraced all the land around the shores of this Bay, and that Skaill was the original head-house thereof, while Sutherquoy proper was an early expansion southwards. The early settlement of that area is well-attested by the fact that it was fully skatted.

On general principles it would seem indeed very unlikely that a Norse chieftain landing in this bay would restrict his ' land-take ' to one side of it only, when the bulk of good arable land was on the other north side. It may not be irrelevant to note that Skaill itself is adjacent to the now-famous prehistoric hamlet of Skara Brae, and though it is improbable that any traces of that settlement were visible when the Norse arrived yet other unknown factors, such as the abundant freshwater supply in the adjacent Loch of Skaill, may have helped to dictate choice of site.

To the foregoing interpretation it may be objected that the parish church is situated at the north-east corner of the bay some half-mile or so distant from Skaill, and from what is noted in Part II the head-house of the tunship or settlement might be expected to have been in that immediate neighbourhood. Assuming—(what is indeed pretty certain)—that the parish church here, as in various other areas, developed out of the ' urisland chapel ' of the chief local magnate at the time, it must also be remembered that between the date of the original settlement and the the erection of urisland chapels in the 11th century three hundred years must have elapsed, during which time many changes might have taken place. Under udal law estates were subject to division among heirs, and by the 11th century Skaill may not have been the residence of the local chieftain at all. In that connection it is significant to recall that in the first *Statistical Account of Scotland* (c. 1790) we are told about the remains of an old structure on the shore of this bay, quite near the church, which was known as the Castle of Snusgar. No evidence of that structure is now to be seen above ground, but in its immediate vicinity there was found in 1858 the greatest hoard of Viking silver ever recovered in Scotland. From the evidence of coins in that hoard it was proved that it had been deposited after the year 945 A.D.—how long exactly it is impossible to tell. There is no record of castles in Orkney before the middle of the 12th century, and it is practically certain that parish churches were not established until a century later, and perhaps later still. And it may be regarded as certain that any man owning a castle in that area would have been the local chieftain at the time. In these circumstances the situation of the parish church is more easily understood. And, as already suggested, the tunship in which it was located being presumably called Sandwick, that name became thereafter the name for the whole parish just as we have seen happened in the case of Orphir, of Firth, and of Stenness.

The foregoing interpretation of course cannot be proved, but it is submitted as the most reasonable hypothesis to explain the following puzzling problems:

1. How the name of a mere bay could be applied as a name for an extensive parish.
2. How a mere quoy-name (Sutherquoy) could have come to include a 'Skaill.'

It should further be noted that while the south side of Skaill Bay with Sutherquoy (including Skaill) is found in the old Rentals as part of South Sandwick, the parish church and all on the north side of the bay were in North Sandwick. As indicated above, there was never (so far as we can tell) any parish church in South Sandwick and if we knew when and why such division into North S. and South S.. was made the above problems might be simplified.

Finally, it may be noted that if the above interpretation be approximately correct we should have here a parallel to what happened in Deerness. There also we have a sandy bay and the adjacent tunship is still known as Sandwick. The chief old house therein is also a Skaill, adjacent to which is the parish church. This parish, however, was given the name of the peninsula it completely occupies—viz. Deerness.

Skaebrae (ske·bre, ske′brə) : Scalbrycht, Scalbra 1500; Scabra, Scabrae 1595; Scabray 1614, 1642; Scabrae 1739. An urisland tunship in North S. Same name as Skelbrae, Sanday, q.v.

Skorwell (skɔr·wəl, skar·wəl) : Scorowell (9d. land) 1492; Scrowell alias fermandikis (9d. land) 1500; absent from 1595 R. where Stove a 9d. land figures in its place, and likewise in later rentals; Skoirvaill, Skoirwall, and Skorvall 1565 (R.E.O.); Scorwall 1642; in the 1739 R. Scorwall appears as a 45d. land embracing the five 9d. lands of Brakos Land (i.e. Brecksland), Sowie Land, Eriksland, Boysland and Stowck Land (i.e. Stove which had pertained to the prebend of St. Duthoc). It would thus appear that the 9d. land of Stove represented the original Skorwell.

Origin of name rather doubtful. Stove lies in a valley running in a north-easterly direction from Skaill Bay, and hence perhaps O.N. *skorvǫllr, -vellir,* field[s] in a 'score' (cut) or valley.

Smerquoy : 1595. Apparently a small unit in Skorwell or North Dyke, and entered as "Ley" (i.e. uncultivated). Name apparently obsolete today: see the name—St. Ola.

Sutherquoy: Southirqwy (7d. land) 1492 pro rege; Suthirquy (17d. land) 1492 pro epo.; " Suthirquoy wes ay to the Kingis scattis as his auld rental bearis XXIIII d. terre" (1500 R.); Southerquoy alias Gorn 7d. land pro rege (1595 R. pro rege); Sutherquoy (1595 R. pro epo.).

The lands included herein were Skaill (8d. land), Gorn (8d. land), Newbigging (5d. land) and Brek (3d. land). 1d. land of Gorn was Stouck or prebend land. See also *sub*. Skaill. ' Southern quoy.'

South Seatter: South Setter 1595; Southsetter 1642; South Seatter 1739. A 3d. land above Yesnaby.

Sowlisyord: Sowliszard 1492; Sowlisyord 1500; Sowiland 1595; Sowieland 1595, 1614, 1642, 1739. One of the Skorwell 9d. lands. See further Botolfsyord *supra*. The first element is certainly a personal name but it is uncertain exactly which: O.N. *Solli?, Sǫlvi?, Sǫrli?*, etc.

Stoddisyord: Stodzorde 1492; Stoddisyord alias Gryndlenth 1500. Not on record later. A 2½d. land in Housegarth tunship. First element pretty certainly a personal name as in the other *-yords*, but no similar O.N. name seems to be on record.

Stove: Stoif 1595, 1642; Stoiff 1614; Stove 1739. A 9d. land in Skorwell (1739 R.) which would appear to represent the original Skorwell, q.v. *supra*. O.N. *stofa*, a room; see also Stove, Sanday.

Tenston: Tenstath 1492; Tenstaith, Tenstayth 1500; Tenstay 1564 (R.E.O.); Tensta 1595; Tensten 1614; Tenstane 1642; Tenston 1739. An urisland tunship in South S. A *-staðir* name with first element no doubt a personal name—exactly which is uncertain: O.N. *Tannr?, Tanni?, Tindr?* etc.

Tronston: Cronstath ('c' for 't') 1492; Tronstaith 1500; Tronstane, Tronstaye 1565 (R.E.O.); Tronstane 1595; Tronston 1614, 1739. A 6d. land in North S. Another O.N. *-staðir* name, with personal name as first element, probably O.N. *Þróndar-staðir*, ' stead' or settlement of a man Thrond.

Trinnigar: Trinniger 1739. A 3d. land in Aithstun; earlier forms lacking. The second element represents no doubt O.N. *garðr*, garth or farm; the first is doubtful, but almost certainly a personal name.

Unigar (un·igər, un·əgər): Owkyngarth 1492; Hawgengarth alias Owgengarth 1500; Vingar (by error for Unigar) 1595; Wniger 1614; Vnigair 1642; Unniger 1739. A 3d. land in South S.

O.N. *Hákonar-garðr*, Hakon's garth or farm. The phonetic development revealed here is interesting. The first of the two 1500 forms is actually the oldest of all, and indicates a pronunciation that was obsolete by 1492, but even in that form the old genitive ending had been dropped. Then by that time also the long O.N. *á* had been rounded to ou and the hard *k* softened to *g*. Presently this guttural went also, and the several phonetic stages may be indicated as follows: Hákonar>Howken>Howgen >Owgen>Owen>Un. It is curious however, how the medial vowel of the final form seems to be a survival of the old genitive inflexion *-ar*.

This Uniger is in South S., but in North S. there is another—usually called North Uniger.

Upsale: Ayth and Upsale 1492; Aith and Upsale, ane uris terre, 1500. This very interesting name is never on record afterwards and seems to be entirely obsolete locally, the site even being forgotten and unknown. The second element represents O.N. *salr*, a hall, of which Rygh (*Indl.* p. 73) writes that it appears in many farm-names " of which the most widely spread is the certainly very old Uppsalir [plural]. . . . So used, *salr* seems to have no other meaning than residence or farm." Magnus Olsen suggests [N.G. X, 161] that so used Uppsalir may represent an an ' upper farm.'

This is the same name as that of the Swedish city of Upsala, the residence of early Swedish kings, and the old centre of Odin-worship in that country.

Voy (voi) : Wawis be West 1492; Wawis be West, Waw 1500; Woster Woy, Eister Woy 1595; Westervoy, Eastervoy 1642, 1739. These formed a 6d. land round the head of the bay in the north-west part of the Loch of Stenness. The early forms Wawis suggest an old plural (*Vágar*) of O.N. *vágr*, a bay or voe, but the later pronunciation points to the old dat. sing. *Vági*. This name occurs also in St. Andrews parish.

Warth : id. 1595, 1614, 1642, 1739. A 3d. quoyland north of Skaebrae. Origin obscure; it seems to indicate O.N. *varða* or *varði*, a beacon, but this farm is on a low plain.

Wasbister (waz·bəstər) : Wasbuster 1492; Wasbustar 1500; Wasbister 1614. A 9d. land tunship between Lochs of Stenness and Harray. O.N. *vaz-bólstaðr*, ' lake-bister ': the same name occurs in Rousay.

Windbreck : id. 1739. A small unit in Easter Voy. See this name—Birsay.

Yesnaby (jɛs·nəbi, -be) : Yeskenibie 1536 (R.E.O.); Zeskynnebea 1589 (R.E.O.); Yeskennabe, Yeskenabe 1595; Yescannabie 1595 (S.C.H.); Zeskinaba 1614; Zescannabie 1642.

A 6d. land on western seaboard. An extremely puzzling name; the final element suggests O.N. *bœr*, a farm-settlement, but this consists of relatively barren wind-swept soil, very exposed to western gales and sea-gust, and is much smaller than the normal Orkney bae-farms. The preceding element is most obscure.

OTHER FARM-NAMES

Appiehouse : three farms so named—in Tenston, Hourston and Quoyloo. See this name—Stenness.

Appietown : now frequently pronounced Appieteen. Farm in Wasbister. See this name—Rendall.

Conziebreck (kon·ji–brɛk) : small farm near Trinnigar. First element of doubtful origin.

Dounby (dun·bi) : a farm in Hourston tunship. A village of this name has sprung up here in the course mainly of the last hundred years. A puzzling name. The ending suggests an O.N. *bœr* farm, but as I have seen no mention of the name in old records its origin must be regarded as doubtful.

Fidge (fɪdʒ) : a farm in Skaebrae tunship. This name is fairly common in Orkney and represents O.N. *fitjar*, pl. of *fit*, meadow.

Gairsty : a farm in Tronston tunship. There must have been an old gairsty hereabouts. See Treb, N. Ronaldsay.

Garricot (gɛr·ikot) : a small farm in Sutherquoy area. A cot-name (O.N. *kot*), but first element uncertain. Cf. Gerraquoy, Birsay.

Gyren (gair·ən) : a small farm on hillside of same name. See Green, Birsay, but reason for name here unknown.

Hooveth (høv·əþ) : a small farm high up in North Dyke area. It is especially regrettable that old forms of this name are completely lacking. It has been pointed out *sub* Benzieclett (supra) how in this West Mainland area a final 't' tends to be aspirated to 'th'; phonetically, therefore, this name might well represent O.N. *hofit*, 'the hof' or heathen temple (or land pertaining thereto). Rygh in his *Indledning*, however, notes that a Norse word *hov* (n.) signifying merely a height or elevation (something 'heaved up') appears in many farm-names in the south of Norway. That word Torp (N.N.O.) regards as really the same as the O.N. *hof* (temple). In these circumstances the real significance of this Orkney farm-name must be left doubtful, but as Rygh says that *hov* as a farm-name is always found in the def. form—*Hovet*, it would appear dangerous to assume any other sense in this case.

It is noteworthy however that in a Sasine of 1785 a farm Hoveth is recorded from Grimeston in Harray.

Howaback : a farm in Hourston. On an old tunship map this name appears as Hobac, and as a small burn flows down here that form might well represent O.N. *ár-bakki*, river-slope or bank, a name that appears repeatedly in Orkney in the form Oback. Cf. however the Fær. place-name *Hóvabakki*, 'high bank or slope.'

Laith : a farm in Tenston on low ground. Origin uncertain.

Lee : a small farm in Hestwall : see Clumlie.

Quean (kwi·ən) : a farm in Newgarth tunship. A def. form of O.N. *kví*, enclosure for animals; most probably nom. sing. *kvín* (kví-in), 'the' quoy.

Queena (kwin·ə) : a farm in Hestwall. See this name—Stenness.

Queenalanga : a farm in Hourston. O.N. *kví-langa*, the long quoy; perhaps in accus. *kvína lǫngu*.

Roundadee : a farm on hillside south of Yesnaby. The second element here is probably O.N. *dý*, a marsh, but if the first signifies 'round' it must be a relatively late formation as No. *rund* is borrowing from L. German.

Scarrataing (skar·ə-teŋ, -tiŋ) : small farm on point projecting into Loch of Harray. The same name is frequently applied to points on the sea-coasts of Orkney—O.N. *skarfa-tangi*, scarf (or cormorant) point.

Skeithva : an old house in Lyking tunship. See next name.

Skethquoy : a farm in Tenston about half a mile from last-mentioned name. The first element in these names is apparently O.N. *skeið*, but the sense in which it is to be understood is obscure; see Skae (Deerness).

Stockan : a farm in North Dyke, and another of the same name near the Mill of Rango. This name occurs also in Norway (Stokken) where it is not always clear how it should be explained. In his *Indledning*, Rygh cites O.N. *stokkr*, a stock, a hewed down tree-stem, and suggests various senses in which that term might be explained in farm-names. The only sense that might perhaps apply in these Orkney names would be—'*rydningsplads*,' a clearing for cultivation, but whether that be the correct sense is doubtful.

Swartland: farm in Skaebrae tunship. O.N. *svart-land*, black land. This farm lies on low flat marshy ground, and the name has reference probably to the dark mossy nature of the soil.

Tufta: farm in Skaebrae tunship. See this name—Harray.

Velyie (vɛl·ji): farm in Sutherquoy. Cf. Velzian—Harray and Rendall.

Vetquoy: farm at northwest corner of Harray Loch and in what was probably the old tunship of Lerely. Perhaps O.N. *feit-kvi*, ' fat-quoy,' i.e. very fertile; cf. Fær. *feitilendi* and the Shet. *Fedaland*.

Via (vi·ə): farm adjacent to the famous Stones of Via. This name does not appear in any old Rental, and the earliest mention of the name I have seen is in Anderson's *Guide to the Highlands and Islands of Scotland* (London 1834) where mention is made of " the Stones of Via, a structure which till now has entirely escaped observation . . . ." The Orkney section of this Guide was apparently contributed by the Rev. Charles Clouston, minister of this parish. In the 1739 R., however, a farm Fea is recorded from this tunship of Hestwall and it may be suggested that Via is a corruption of that name—O.N. *fjall*, a hill.

Vola: farm in Skaebrae. Origin of name uncertain.

Yuildadee (jøld·ədi): also spelt Yeldadee (jɛld·ədi): farm on hillslope south of Sutherquoy. Origin uncertain. The last element would appear to be O.N. *dý*, a marsh. Cf. Yeldabreck, Birsay, and Yeldavill, Harray.

STROMNESS

Stromnesse 1483 (R.E.O.); Stromnes and Stromness indiscriminately thereafter. O.N. *straum[s]-nes*, i.e. the ness projecting into the rapid 'stream' or tideway of Hoy Sound.

Strangely enough this now familiar name is never mentioned in the O. Saga, and but rarely in old records of any sort. In the early Rentals moreover the farms in a considerable area of the parish are specified merely by reference to their skattable values and only occasionally by their individual names. Even the skatting itself is difficult to assess with certainty. In the north of the parish Kirbister was a 6d. land and the neighbouring Redland a 3d. land—together making up a half-urisland, which, as we may observe on Mackenzie's Charts (1750) was enclosed by a common hill-dyke. On the south side of Redland Hill the chart shows another hill-dyke round the cultivated lands of Quholm, Bowbrek, Garth and Lee. These together formed a second half-urisland, but there our difficulties begin.

The remainder of the old arable lands of this parish lay along a coastal arc some four miles in length and roughly a mile in width between the hills and sea, stretching from the Bridge of Waithe in the east to the Black Craig in the west. Midway along this belt is Stromness Harbour, the old Hamnavoe, one of the safest and most sheltered havens for small vessels to be found in Orkney. The portion of this coastal belt west of Hamnavoe was apparently all termed originally Stromness, but by the date of our earliest Rentals with one small exception it appears as two tunships—Inner and Outer Stromness, each skatted as two urislands. They were of different capital value, however, by that time, for while Inner Stromness lands were valued at 3 marks per pennyland Outer Stromness lands were 4 marks per pennyland. It is uncertain but improbable that there ever was a tunship wall dividing these two tunships.

The small 'exception' just referred to is puzzling. Between Hamnavoe and Inner Stromness, on the original *Straumnes* itself, was a small ½d. quoyland farm called Ness which was not (as might have been expected) included in the adjacent two urislands (36d. land) of Inner Stromness. For such a very curious omission or exclusion there must have been some very pertinent reason, obscure though that may now be.

The coastal belt east of Hamnavoe presents further difficulty. It now goes under the inclusive name of Cairston, and in the Rentals some of the lands therein were erroneously termed quoylands from the fact that they paid no skat. In reality, however, they were bordlands, parts of the private 'Auld Earldom' estate, and it was on that account they were unskatted. It is now impracticable to tell how much of this area was actually bordland, but altogether it was assessed as 26½ pennylands, i.e. just ½d. less than 1½ urislands. Was the odd ½d. land that of Ness on the west shore of Hamnavoe? If so, then the whole parish would have included 6½ urislands with a few quoylands in addition.

Whatever the true interpretation of those facts may be, the excellent harbour of Hamnavoe itself cannot be left out of account. Any Norse settler arriving in this part of Orkney would inevitably have chosen this haven in which to anchor or lay up his ship, and it is probable that for his 'land-take' or settlement he would have taken all the land on each side of the harbour. In that case we should see here yet another of those large original settlement-units referred to so often already, in this case an estate of such size as to be later skatted as no less than 5½ urislands, with a magnificent harbour in its midst. The names Inner and Outer Stromness imply that division of an earlier 'Stromness' had taken place at some time, and the fact that Ness was not incorporated with Inner Stromness might imply that the harbour itself went with the coastal belt to the east. That that portion was an important estate later is obvious from the fact that in the 12th century a castle had been erected thereon at the Bu of Cairston.

A further problem arises in connection with the parish church. The present parish church of St. Peter is situated in the town of Stromness, but it is said to date only from 1814. The former parish church of St. Peter stood at the graveyard on the shore of Hoy Sound in the tunship of Inner-town, adjacent to remains of what is traditionally called "Monker House," the early history of which is completely unknown. In a document of 1545/6 (R.E.O., 229), however, we read of Outer Stromness "in the parichin of the haly cros," which must be a reference to a still earlier parish church dedicated not to St .Peter but to the Holy Cross. As to the location of that church we may have a hint in the 1627 Report of Stromness printed in Peterkin's Rentals. There it is stated that the vicar's residence used to be "in Conquescoy in Kairstane, qlk wes ye viccars maniss and his glibbe." That is quite close to the Bu of Cairston, where the O. Saga records that a castle stood in the 12th century, and as Mr. Clouston argued in his *History of Orkney* there can be no reasonable doubt that the original parish church of the Holy Rood likewise stood in that immediate vicinity.

Such a conclusion would be entirely in line with what we know happened in other parishes in Orkney. It is pointed out in Part II how in every case that can be checked it was the private or urisland chapel of the leading family in each area that was raised to the dignity of a parish church when these were established. And though we do not know the precise date when a parish organisation was introduced here in Orkney there is every probability that the leading family at the time would have been the descendants of the leading family in the 12th century. And that the owner of a castle was the leading man or local chieftain in the Stromness area then, can hardly admit of doubt.

Now as we have previously noted, it was either the name of the tunship where the parish church stood, or the name of the church itself, that normally became the parish name. Why then it may be asked is not the parish called Cairston or Holy Cross parish (as in Westray and Sanday)? From the document quoted above we know that it actually was called for a time the parish of the Holy Cross. When and why then, was the name changed to Stromness? A partial answer to the first question would be—after the parish church was established in the tunship of Inner Stromness, but exactly when that took place we do not know, nor do we know the reason for the change. Here we are reduced to mere conjectures, but it may be surmised that a clue may be found in the Monastery remains referred to above. Apart from the name there is nothing at present visible, nor any records, to confirm that a monastery* once stood there in Innertown. On the other hand there is nothing to disprove it, and one can point to the monastic ruins discovered on the Brough of Birsay where no such monastery was previously ever suspected. And if there was a monastic establishment here in Stromness it may be suggested that it was the monastic church at that site which became the new parish church for the area after the Holy Cross church in Cairston had fallen into decay.

---

* "A fragment of a bronze mounting with a Celtic pattern inlaid in gold, now in the National Museum, was found in the Monker Green [here], and, near the same place, the terminal portion of a Celtic brooch."
 A.M. *Inventory*, p. 321.

Finally, it may perhaps not be irrelevant to suggest that some memory of the old church in Cairston is implicit in the much later name of the " Cairston Presbytery."

RENTAL NAMES

Bea: 1500. A 3d. land in Innertown. Name now obsolete. O.N. *bœr*, a farm; see Bae names—Part III.

Bowbreck: Bowbrak 1492; Bowbrek 1500, 1614, 1642; Bowbreck 1595, 1739. A 2d. land in Whome area. Name obsolete; first element of uncertain origin.

Bowbustirland: see Howbusterland.

Breckness: Breknes 1545, 1565 (R.E.O.); Breckness 1642, 1739. In Outertun— probably the chief farm in that tunship—situated on a low headland where remains of a broch are still to be seen. Old skattable value not recorded.

This name undoubtedly has reference to the skerry called Braga (bra:gə) which lies about 300 yards offshore at this point, and on which billows from the open Atlantic are constantly breaking, at times with terrific violence. It is an interesting name which occurs again applied to a reef off Costa Head. According to the *North Sea Pilot* the name also appears in Shetland where there is a Braga Ness (= our Breckness) and a Braga Rock (twice). The word is derived from (or cognate with) O.N. *brak*, crash, noise, uproar, etc.; No. *braak*, Dan. *brag*, id. Applied as a name for a rock or skerry on which waves break the term might perhaps be interpreted as 'the crasher' or 'the tumultuous one.' Cf. similar terms in *The Place-Names of Shetland*, p. 133.

Brinnigar: Brownigair 1595; Browneger 1614; Brounigair 1642; Brinniger 1739. In 1595 entered as 2½d. land in Innertun. It later became the minister's glebe. The present (and 1739) form might suggest that burning had occurred here, but the earlier spellings render that doubtful. Brown, however, was a common surname in Stromness area in the 16th century, and that may have influenced the spelling.

Brettabreck (brɛt·əbrɛk): Brittabreck 1739. A farm in Kirbister tunship. See this name—Rendall.

Bu of Cairston: see Cairston.

Cairston (kɛr·stən, kjɛr·stən): *fyrir Kiarreksstaupum*. O. Saga; Bull of Kerstane 1492; ... Karstane 1500; ... Kairstane 1595; Cairstane, Cairsten 1614. In 1500 entered as 14d. land, but several small 'onsetts' were included therein; by itself, according to Mr. Storer Clouston, the Bu was probably a 9d. land or ½ urisland. An O.N. *staðir* name, signifying the stead or settlement of a man *Kjarrekr*, an otherwise unrecorded name. For Bu—see Part III.

Citadel: Cittadeall 1739. Farm in Innertown. Its early history is quite unknown.

Clouk (kløk): Cloukland 1492; absent from later rentals until 1739 where Cluck appears as a 6d. land in Inner Stromness. The Beatons of Clouk were one of the most important local families c. 1600.

Origin of name obscure, but it appears again in Rousay and Evie. The pronunciation indicates that it is a different name from Cloke (Birsay).

NAME MATERIAL

Cloviger (klov·igər, klou·əgər) : id. 1739. A farm in Kirbister tunship. Perhaps O.N. *Klaufa-garðr,* garth or farm of a man Klaufi, but the first element may represent O.N. *klauf* (or *klofa* or *klofi*)—a cleft; in Færoe *klovi* is used (in place-names) of a cattle-track.

Creya: Crega (1619 Sas.); Crea 1739. A small farm in Outertun. There is another farm of this name in the Kirbister area. See this name—Orphir.

Dale (del, dil) : a farm in Inner Stromness which does not appear in the Rentals. Strangely enough there was another of this name in Outer Stromness which is similarly absent from the Rentals but well attested in sasine records : " one penny udal land in the Uttertoun of Stromness in that part thereof called the sex penny land of Deall . . ." (Sas. of 25/11/1619); " within the 6d. land of Deall . . . in the Uttertoun " (Sas. of 21/6/1620). There is no farm of that name there today, but on the 6-inch O.S. map part of the tunship is marked ' Deal.'

In neither tunship is there any very noticeable dale or valley to explain the name, which may thus have to be classed with Dale in Evie and Brendale and Ervadale in Rousay, q.v.

Deasman: 1739; Deaman 1764. A small unit in Innertun.

Don: Done 1588 (R.E.O.); Done, Ovirdone and Nethirdone (1617 Sas.); Don 1739. A small farm in Outer Stromness near the Black Craig. Origin of name obscure; if it were not that the same name seems to occur twice in Norway (where it is also unexplained, G.N. xv) one might be disposed to suspect Celtic origin, all the more so in view of the name of the adjacent farm—Straither q.v.

Fea: Waster Fea 1739. Farm in Outertun. O.N. *fjall,* hill.

Fealquoy (fjal·kwi) : Fewqui 1490 (R.E.O.); Feaquoy, Fealquoy (1617 Sas.); " the 6d. land of Fealtquoy in Uttertoun " (1620 Sas.); Fealquoy 1739. In Outer Stromness near Black Craig. O.N. *fjall-kvi,* hill-quoy.

Feawell: Feuell 1500; Fewall 1595; Fewell 1614, 1642; Feavell 1739. A 2d. land in Cairston. Origin uncertain; see this name—Birsay.

Forse: 1739. Small unit in Cairston . O.N. *fors,* waterfall.

Garson: Garisend 1492; Garsend 1500; Garsand 1595; Garson 1739. A 1d. so-called quoyland in Cairston. O.N. *garðs-endi,* ' dyke-end ' : there was an old gairsty in this area: see Treb, North Ronaldsay.

Garth: id. 1492, 1500, 1595; Gerth 1614; Gairth 1642, 1739. A 1½d. land in the Whome district. O.N. *garðr,* a farm.

Hamiger: Hammager 1587 Charter (R.M.S.); Hamigair 1595; Hameger 1614; Hammagair 1642. A ½d. land on east shore of Hamnavoe, or Hamlavoe (Stromness Harbour). O.N. *hamna[r]garðr,* haven-garth.

How: id. all rentals. A 5d. land in Cairston . So named from a prominent (probably broch) mound : O.N. *haugr,* mound.

Howbustirland: 1492. A 3d. land in Inner Stromness which appears in the 1500 R. as Bowbustirland. Whichever may have been the correct form it is now obsolete and the site unknown.

Kirbister: Kirkbustir 1492; Kirkbustare 1500; Kirbustare 1595; Kirbister 1614. An old 6d. land tunship bordering on Sandwick parish. O.N. *kirkju-bólsta˙r,* kirk-bister or farm.

Langskaill: Langskail 1739. A farm in Kirbister tunship. O.N. *langi-skáli*, long hall.

Leager: " . . . the 6d. land called Leagair in the Utertoun" (Sas. 1620); Leager 1739. Probably O.N. *hlíðar-garðr*, slope-garth or farm.

Lee: Lyth 1500; Lee 1595; Lie 1614, 1642. A 1½d. land in Whome area which pertained to St. Duthus stouk. O.N. *hlíð*, slope.

Lerquoy: Leirquhy 1554 (R.E.O.); Larquoy 1739. Small farm in Outertun. O.N. *leir-kví*, clay quoy.

Mousland (mus·lən): Nethir Mousland 1492, 1500, 1642, 1739; —Musland 1614; Over Mousland 1642, 1739. These two units are situated up in the hill near the Black Craig, and quite separate from Outer Stromness tunship. They formed apparently 1½d. lands of quoyland. The name is of special interest as being a third example of the name previously noted in Westray and Sandwick (q.v.).

Pow (pou): Poll 1500; Pow 1739. This farm-name (O.N. *pollr*, a pool) raises the problem of the boundary between Inner and Outer Stromness. In the 1500 R. Poll appears as a 2d. land in Outer Stromness, but in the 1492 R. a man named Will Pole, and afterwards Will Pow (presumably the same) is on record as owning land in Inner Stromness, and in the 1739 R. Pow appears as part of Inner Stromness. Today, there are two farms—Pow and Upper Pow—both of which seem to be in the Innertun.

That there was no visible boundary between the two tunships would seem to appear from a surviving record (Kirkwall S.C.H.) of a planking (legal division of lands) of Inner Stromness in 1765. " . . . And as there were some parts of the arable or corn lands of Uttertown intermixed by runrig with some parts of the lands of Innertown we have divided the same, and *the one town from the other* by a line which runs from the March Stone at the goe of Stinniger and upward to the March Stone at the west corner of Pressquoy [probably Priests' Quoy. H.M.], and from that in a crook eastward to the top of the Green Hillock, and from thence upwards through the middle of the Green Gate leading up to John Stout's house called Gentle Junes (?James)." It is doubtful whether that boundary line can be traced today.

Pultisquoy: Pultisquy 1492; Pultisquoy 1500, 1595, 1614, 1642. A quoy in Outertun. First element obscure.

Quoys: Quyis 1492. A 2d. land in Innertun. Not recorded in later rentals. but reappears in a sasine of 1624. O.N. *kvíar*, quoys.

Redland: Raland 1492, 1500; Redland 1595 and always thereafter. A 3d. land bordering on Loch of Stenness. In view of the two earliest spellings the origin of the name must be regarded as uncertain, but most probably it has had the same history as Redland (Evie) q.v.

Seatter (set·ər): Cetyr 1492; Satir, Sattir, Setter 1500; Seter 1595, 1642; Setter 1642. An old 1d. quoyland adjacent to Redland. O.N. *setr*, a farm, homestead.

Stenigar: Stanagarth 1492; Stanagar 1500; Stannagarth 1583 (R.E.O.). A ½d. land in Innertun. The name is still remembered, and derives from O.N. *steina-garðr*, 'stones farm'; two standing-stones are still to be seen here about 200 yards from the beach.

NAME MATERIAL 165

Straither (streð·ər): Strethin [the final -in probably by error for -er] 1620 Heart's Sas.); John Byning in Straith 1625 (id.); Strather 1739. A small farm in Outer Stromness near the Black Craig and adjacent to a farm —Don.

An interesting but puzzling name with which may be compared that of a notable spring-well in a marshy valley near Skaill in Egilsay—the Well of Struith (strøð). In his *Concise Dict. of English Place-Names* Ekwall cites O. Eng. *strōd, strop* and defines as "marshy land overgrown with brushwood," and he adds " a derivative *strother* with the same or a similar meaning is also found in place-names." In the *Chief Elements used in English Place-Names* (Eng. Pl.-N. Soc.) it is stated that O.E. *strōd, stroð* is "only known from charter material," and a M.E. *strother*, 'marsh' is cited as a derivative of *strōd* and said to be confined to the north country (of England).

In his *Celtic Place-Names of Scotland*, however, Watson refers to the Celtic *sruth*, a stream or current, and cites a derivative *sruthair* which is common in Ireland and appears in Struthers and Anstruther in Fife, Struther in Lanark, etc. Then he adds: "The streams are tiny in all cases."

Now it is noteworthy that a small stream or burnlet flows down past this farm of Straither, and that fact together with the improbability of an O.E. or M.E. word appearing in Orkney place-names would seem to make a Celtic origin inescapable. And that being so, the name of the adjacent farm Don may also be significant. It may be noted finally that both farms lie on the very outskirts of this Stromness settlement.

Whome (hwom): Ovirquham 1492; Ovirquhame 1500; Overwhom 1595; Quhoam 1595; Whome 1642. Overwhome by itself was a 4d. land. O.N. *hvammr*, a rounded valley; in S.W. Norway the modern form is Homm.

OTHER FARM-NAMES

Arian (ɛr·iən): Earying (1621 Sas.); Airing (1624 Sas.). A farm in Kirbister area. This farm (at the opposite side of the parish from Straither) has its name from a definite Celtic source. In the Flatey-book MS. of the *Ork. Saga* a place in Caithness is termed *Asgrims-œrgin*, a name which in the old Danish translation appears as *Asgrims-erg*. There is general agreement that *erg* is a Norse borrowing from Old Celtic *airge*, a shieling —(Mod. Gael. *airigh*). The form *œrgin* is probably a Norse neuter pl. with def. article *-in* postfixed, the whole compound thus signifying 'the shielings of Asgrim.' In that case the Celtic term had been so fully incorporated into Norse that it could be treated grammatically as a Norse word.

There can be little doubt that in Arian we meet again that adopted Celtic name, and it may be added that in the Rousay hills is a ruinous old structure still known as The Styes o' Steenie-iron, in which the final element (now assimilated to the familiar Eng. word) is almost certainly the same. Steenie- probably refers to a standing-stone which is on record from that vicinity. (See further my *Place-Names of Rousay*). Cf. also Airy and Airafea (Sanday, Stronsay, Westray).

Brockan: Broaken 1679; Brockine 1694; Brocken 1764. A farm in Outer Stromness on a steep slope. See this name Rendall, Birsay, etc.

Clouster (klust·ər): a farm in Cairston. See this name—Westray.

Congesquoy: Conquescoy 1627. A farm near Bu of Cairston. Origin rather uncertain, but the first element is probably the Scots legal term 'conquest'—in the sense of 'acquired' or 'purchased' (by one of the earls).

Croval: a farm in Innertun. Origin obscure: the name occurs also in Sandwick.

Kingshouse: a farm in Outertun. Origin uncertain: cf. this name—Harray.

Leafea (li·fiə): a farm in Inner Stromness. Origin uncertain.

Miffia: a farm in Outertun. See Mithvie, Rousay.

Sourpow: Sorepool and Sorpuill (1621). A vanished farm in Kirbister. See this name—Orphir.

Stank: id. 1580 (R.E.O.); 1628 Sas. A farm in Outertun. Probably the Scots term *stank*, stagnant pool, etc.

Stove: id. (Sas. 1619, 1628). In Outertun: name obsolete here. See Stove, Sanday.

Weardith (wɛrd·əþ): a farm on north slope of Redland Hill. Origin obscure.

Whomslie (hwoms·li): a farm in Whome valley. O.N. *hvamms-hlið*, valley-slope.' It is rather curious to note another Lee in this valley: see Lee *supra*.

South Isles

BURRAY

Burray 1492, 1500, 1595; Borowray (Fordun, c. 1375); Borgaröi (Claussön, 16th century). O.N. *borga[r]-ey*, 'broch[s]-isle'; so named from the presence of two brochs.

This island does not figure in the *O. Saga*, and the early rentals are unusually defective in regard to it, even the spelling of the few names mentioned being very unreliable. In 1494 King James IV made a grant of the isle to the Orkney Bishop Andrew, and as it was later held as a feu from the bishopric the individual farms do not appear in the 1595 R. at all. It had been originally skatted as 2 urislands though the skatted units specified in the rentals did not amount to quite that figure. Very little is known of the early history of the island.

In the 1492 and 1500 R. the main island divisions were an Overtoun (or Uppertun), a Nethertoun, and The Bu. In the 1627 Report of the island the Overtoun is termed the North Toune and the Nethertoun becomes the South Toune. The hilly western area of the island was still apparently uncultivated.

RENTAL FARM-NAMES

[The 1627 references *infra* are to a very ill-spelt list of names mentioned in a Report of Burray printed in Peterkin's Rentals.]

[The] Bu: Le Bow 1492, 1500. A 9d. land bordland farm. See Bu (Part III). This was the residence of a famous family—The Stewarts of Burray—whose mansion here was one of the greatest houses in Orkney in the 17th and 18th cent. A very interesting Inventory of its contents appeared in P.O.A.S. XII.

Gilbroch: 1500; Gilbroth 1492. An obsolete name. If the 1500 form is correct it might indicate a farm adjacent to one of the old brochs. There is a farm today—Gillietrang, but that is an obscure name also, and the farm is not near either broch.

Girsay Schottis: 1500; Grosray Schotts 1492. That name is quite obsolete and nothing can be said as to the meaning of the name or the site of the farm. No similar name occurs in Orkney.

Housebreak: id. 1627. A 1d. land in the South-tun. O.N. *hús-brekka*, 'house-slope.'

Leith: id. 1627. A 1d. land near east point of island. Probably represents a Scots spelling of O.N. *hlið*, slope.

Lurdy: Lerdie 1627. A 1d. land in the South-tun. Origin obscure.

Pole: a farm in the South-tun; probably represents the 3d. land of 'Poolsherp' of 1627. For the first element of that name cf. Pow (Sandwick), and for the second see Skerpie.

Quoys: Quyis 1492, 1500. A 1d. land in North-tun. O.N. *kvíar*, 'quoys.'

Taftnica (taftnɪk·ǝ): Tefaiecca (sic) 1627. A 2d. land in South-tun. First element O.N. *topt*, site of a house; second obscure.

Skerpie: probably the 'Sherpe' of 1627—a 1d. land in South-tun. This seems to be a name that occurs also in Norway where Magnus Olsen (G.N. X, 74) postulates an O.N. *skerpa as the source. That word he would regard as a deriv. of the O.N. adj. *skarpr*—used of dry shallow soil. That adj is commonly employed in Orkney in the form *skarpy* for poor shallow soil: e.g. 'poor skarpy brecks.'

The addition of '-sherp' to 'Pool-' (v. Pole) is curious, but might point to a division of a larger unit.

Weddell (wedl): Wedaaie 1492; Wedale 1500. Farm still today in north of island on shore of the narrow Weddell Sound between Burray and Glims Holm. That sound (acc. to *North Sea Pilot*) "is shallow and foul, weeds appearing at low-water springs nearly all the way across." Hence the name—O.N. *vaðill*, a wading-place or ford. Cf. Wald (Firth).

Westermeil: Westermuill 1627. A farm on south shore of island near a sandy beach. O.N. *vestr-melr*, 'west-sand.' There may have been a farm *Melr* from which this farm lay to the west, but that is uncertain.

Yeldabrek: Zeldbrak 1492; Yeldabrek and Yelbrek 1500. A vanished farm in the north of the island. Name obsolete; cf. this name—Birsay.

Though early forms of Burray farm-names are relatively few and unreliable, a complete list of names from 1750 is by lucky chance available. About three years previously the Jacobite Sir James Stewart of Burray had died in prison in London, and his property had passed to a relative the Earl of Galloway, who had as factor a well-known Orkney gentleman-farmer —Patrick Fea of Airy. Fea's rental of the Stewart estates in Burray, Flotta and South Ronaldsay, etc., is still preserved in Kirkwall, and it may be of some interest to enter here a list of all the Burray holdings at that date so as to give a comprehensive picture of the farm-names in one small island two centuries ago. The skattable liability of each is added.

BURRAY FARMS—1750.

Cott	1d. land	Tumol of Milldam	—
Housebrake	2d. ,,	Tumol of Holy Graves	—
Leith	1d. ,,	Tumol of Whitestane	—
Fea	1d. ,,	Trofer or Tropher	1d. land
Homes	1d. ,,	Tofts	1d. ,,
Poll or Poall	1d. ,,	Lodgicks	1d. ,,
Scarpa	1d. ,,	Gillytrang	1d. ,,
Loordie	1d. ,,	Tumol of Crapygoe	—
Toftnicca	1½d. ,,	Bruntbigging	1d. land
Sellyland	1½d. ,,	Warebanks	1d. ,,
Westermeall	2d. ,,	Woddel	4d. ,,
Tumol of Banks	—	Newbigging	1½d. ,,
Tumol of Ourigar	—	Linkshouse	1½d. ,,
Tumol of Bruntland	—		
		Total	27d. land

It will be noted that the total of these amounted to only 27d. land. The missing 9d. land was the Bu of Burray—the manorplace of the estate (termed in the rental by the Scots name—The Mains)—which was not rented but farmed on behalf of the proprietor. The island thus was a 2-urisland unit.

It will also be noted that places termed 'tumols' were unskatted. For that term—see further Part II.

SOUTH RONALDSAY

Rognvalldzey, Raugnvallzey (*O. Saga*); Roghnals œy, Rognaldz œy 1329 (D.N. II. No. 170); Ronalsay 1500, 1595; South Ronaldsa 1570 (R.E.O.). The addition of 'South' became necessary to distinguish this island from the former Rinansey—the present North Ronaldsay—which name was being assimilated to Ronaldsay also. O.N. *Rǫgnvalds-ey*, the island of Rognvald, a man of whom nothing else is known.

The relatively frequent references in the *O. Saga* to this island were probably due to the fact that it is the nearest of the larger Orkney islands to the Mainland of Scotland, and was thus on the main line of communication. It is an isle of low hills, the arable lands for the most part lying in numerous valleys or lowlands round the coast. It is difficult to distinguish the larger original settlements, but altogether the isle contained some eight scatted urislands together with a rather unusually large number of unskatted quoylands.

A document of 1329 written in Kirkwall in the current Norse language of the time is of exceptional interest in regard to this island. It deals with a sale of lands which had belonged to the family of a Norwegian magnate Herra Thore Hakonsson (a former Chancellor of Norway), one of whose daughters Elin had been married to Herra Erling Vidkunnsson, the Drottseti or regent of Norway at that time. Nothing is known of any earlier connection of that family with Orkney apart from the fact that Thore and his wife had accompanied the Maid of Norway on her ill-fated journey over here, and were no doubt present at her death in 1290.

For us the real interest of the document lies in the early forms of the farm-names mentioned, which were as follows: *Stufum, Kui kobba, Klœte, Þordar eckru, Borgh, Leika kwi, Lið, Haughs œiði* and *Petlandz skœr*—all of which, with the exception of the second, can still be easily recognised today.

South Ronaldsay was divided into two parishes named St. Peter's and St. Mary's after the respective parish churches, the former of which is in the old Paplay district and the latter at Burwick where there was an earlier chapel dedicated to St. Colm.

RENTAL NAMES

Aikers (ɛk·ərz) : Akirrs 1492; Akirishous 1500; Acrys 1545; Acres 1564 (R.E.O.); Akerhous 1595. A 6d. land tunship in the middle of the island. See this name—St. Andrews.

Arneip : Arneip viz. Burvik 1500; Arneip viz. Burwick 1595. Not on record later, and name obsolete. Second element obviously O.N. *gnipa*, No. *nipa*, a steep precipitous hill or headland. Here the reference must be to the high cliffs where the hill immediately west of Burwick farm falls down abruptly to the sea. The first element pretty certainly refers to eag'es— O.N. *arna-gnipa*, 'eagles' peak or headland.' That eagles used to frequent this island is evident also from the following quotation from Low's *Tour* (1774) : " In my way over the hills [north of Sandwick—a couple of miles farther north along this coastline. H.M.] saw some tumuli; the old men called them Earny couligs, the meaning of which they could not explain." 'Earny' of course refers to ernes or eagles : hence 'Eagle-hillocks.'

Atla : "Ane fd.ing land called Atla lyand be south the burn [in Widewall] " 1595. Not on record again, and name apparently obsolete. It is probable that the 't' is (as so often occurred) a misreading of 'c,' in which case the name may have been the same as the first element of Hackland (Rendall) q.v.

Barswick (bars·ək) : Barzvic, Barŏzvik (*O. Saga*); Braswic 1492; Berswick 1584; Barswick 1627.
A 6d. land adjacent to a precipitous headland called Barth Head. O.N. *barðs-vik*, from *barð*, verge, brim; used of precipitous hill-sides. In Færoe and Shetland the name is applied, as here, to high, precipitous, abrupt headlands.

Blanster: Blosetter 1492; Blanksetter 1500, 1595; Blansetter 1644, 1668. A former 6d. land and a farm still. The second element is O.N. *setr*, a homestead, farm, but the first is uncertain owing to the various spellings. The 1492 form is manifestly wrong, and in 'Blanksetter' it is doubtful whether the 'k' is not an intrusion; that sound is certainly never heard today. If then we could assume that the present form is nearer the original we could suggest a personal name Blann or Blian. Two Orkney men so named are on record in the *O. Saga*, but the name is apparently not Norse, and Lind queries it as being a Celtic name.

Brough (brox): Borgh 1329; Burch 1492; Burgh 1500; Brought 1595. Together with Tarland it formed 7¾d. land in 1492. This farm is on the south point of the island, and the farmhouse stands at the site of an old broch. O.N. *borg*, a fortification.

Burrowland: 1492, 1500; Brunaland and Brualand 1595. This went along with Stewis to form a 3d. land in 1500. The name has now been corrupted to Braeland. Origin obscure; the first element suggests an O.N. *borg* (see last name) but there is no record of any old broch here. See next name also.

Burrowell: 1492; Burwell 1500; Burwall 1595. The name has now been corrupted to Broll. In the rentals this went along with Linkletter and Windwick to make up a 9d. land or half-urisland. Origin again obscure as in the preceding name; Burrowland and Burrowell are about 1½ miles apart—on the east side of the island.

Burwick (bʌr·ək) : Burwic 1492; Burvik 1500; Burwick 1595. A 6d. land at head of a bay on the south-west corner of the island, looking out on the Pentland Firth. O.N. *borgar-vík*, 'broch-bay'—so named from the broch at Brough half a mile from the head of this bay which is practically the only landing-place on this wild coast. It was from here that in olden days there was a ferry service over to Caithness.
In the 1500 and the 1595 R. Arneip q.v. is an alternative name for this farm.

Cara: a Mawnis Cawra (having surname from this farm) appears on record as a roithman in 1514 (R.E.O. 87); Caro 1563 (R.E.O.); Cara 1595. This is absent from earlier rentals, and in the 1595 R. we read : "Of old not contained in the rental and lately scatit, being 1d. land." The farm is at the north-east corner of the isle, and the Cromarties of Cara were one of the leading Orkney families in the 16th century. Origin of name obscure, but cf. Garay *infra*.

Cleat (klet): i Klæte 1329; Clate 1492; Clait 1500; Cleat 1595. This went along with Isbuster in the rentals to form 6d. land. See this name— Sanday.

NAME MATERIAL 171

Cletts: Clettis and Cletts 1595. A 2d. quoyland near Paplay district. Apparently O.N. *klettar*, plur. of *klettr*, rock.

Coulls (kuls): Caulis and Coulls 1595; as surname—Cowlis 1545 (R.E.O.); Cowllis 1589. A 1¼d. land in Sandwick district. Probably a Scots plur. of O.N. *kula*, a knob, protuberance; in Shetland that word is used of a rounded hill or sunken rock, etc.

Garay: (1) id. 1492, 1500; Gara 1595. A ½d. land in Grimness, next in rental to Cara.
(2) Garaye 1557 (R.E.O.); Cara 1595. A ½d. land in Sandwick.
The former seems no longer existent, but the latter is a farm still today, commonly now spelt Gairy. Origin rather uncertain; from the various spellings one might suspect that Cara *supra* and Garay are one and the same name. Cf. Gara (Orphir).

Garth: Gart and Garth 1595; Gairth 1696. A quoyland adjacent to St. Margaret's Hope. O.N. *garðr*, a farm.

Gossigar: Gossakir 1492, 1500; Gossagar, Gossagair 1595. A 6d. land in South Parish. The earliest spellings seem to indicate that the second element is O.N. *akr*, as in Thurrigar, (arable land) rather than *garðr*; the first element in the name of an early settlement of this character points to the personal O.N. name Gási: hence—the 'farm or fields of a man Gási.'

Grimness: Grymnes 1492, 1500; Grymness 1595. A 6d. land tunship in north-east of island. O.N. *Grims-nes*, so named after a man Grímr.

Grutha (gruð·ə): Grudo 1492, 1500; Gruthay and Grutha 1595. A 3d. land at beach adjacent to the old Grimness tunship. Probably O.N. *grjót-haugr*, rocky or stony mound.

Herston: Harthstath, Harstath 1492; Herstath 1500; Harstane and Herstane 1595. A 3d. land tunship on a peninsula by itself. O.N. *Harðar-staðir*, the stead or settlement of a man Hǫrðr.
This name is interesting as the only instance of a *staðir*-name in the South Isles of Orkney. In Part III it is pointed out that such names seem to imply a re-settlement of older estates by some earl (probably Thorfinn Sigurdson) for security reasons, and if that be the case the original name of this may perhaps still be discerned in the name of a small farm thereon—viz. Sabay (O.N. *sœ-bœr*). Elsewhere I have suggested that as this peninsula projects out into the main entrance channel to Scapa Flow from the south a trusted retainer (?Hǫrðr) might have been settled here to give timely warning of any hostile force approaching.

Holland: id. all rentals. The skatting varies slightly in different rentals, but it was probably an original half-urisland, on the highest ground in the south part of the island where it is situated. See this name—North Ronaldsay.

Hottit: a small farm today in Grimness. In the 1595 R. is entered a ¼d. land Howcost which is pretty certainly the same farm ('c' by mistake for 't,' and an 'f' having been read as a long 's'). The name is interesting but obscure. It might conceivably represent an O.N. *hoftopt* (temple-site), or it might equally well be the same as Howatoft—a name in the neighbouring tunship of Paplay—a not uncommon name in Orkney (q.v. North Ronaldsay).

Hoxa: a Haugahæide (O. Saga*); a Haughs æiðe 1329; Hoxa 1492; Hoxay 1500; Hoksay 1595. A 6d. land in north-west of island. O.N. *haugs-eið*, mound-isthmus.

The peninsula of Hoxa is joined to the rest of South Ronaldsay by a low isthmus on which is a large broch-mound prominent from afar. Here it probably was that Earl Thorfinn Skull-cleaver was 'howe-laid.' The 6d. land settlement embraced lands on each side of the isthmus.

Inyequoy: Inzequoy 1492; Inyequoy 1500. Not on record later, but it was a ¼d. land situated in the South parish. In Fea's Rental of 1750 a Quoy Inga is grouped with Masseter and may be the same place. Origin of first element uncertain.

Isbister (aiz·bestər): Estirbuster 1492; Ystabustare 1500; Isbuster 1595; Iysbuster 1627. Formed with Cleat a 6d. land in the south-east corner of the island. A *bólstaðr* name of which the first element is uncertain. Isbister chances to be both the most easterly and the outermost settlement in this area, and from the diversity in the two earliest forms of the name it is difficult to tell whether the name represents O.N. *eystri-bólstaðr* (more-easterly-bister) or *yzti-bólstaðr* (outermost-bister): perhaps both names may have been in use. Cf. Isbister—Rendall and Birsay.

Kirk: id.—all rentals. A 3d. land on east side of island. It was udal land as late as 1601 (U.B.), and as there is no record of any church or chapel here the name is decidedly puzzling.

Leoquoy: Leika kwi 1329; Leoquoy 1500; Lykway 1584; Lyaquoy, Lykquoy 1627. A ½d. land in Brough neighbourhood. It was not kirkland in 1329, but by 1500 it is stated to have been 'gevin to Sanct Magnus altar.' O.N. *leika-kvi*, sports-quoy.

Liddell: Liddale 1492, 1500; Luddale, Luddle 1595. A 6d. land in south end of island. The same name occurs in Norway where it is interpreted as O.N. *hlíðar-dalr*, 'slope valley,' a term that suits the local topography here also.

Linklater: Lyntlett 1492; Linclet 1500; Linklet 1595; Linkletter 1627. This went along with Burwell and Windwick to make up a 9d. land or half-urisland. See Linklet—N. Ronaldsay.

Lythe: a small farm of today near Osquoy which (it may be deemed certain) represents the rental Qwylith 1492; Quyleith 1500; Quoyleith 1595. In the rentals that unit was grouped together with Quoybrown and Ossaquoy as forming 1d. land, and it almost certainly represents the Lið of the 1329 sale of lands referred to above. That item was stated to be " xx skillinga kaup," i.e. 1½ marks of land or probably ½ pennyland: four marks = 1 pennyland was the commonest equation in Orkney. O.N. *hlíð*, slope.

Lythes (laiðz): Lythis 1492, 1500; Lyths 1595. A 6d. land on east side of island. O.N. *hlíðar*, pl. of *hlíð*, slope.

---

* In the Index to his edition of the O. Saga Prof. S. Nordal has " Haugaheiðr (*rettere* Haugseið) . . ." The correction may partly be superfluous. On this isthmus there are actually two mounds (v. *A.M. Inventory*), and the name may thus have been originally not Haugseið but Hauga-eið, 'the isthmus of *mounds*,' though in later times it certainly had the singular prefix.

Masseter: Morsetter 1492, 1500; Marsetter 1595, 1627. A 6d. land in St. Mary's Parish. See Mosseter (Rendall). The vowel change here is peculiar.

Osquoy (ɔs·kwai): Ossaquy 1492; Ossaquoy 1500, 1595. See Lythe *supra*. The same name appears in the 1595 R. for the present Essaquoy (Rousay). First element obscure, though it might phonetically represent *Asu-*, genitive of O.N. personal female name—*Asa*.

Paplay: id. all rentals. A half-urisland tunship on east side of island where the parish church of St. Peter stands. This is another memorial of the early Celtic mission in Orkney. See Paplay (Holm).

Pool (pøl): Puile 1595. A ½ farthing land in Grimness. See Pow (Sandwick).

Quoyangry: Quoyanyie, Quoyaurie 1595. A ¼d. land in St. Mary's. This name occurs in several parts of Orkney, but no satisfactory derivation for the second (qualifying) element can be suggested.

Quoybanks: 1595. A ¼d. land in Grimness. Second element O.N. *bakkar*, banks, slopes.

Quoybrown: Qwybrown 1492; Quoybroune 1500; Quoybrownaris 1595. See Lythe *supra*. Probably the present Quoyboon or Quoybond. Origin obscure.

Quoybutton: Qwybuttin 1492; Quoybuttin 1500, 1595. A ¼d. land, probably in Hoxa area. The name seems extinct.

Quoycuise: 1595. A 1d. land in Kirk area in east of island.

Quoydoun: 1595. A ½ farthing land

Quoyleith: Qwylith 1492; Quyleith 1500; Quoyleith and Quoylyth 1595. See Lythe *supra*.

Quoys: (1) Quyis 1492; Quoyis 1500; Quoys 1595. A ¼d. land in Grimness.
(2) There was another Quoys (Quyis 1562 R.E.O.) in the Holland tunship, and there were probably others. O.N. *kvíar*, 'quoys.'

Quoysharps: Qwyscharpis 1492, 1500; Quoyscarpis 1595. A 1d. land; site unknown, but apparently near Cara. Origin uncertain.

Quoyschorsetter: 1595. A 1d. land now called Quoyhosseter. Origin uncertain.

Quoysmiddie: 1595; Smiddie 1595. In Grimness—now called Smiddy.

Ramsquoy: 1627; Rampsquhy 1584. Site unknown, but probably in Thurrigar. First element probably O.N. personal name Hramn.

Ronsvoe: *i Rögnvaldz-vag* (*Hakon Saga*); Ronaldiswaw 1492; Ronaldiswo 1500; Ronaldisvo 1595; Ronsvoe 1750. The old half-urisland tunship at head of bay of same name—the present St. Margaret's Hope (bay). Here King Hakon Hakonsson's fleet lay at anchor for some days in August, 1263, when an eclipse of the sun caused foreboding.

The Rognvald whose name appears here has probably been the man after whom the whole island has its name, in which case this tunship no doubt would have been the earliest Norse settlement in the island.

Sandwick: id. 1492, 1595; Sandwik 1500. A tunship nestling round a fine sandy bay on the west side of the island. Though entered in the rentals as a 9¼d. land it has probably been skatted originally as a 9d. land or half-urisland. O.N. *sand-vík*, sandy-bay.

Serrigar: Southirgarth 1492; Suthirgarth 1500; Surregair 1595. A 2d. land immediately south of Sandwick of which it may perhaps have been a part. O.N. *suðr-garðr*, south farm.

Sorquoy: Sourqwy 1492; Sourquoy 1500; Sorquoy 1595. A 2d. land in Paplay area on east side of island. O.N. *saur-kví*, 'muddy quoy'; *saurr*, mud, filth, etc., was a common place-name term, conveying the idea of muddy, boggy, dirty ground.

Suckquoy (suˑkwi): Souquoy, Sowquoy 1595. A ½d. land in Sandwick area. There is much confusion in surviving records of this name. In the Rental of the Provostrie of Orkney (1584) we find: "Item, the half-penny land of Sarkwoy payis yeirlie 4 meillis malt." In Craven's *Church Life in S. Ry.*, p. 11, we read of the ½d. land "'callit Surquoy,' and of 'the town of Surquoy and Shire (sic) of Sandwick'; and on p. 112 of a lawrightman 'Jon Ritchie [Surkquoy?]'" in Sandwick.

From these forms it would seem to emerge that this name was originally the same as the previous—Sorquoy.

Stensigar: Stansgair and Stensagair 1595. A ½d. land in Sandwiɔk tunship. The name has reference to a standing stone there: hence the same name as Stanger (Birsay)—O.N. *steins-garðr*.

Stews (stuːz): *i Stufum* (1329); Stowis 1492, 1500, 1595. This went along with Burrowland in the rentals to form a 3d. land.

O.N. *stúfar*, pl. of *stúfr*, a stump. That this, at first sight ridiculous, name for a farm is to be interpreted in the literal sense of 'stumps' can be confirmed in a surprising manner. In 1774 the Rev. George Low, minister of Birsay, and one of the most gifted divines ever stationed in Orkney, was encouraged to make a general scientific research tour through Orkney and Shetland. His account of the tour had to await publication for over a century, but it was at last published in Kirkwall in 1879, and proved to be an invaluable record. He visited S. Ronaldsay, and on page 25 of the printed *Tour* we read: "Left this spot, and, May 12th, proceeded southward by Stowse Head; observed on the hill the remains of a triangular monument, but very much defaced, and two of the stones broke to the *stumps*. Tradition is there none as to the reason, etc., of its erection. The figure with a plan follows together with the foregoing."

The figure and plan referred to are reproduced in the book, and show one tall standing-stone and two short stumps of stones arranged as he indicated in a triangle. From the recent A. M. Inventory it appears that only the tall stone is now to be seen, and thus, but for Low's record, the name Stews would remain a perennial puzzle.

The name, however, is interesting in another way because from the recorded forms we are able to narrow down the period at which the Scots/English plural ending displaced the old Norse grammatical forms.

Tarland: id. 1492, 1500, 1595, 1627. Linked with Brugh as a 7¾d. land. Origin uncertain.

Thurrigar (þʌrˑigər): *i þordar eckru* 1329; Thurrok, Thurdrak 1492; Thurdrakir 1500; Thordorkore 1509 (R.E.O.); Thurrekirk 1595; Thuiragar,

Thurregar 1627. A 6d. land in St. Mary's parish. O.N. *þórðar-akr* (? *-ekra*), 'Thord's acre' (farm).

Toung: 1595. A ¼d. land—site unknown.

Widewall: i Viþivagi (O. Saga); Widewaw, Wydewaw 1492; Wydwall 1500; Wydewall 1595. A 12d. land tunship abutting on a bay of the same name and divided into two 6d. lands—'benorth' and 'besouth the burn.' Formally, the Saga form suggests O.N. *viðivágr*, 'willow-bay' (Eng. *withy*), and the low-lying boggy ground here might well have had a growth of willow scrub in early Norse times. As it happens, however, two other words may be thought of as forming the first element—O.N. *viðr*, adj. wide, and *viðr*, tree, timber, wood. Though this bay has a relatively narrow entrance it opens out 'wide' inside, and at one part of the beach at low water can be seen the remains of a submerged forest—tree-roots embedded in the sand. Origin therefore rather uncertain.

Windwick (wɪn·ək): Windwit ('t' an obvious mistake for 'k') 1492; Wyndwick 1500, Windwick 1595. Together with Linkletter and Burwell this made a 9d. land or half-urisland. This tunship nestles at the bottom of a picturesque and secluded valley at the head of a bay of the same name on the east side of the island. The bay itself is open to the eastern ocean, but cannot be deemed any windier than others unless perhaps it may have been thought specially liable to gusts coming down from the higher ground around the bay. The tunship itself, however, is a sheltered fertile area, and it is probable that the first element of the name represents not 'wind' but O.N. *vin*, pasture—as in Vinbreck, Vinquoy, etc.

Yorbrandis: 1500, 1595. An old 1d. land in Grimness which undoubtedly is the farm known today as Brance. An interesting but obscure name; it was quoyland (i.e. unskatted), and it may be surmised that formerly the name may have been "Quoy-Yorbrandis" i.e. the quoy of a man Y——. In Lind, however, no O.N. name is cited from which such a personal name could be derived. Origin therefore obscure.

OTHER FARM-NAMES

Biggings: there are at least two farms of this name; cf. Biggings (St. Andrews).

Brain: a small farm in Paplay. In Fea's Rental of 1750 a farm in Paplay appears called Brae, which may be the same. Origin obscure.

Brance: see Yorbrandis *supra*.

Brecks: a farm in Grimness. This was the place from which Sir James Sinclair of Brecks (of Summerdale fame) took his territorial designation. (See article on this man by Mr. Clouston, P.O.A.S. XV). O.N. *brekkur*, slopes.

Bu: there was a Bu of Hoxa and one of Linkletter. See Bu—Part III.

England: id. 1677 Perambulation. A 1d. land in Paplay. If this is a genuine native Orkney name (which is doubtful) it could represent—O.N. *eng-land*, meadow-land.

Farewell: a farm in Ronsvoe tunship. See Faraclett (Rousay).

Flaws (fla:z): there were two or more farms of this name in Holland tunship. See Flaws (Holm and Paplay).

Gerraquoy: a farm in Grimness. Same name found in Birsay, q.v.

Gimps: a farm in Grimness. It is specially regrettable that older forms of this name are lacking, as it would appear to contain a stem *Gims-* which is found in several Norw. farm-names but is unexplained.

Grindilla: in St. Mary's parish—near Tomison's School. See this name—Sanday.

Halcro (ha·kro): Halcro *alias* Holland 1545 (R.E.O.). A farm in Holland tunship which never appears in the older rentals. The Halcrow family were one of the most outstanding Orkney families in the 16th and 17th centuries, and an Andrew Halcro of that ilk is on record from the early 16th century, but there is reason to believe that their original seat was elsewhere than at this modest farm in Holland-tun. The whole problem is discussed at length in a paper by Mr. Clouston—"The Origin of the Halcros" P.O.A.S. XI.

The origin of the farm-name is also obscure. There is another farm Halcro in Caithness, but no connection between the two is known. One of this family—Malcolm Halcro, Archdeacon of Orkney, appears in the records of St. Andrews University in 1512 as "Malcolm Hawcrowk," and the 'l' is never sounded in the local pronunciation.

Hallbreck (ha·brɛk): small farm in Grimness. See Haabreck—Westray.

Hools (hulz): a farm on hillslope south of Widewall. Apparently a plural form of Hool. q.v.—Sanday.

How: farm in Hoxa tunship taking name from the mound referred to in the name Hoxa itself.

Howatoft: farm in Paplay tunship. See this name—North Ronaldsay.

Knockhall (nokha·): farm on east shore of St. Margaret's Hope (the bay). I have no early record of this name (prior to 1750) which is pretty certainly not of Norse origin. But in Alex. Bugge's *Den Norske Trœlasthandels Historie*, II. (p. 143) it is noted that among the Orkney vessels touching at Ryfylke in Norway in 1568 was one skippered by a "John Ronaldso of Kuichuall." Bugge regarded that name as Kirkwall, but that town was familiar in those days to people in West Norway and would be unlikely to be so misspelt, and the skipper's surname would suggest he was a South Ronaldsay man. If that be so the name in question could hardly be other than that of this farm which is on the shore of the bay from which his own surname was ultimately derived—the old Rognvald's Voe or Ronsvoe. Origin of name is however obscure.

Midtown: there were two farms so named—in Hoxa and Herston.

Oback (o·bək): small farm in Widewall on a burn-bank. See this name—Orphir.

Olad: there is an Olad and an Olad Burn in Widewall; and an Olad in Sandwick tunship which appears in Fea's 1750 Rental as Quoy Olet. In these names the first element has reference to a burn or stream—O.N. *á*—and with the second element one may compare Lettan (Sanday) q.v. The name thus probably signified 'burn-flat,' i.e. flat ground on the banks of a burn.

Quindry (kwɪn·dri): a small unit at north corner of Widewall Bay. Origin uncertain.

Quoyloo: in St. Mary's Parish—near Tomison's School. See this name—Sandwick.

Rockerskaill: a small farm in Aikers tunship. Interesting as being the only -skaill name in the South Isles. Nothing is known of its history, and it seems to be different altogether from the old type of chieftain's seat of Saga days. First element quite obscure.

Roeberry: now one of the largest farms in the island, but appearing in none of the early rentals. In Fea's Rental of 1750, however, is entered the ¾ farthing land of Roeber in Hoxa tunship which must be an earlier form of this name. The origin is fairly clearly O.N. *rauða-berg*, red rock or headland, but such a name is usually applied to a bolder headland than can be seen here.

Schusan (sku·ʃən): a farm in Widewall. Origin obscure.

Scows of Cara: Fea's Rental, 1750. Evidently near Cara and now probably part of Berridale farm. Cf. Scows (Orphir).

Sabay: see under Herston. For the name—see Sabay (St. Andrews).

Weems: a farm in Paplay. Origin obscure.

Whistlebere: a farm in Paplay. See this name—Sanday.

Windbreck: farm near St. Margaret's Hope. The Sutherlands of Windbreck were an important Orkney family in the 17th/18th century. See this name—Birsay.

HOY AND GRAEMSAY

Haey (*O. Saga*); Hoye 1492; Hoy 1500 and later. O.N. *há-ey*, 'high island.'

After the Mainland, Hoy is the largest of the Orkney Isles, measuring about 12 miles in length from north to south, by 5 or 6 miles in average breadth, and its Ward Hill (1564 ft.) is the highest of all in the group. At the southern end of the isle an attenuated isthmus, only a few yards in width and two or three feet above high water level, connects a peninsula known as South Walls with the rest of the island, and between these two very unequal portions is a narrow four-mile-long bay or fjord called Longhope, an anchorage much frequented in olden times by sailing vessels awaiting a favourable wind or tide to pass through the Pentland Firth.

Besides Longhope there are so many other bays or voes in the southern part of the island that this area in Norse days was termed *Vágaland*—land of voes. Before the end of the 14th century, however, a form Wawis was coming into use, which was simply a scotticising of O.N. *vágar*, bays or voes. Wawis in turn was later anglicised or 'corrected' to Walls, presumably on the erroneous assumption that Wawis was a Scots form like 'wa's' or 'waw's' (walls e.g. of a house). Fortunately in this case the old local pronunciation has been tenacious enough to survive, and, except on paper and by people over-anxious to 'speak proper,' the name is still normally referred to as Waas (wa:s).

The island is divided into two parishes—Hoy and Walls, of which the latter is much the larger, embracing not only the peninsula of South Walls but more than half of the larger block also. That parish division must have been made before Norse ceased to be spoken, because the dividing line across the island reaches the west coast close to the mouth of a stream called the Summer Burn, while the valley of that burn is known as The Summer of Hoy. That word Summer undoubtedly represents an O.N. *sunnmære*, 'south border,' in other words—the south border of Hoy parish. It is of interest in that connection to note that Summerdale—the site of a famous Orkney battle in 1529—is on the border between the parishes of Stenness and Orphir.

By far the greater part of Hoy consists of hilly ground, and only a very few patches around the shores of the larger island block admit of cultivation. In the *parish* of Hoy the arable lands are confined to a fringe at the north-east corner and the small valley of Rackwick on the west coast opening out to the Atlantic. The whole of these arable lands amounted in old days only to 51 pennylands, all bordland or private property of the 'Auld Earldom,' and as such skat-free. The chief farm was the Bu of Hoy, adjacent to the houses of which was the parish church.

In the *O. Saga* Hoy is mentioned only three times, twice casually as the residence of a man Jón Væ013ngr who lived "í Haey á Upplandi." The 'Upland' must imply the northern hilly end of the island, all of which (as just noted) was private earldom estate. As Jón was a *gœðingr* of Earl Paul, Mr. Storer Clouston (*Orkney Parishes*) reasonably assumed that he must have been the local chieftain, endowed with a bordland 'fief' and resident at the earl's Bu of Hoy.

Jón's second name (or nickname?) raises a curious problem. The *Saga* tells also of another Jón Væ ngr—a nephew of the former. O.N. *væ ngr* meant a 'wing,' and, whether by coincidence or otherwise, on the north shore of South Walls abutting on Longhope a small district is still known as The Wing. O.N. *væ ngr* is a rare element in Norwegian place-names, and in personal names (so far as I know) confined to these two—uncle and nephew. It is thus difficult to disassociate these personal names from that of the place-name in the island where at least one of them lived; but on the other hand no other man is on record in the *Saga* with a territorial se ond-name—that is—without a preposition 'in' or 'at' in front. We are thus completely in the dark as to whether the personal name was derived from the place-name, or *vice-versa*, or whether the two are entirely unconnected.

RENTAL NAMES

Beneith the Hill: 1492; Beneth the hill 1500, 1595. An urisland tunship, but there is some confusion in the rental entries here. The 1492 R. included Rackwick, a 3d. land, as part of this urisland (18d. land), while in the 1500 R. it is regarded as an urisland apart from Rackwick. In the 1595 and 1614 rentals Beneath the Hill is entered as 15d. land only, but in the Report of Hoy (1627) it again appears as an urisland by itself.

The name though obsolete is interesting as an obvious translation of a Norse prepositional name—O.N. *undir fjalli*, ' under the hill.' Cf. the district name in Birsay—Above (or Abune) the Hill; see *sub* Bea (Birsay).

Braebuster: Brabustir 1492; Brabustir abone the hill 1500; Brabustare 1595; Brabuster 1614. An urisland tunship on north coast of Hoy. From the 1500 entry it would appear that this isolated district and the Beneath the Hill district had their names given relative to the hill ridge dividing them from each other—not to the great Ward Hill which dominates both. O.N. *breiði-bólstaðr*, ' broad-bister ' or farm.

Bu of Hoy: Bull of Hoye 1492; The Bull 1500; Bull of Hoy 1595 etc. A 9d. land or half-urisland farm. See Bu—Part III.

Claisbreck: 1492; Glasbrek 1500; Cleshbreck 1739. Name obsolete. This was a 1d. land "umbesett" attached to the Bu. First element obscure. For umbesetts see *sub* Bu—Part III.

Dale: 1500; Dell 1614; Daill 1739. A small "outbrek" of Braebuster.

Garson: Garsent 1492, 1500. Another "umbesett" of the Bu. See this name—Sandwick.

Hallay: 1500; Halla 1614; Halay 1642. An "outbrek" of Braebuster. See this name—Deerness.

Quoys: Quyis 1492, 1500; Quoys 1739. Another "umbesett" of the Bu.

Rackwick: Rakwic 1492; Rakwik 1500; Rackwick 1595. A 3d. land tunship in a valley by itself on the west coast of Hoy. See this name—Westray.

Selwick: Sellewit (sic) 1492; Sellaviok 1500; Salwick 1739. Another "umbesett' of the Bu. O.N. *sela-vik*, bay of seals.

Warbuster: Werbuster beneth the hill (a 9d. land) 1500; Warbister 1627; Warbuster by South the Burne (9d. land), and W—— by North the Burne (9d. land) 1739. The dual Warbuster of 1739 had replaced the former Beneath the Hill; the burn referred to was evidently that running down from Orgill. The name is now obsolete. O.N. *varð-bólstaðr*, ' beaconbister '; the varð- (O.N. *varði* or *varða*) in this case must have had reference to the Ward Hill of Hoy which towers up precipitously immediately behind the tunship.

OTHER FARM-NAMES

Lythe: a small cot. O.N. *hlíð*, slope.

Lythend: another small unit—which calls for inclusion here as a reminder of the famous Icelandic *Hlíðar-endi* (slope-end), the home of the immortal Gunnarr.

Murra: a small cot in Braebuster area. Perhaps O.N. *mýrar*, mires.

Orgill: the chief farm now in Hoy, on the banks of a stream. O.N. *ár-gil*, stream-gill or valley.

Slack: a small cot in Braebuster area. O.N. *slakki*, a saddle or gap in a hill-ridge.

GRAEMSAY

Grímsey (*O. Saga*); Grymesaye 1492; Grymesay 1500, 1595; Gremsay 1614; Gramesay 1642. O.N. *Grims-ey*, isle of a man Grímr. The present spelling and pronunciation is no doubt due to confusion with the local surname Graeme.

This small isle in Hoy Sound between Hoy and Stromness is twice mentioned in the *O. Saga*, but without reference to any resident there. Ecclesiastically it seems originally to have been attached to Stromness parish, becoming conjoined with Hoy only in the 16th century. It contained two urislands of skatland together with some ¾d. land quoyland.

RENTAL NAMES

Bu: The Bull of Grymesay 1595. This was not a genuine old Bu, but a new name given to Sandisend after James Stewart, an illegitimate son of Earl Robert Stewart obtained a feu of parts of the island in the latter half of the 16th century.

Corrigill: Corgill 1492, 1595; Corgillis 1500. A 6d. land in southwest of island. A puzzling name, as there is no noticeable 'gill' here, and complicated by the fact that the neighbouring Suthirgarth of 1500 R. appears as Suthergill in 1595 R. That name appears later on Mackenzie's Chart of 1750 as Surrigal—a name covering the southern part of the isle. Cf. Corrigill (Harray).

Oute apoune the Ile: 1492; Outohoy 1500; Out-upoun-the-yle 1595. A 4d. land in the north-west and higher part of the island. This is a curious and interesting name—now obsolete—which would seem to be a translation of an O.N. *úti uppá ey*, lit. out upon the isle.

Quoys: Quyis 1492; Quoyis 1595. A 3d. land. O.N. *kvíar*, quoys.

Sandisend: 1492, 1500; The Bull of Grymesay 1595. A 9d. land. The name now in use is Sandside—a corruption similar to what has occurred in Sandside (Deerness) q.v. The farm-houses of Sandside stand at the east end of a sandy beach.

Suthirgarth: Suyirgarth 1492; Suthirgarth 1500; Suthergill 1595. A 12½d. land in south of island. Name now obsolete. The earlier forms plainly indicate O.N. *suðr-garðr*, 'south garth' or farm, and the change to Suthergill in 1595 is inexplicable unless it was due to confusion with the neighbouring Corrigill.

OTHER FARM-NAMES

Cletts: this like each of the others immediately following is only a small farm or croft. See this name—South Ronaldsay.

Garson: see this name—Sandwick.

Gorn: see Goir—Sanday.

Quoynaknap (kwin·ə-[k]nap··): O.N. *kvin-á-knappi*, the quoy on a hillock or 'knob': *knappr*, knob or hillock, etc.

Ramray: origin quite obscure.

Veval (ve·vəl): origin uncertain; cf. Feaval—Birsay.

Windbrake: cf. this name—Birsay.

Windywalls (wɪnd·iwaz): see this name—Eday.

WALLS

Vagaland (*O. Saga*); Wawis 1492; Wallis 1500, 1595, 1614, 1642; Wais 1568 (R.E.O.); Walls 1595. O.N. *vága-land*, land of bays (voes). See further *sub* Hoy parish name.

Air: 1601 (U.B.), 1739. This appears among the udal lands in the Uthell Buik of 1601, but is not on record previously. The house of Air is shown on Mackenzie's Chart (1750) situated at the beach or 'ayre' (still so named) near where the public road bends inland a little to the east of North Ness on the north side of Longhope. O.N. *eyrr*, gravelly beach.

The Sandisons of Air were an important local family in the 17th century, but the lands of Air would seem to have been part of the old Seater (North and South) tunship.

Aith: 1492 and all later rentals. An urisland tunship on east side of the isthmus connecting South Walls with the rest of the island. The headhouse of this tunship was (and is) called The Bu of Aith. O.N. *eið*, isthmus.

Aithsdale: Aithsdaill 1739. Part of Aith tunship. O.N. *eiðs-dalr*, isthmus dale or valley.

Booth: 1739. A small unit in Osmundwall tunship. This is on the shore of O―― bay, where there has probably been a merchant's booth in former days. The bay was a calling-place frequently for ships.

Brims: Brymnes 1492, 1500; Brims 1595; Brimbis 1614; Brimnis 1627. A 4½d. land on peninsula of same name jutting out into the Pentland Firth. O.N. *brim-nes*, surf-ness.

Dykeside: 1739. A small unit in Osmundwall.

Fea: Mosound alias Fea 1595; Moasound alias Fia 1614; Fiah 1627; Fea 1739. An alternative name for the former Mosound: a 9d. land tunship stretching from Longhope upwards to the hilly ridge of S. Walls. O.N. *fjall*, hill.

Flawsquoy: 1739. A small unit in Osmundwall. See Flaws—Holm and Paplay.

Green: 1739. A small unit in Osmundwall.

Gyer: 1739. A small unit in Osmundwall. See Gyre—Orphir.

Kirbister: Kirbustir 1492; Kirkbustar 1500; Kirbuster 1595; Kirbister 1614. A half-urisland or 9d. land tunship on south shore of Longhope near its entrance. O.N. *kirkju-bólstaðr*, 'kirk-bister' or farm.

Lyness: Lynais 1627. Not entered in any rental, but appears in the 1627 Report of Walls as 1d. land (udal): probably part of Thurvo tunship. In the absence of older forms of this now famous name its origin must be regarded as somewhat uncertain, but it may represent O.N. *hlíðar-nes*, slope's ness.

NAME MATERIAL 183

Manclett: 1492, 1595, 1627; Manclet 1500. A 2d. land at upper end of Longhope. The second element is clearly O.N. *klettr*, rock (cf. Cleat, Sanday), but in the absence of still earlier forms the first element is obscure. Man- appears occasionally as first element in Norwegian names where its meaning has not been explained, and the fact that a Celtic term—Welsh *maen*—means a stone, and is thus practically synonymous with *klettr*, can hardly be other than a coincidence.

Melsetter: Melsettir 1492; Melsetter 1500 and later. A 3d. land tunship at head of Longhope. The House of Melsetter was the residence of the island lairds the Moodies of Melsetter throughout the 17th and 18th centuries.

Second element of name O.N. *setr*, a homestead, etc., and the first is probably O.N. *melr*, sand, but as Melsetter is in a sense the meeting-place of North and South Walls the O.N. *meðal-* 'mid' cannot be wholly ruled out.

Mosound: Meosomid (-mi- no doubt a misreading of -un-) 1492; Meosound (in Peterkin wrongly transcribed Meesound) 1500; Mosound alias Fea 1593; Moasound alias Fia 1614. A 9d. land in South Walls adjacent to the narrowest part of Longhope. O.N. *mjá-sund*, or *mjó-sund*, 'narrow sound.'

Osmundwall (ɔuz·nə): *til Asmundarvags* (O. Saga); Osmondwaw 1492; Osmoundwall 1500; Osmondwall 1595, 1614, 1627, 1739.

An urisland tunship in east of South Walls adjacent to the bay so named in olden days but now called Kirkhope (from the parish church there). O.N. *Ásmundar-vágr*, Asmund's bay or voe. Here it was that King Olaf Tryggvason came on the Orkney Earl Sigurd the Stout unexpectedly in 995 A.D. and forced him to accept Christianity.

Quoyneipsetter: 1500; Quyneipsettir 1492. Absent from rentals after 1500. A small ½ mark land on north shore of Longhope, but exact location unknown. It may have been part of the old Seater tunship, and if so the -neip- (O.N. *gnipa*) may have referred to the point now known as North Ness.

Quoys: Quyis 1492, 1500; Quoys 1500; Quoyis 1595, 1614. A 4½ markland somewhere on north side of Longhope.

Rysa (rais·ə): Ryssay Sound 1492; Ryssa 1500; Rysay 1595; Ryssay 1614, 1627. A ½d. land on mainland of Walls opposite the island of Rysa.
O.N. *hris-ey*, brushwood isle. Cf. Scots *rice*.

Scarton: Scartoun 1739. A small unit in Osmundwall. Origin uncertain.

Seater: South Settir and North S—— 1492, 1500; N—— and S—— Setter 1595 and later. Each was 1½d. land near North Ness in N. Walls. O.N. *setr*, homestead, etc.

Snelsetter (snɛl·setər, snɛl·stər): 1614, 1627. A farm on south coast of South Walls bordering the Pentland Firth. Not in early rentals, and may have been part of Osmundwall tunship.

First element obscure, but probably a personal name: ? O.N. Snjallr.

Thurvo: Thurwaw 1492; Thurwo 1500; Thurvoe 1595; Thurvo 1614. A 3d. land tunship—so named from adjacent bay north of Lyness. Second element O.N. *vágr*, bay or voe, and the first certainly a personal name though exactly which is uncertain: ? O.N. *Þórir*. It is possible but improbable that the old god's name *Þórr* is indicated.

Wards: Warthis 1492, 1500; Wards 1595, 1739. A 9d. land tunship on ridge of South Walls where there has evidently been an old beacon-site, though the plural form is puzzling. O.N. *varða* (or *varði*), a beacon.

OTHER-FARM-NAMES

Bu of Aith: see Aith.

Cantick: a farm on north shore of Cantick Head, the east tip of South Walls. That bold promontory between Kirkhope (*Ásmundarvágr*) and the Pentland Firth must have been a notable landmark for seafarers from earliest times; but the earliest recorded form of the name I have seen is on Blæuw's map (1650), where immediately south of the headland appears—" " Cantop tyde." The spellings on that map are however quite untrustworthy, and there can be little doubt that " Cantick head " which appears on Mackenzie's Charts (1750) is the correct rendering of the name, and is that which is still in use today.

The origin of the name is quite obscure, and though the first syllable of the word bears the stress one cannot but think of the Gaelic *ceann*, head or headland.

Chuccaby (tʃʌk·əbi): a small farm today, and it is uncertain whether it formed part of Kirbister or of Wards or of Mosound tunship. The name, however, is of great interest, representing almost certainly O.N. *þjukkvibœr* (thick bae or farm), which, in the form *þykkvi-bœr*, occurs as a farm-name in Iceland though the precise interpretation of 'thick' is rather doubtful. Cf. however Thickbigging (Firth).

Eastbister: there are two *bólstaðr*-farms in South Walls—this and West-bister—both on the southern slope to the Pentland Firth.

Garson: a farm near the Bu of Aith.: See this name—Sandwick.

Haybrake: a farm adjacent to Lyness. O.N. *hey-brekka*, hay-slope.

Hurliness: a small farm on a little ness at upper end of Longhope. First element obscure.

Lythes: a farm on south shore of Longhope between Aith and South Ness. See this name—Hoy.

Ore: a farm near mouth of the Burn of Ore which flows into a bay called the Bay of Ore immediately south of Lyness. A rather puzzling name. *Or-* frequently occurs as the first element of an Orkney name—e.g. Orgill, and represents O.N. *ár*, the genit. case of *á*, a stream or river. One might therefore regard Ore as the first-element survivor of a compound name of which the second had been dropped. A possible (and perhaps better) alternative is to regard it as O.N. *ár* the nom. pl. of *á* = 'rivers,' 'burns,' 'streams.' Though the Ore burn is a single stream for most of its course it is fed by two streamlets which unite some two miles or so up from the mouth. Neither of these feeder burns is more than about a mile in length, but their existence does give grounds for regarding Ore as a plural = 'burns.'

Orraquoy: a small neighbouring farm to Ore, of which it has probably been a quoy-outbreak.

Scows: small farm near Ore. See this name—Orphir.

Skerp: small farm in Aith. Cf. Skerpie (Burray).

Swartland: small farm in mossy ground near Seatter. See this name—Sandwick.

Wing: See *sub* Hoy parish-name.

FARA

Faray 1492, 1500; Farra 1595; Faray 1642. Pharay, Phara 1500, 1614.

A small isle adjacent to Walls of which parish it formed a part. It was skatted as a 3d. land.

Same name as Faray in North Isles, q.v.

FLOTTA

Flatey (Fordun) c. 1375; Flottay 1595, 1617, 1627; Flotta 1617. O.N. *flat-ey*, 'flat isle,' no doubt as contrasted with its big neighbour Hoy. (It is rather curious to note how South Isles people usually pronounce the old -ey (island) ending as 'a,' e.g. Cava, Flotta, Fara, Rysa, Hunda, Swona, whereas in the North Isles the ending is always (with the exception of Papa)—'ay' (e), e.g. Rousay, Stronsay, Sanday, Faray, etc.).

This island was part of the old bishopric estate, and along with Burray was early feued out as a whole, so that its individual farms do not appear in any of the early rentals. From the Report of Walls and Flotta (1627) printed in Peterkin's *Rentals* it appears that it had been skatted as a half-urisland, and it formed part of the parish of Walls and Flotta.

In the 18th century, however, it was part of the Stewart of Burray estate, and in Fea's Rental of c. 1750 is a list of all the island farms at that time together with their relative valuation. The list is here appended, and the present-day forms of names are added in brackets.

Affall (Aval) : 2 marks, i.e. presumably 'marklands.' We do not know the number of marks per 1d. land in Flotta, nor the actual number of cowsworths in a markland; but from the relative rents charged it would appear that 1 markland = 3 or 4 cowsworths. Origin of name obscure.

Arp (id.) : a 'cottary.' Origin obscure, but the same name occurs in Burray as that of a small farm of today.

Banks (id.) : a cottary. O.N. *bakkar*, banks; this is near the shore-banks.

Blackywall and Bleckawell (Blackawall) : 2 marks. Probably a hybrid name, which, if we can trust the second 1750 form, might represent 'bleaching-well,' from O.N. *bleikja*, to bleach.

Blomer (Blowmuir) : 2 marks. See this name—Holm.

Brunthouse (id.) : 5 cowsworth. Cf. Bruntbigging (Burray) and Brinnigar (Stromness)

Cameral Joy (Cameraljoy) : a cottary. An unusual name—probably of Scots origin, but meaning quite obscure.

Castlewell (Castlewall) : a cottary. Reason for name unknown.

Croenest (Crows nest) : 2 marks.

Curries House (Curries) : a tumail.

Dam (id.) : 2 cowsworth.

Garson (id.) : 2 marks. See this name—Sandwick.

Hallywell, Hellywell : 2 marks. Site unknown, but the name suggests a holy well. There is one reputed holy well in Flotta—Winster's Well, but that is situated on an uncultivated headland—The Roan.

How : Over How, 1 mark; Neither How, 2 marks. The latter seems no longer existent, but there is still an Over (or Upper) How.

Lingo (Lingall): a cottary. Origin obscure.

Mains, The: now again reverted to its original name—The Bu.

Quoyness (id.): a cottary.

Seraquoy, Soroquoy (Serraquoy): 5 cowsworth. Cf. Sorquoy—S. Ronaldsay.

Shut Behind: an unrentalled croft—site unknown. The origin of this Scots name is quite uncertain, but the place may perhaps have been situated near a 'grind' or gateway in an old hill-dyke which was expected to be 'shut behind' one on passing through.

Standing Stone (Stannanstone): a cottary.

Urback (Orback): 1 mark. O.N. *ár-bakki*, burn-bank.

Watriehall: a cottary.

Whanclet (id.): 2 marks. Second element O.N. *klettr*, rock, and the first uncertain, but might well represent O.N. *hvǫnn*, angelica.

Whome (id.): 5 cowsworth. See this name—Stromness.

Windbrake (id.): 2 marks. See this name—Birsay.

Part II.---Farm Background

A. SKATS AND RENTS

FOR practically everything that can now be learned about Orkney farms in mediæval times we are dependent on information contained in, or to be deduced from, a group of old rentals, the more important of which were printed by Alexander Peterkin in an invaluable compilation* last century. The rentals included in that volume are as follows:—

1. Lord Sinclair's Rental of the Earldom lands in Orkney. Drawn up in 1502, 1503 and 1504, and represents conditions c. 1500 A.D. The only known MS. of this rental (that from which Peterkin's copy was printed) is now preserved in H.M. General Register House, Edinburgh. Though incomplete it is by far the most valuable in Peterkin's collection in regard to the light it sheds on farming conditions in Orkney. In this present work it is usually referred to as the 1500 R.
2. Rental of the Earldom and Bishopric lands in Orkney. 1595.
3. Bishop Law's Rental of Bishopric lands. 1614.
4. A Rental of Bishopric lands. 1642.
5. Another Rental of the same—undated, but from about the same date.
6. Bishopric Compt Book. 1739.

Besides the above rentals a certain number of others survive in manuscript from the 17th and 18th centuries. These add little to the information contained in the earlier, but among Mackenzie's MSS. preserved in Balfour Castle in Shapansay is a copy by that indefatigable 18th century antiquary of the earliest rental of all. In a prefatory note Mackenzie has written: "The most ancient Rental of all . . . by Henry, Lord Sinclair, now in the hands of Fea of Clestran who found it amongst the papers of the Buchanans of Sound, sometime Stewards of Orkney, and of which the following is an exact copy."

That rental had been compiled in 1492, but nothing whatever is known of its fate since Mackenzie's time, and only his copy survives to tell anything of its contents. In regard to the Earldom part of the county it is very similar to the 1500 R., but it is even more incomplete. It includes, however, much of the Bishopric part as well, and is thus of special interest. Frequently also it enables one to check the entries in the 1500 R., and in addition provides the earliest recorded forms of many farm-names. But it has to be remembered that the present MS. is only an 18th century copy, and the copyist was probably not particularly concerned with the exact spelling of names, many of which indeed might have been hard to decipher.

* *Rentals of the Ancient Earldom and Bishoprick of Orkney, &c.* Collected by Alexander Peterkin, Sheriff-Substitute of Orkney. Edinburgh, 1820.

CLASSIFICATION OF LANDS

In the earliest Norse period, prior to the establishment of the earldom in the early 10th century, all lands in Orkney were presumably udal—owned by descendants or successors of the original settlers. By the time of our earliest rentals, however, vast changes had taken place, and we find therein a most complicated and confusing picture from which the following different classes of land emerge:

1. UDAL LAND—owned by descendants or successors of original settlers.
2. EARLDOM LANDS of various kinds:

    (a) Bordlands. These lands were exempt from skat and can thus be fairly safely assumed to represent the earliest private estates of the earldom. Nothing is known as to how they were acquired, but as they included large portions of Sanday and of Westray and one estate at least in Stronsay—all in the North Isles of Orkney nearest to Norway, and part of the south end of South Ronaldsay (nearest to Caithness) in the South Isles, their acquisition or seizure may have been partly based on strategic reasons. On the Mainland there was a large bordland estate lying around the Bay of Houton in Orphir where there was one of the chief earldom halls; at Cairston near Stromness there was another bordland, and in Birsay, as is well known, the great Earl Thorfinn had his principal residence. Our rentals, however, show that much of Birsay was Bishopric property —a fact which must imply that these lands represented an early endowment of the Orkney see—most probably by Thorfinn himself.

    (b) Auld Earldom lands. Some of this class may also have been bordland, but as at least some of these lands were liable for skat they probably represented acquisitions by earls at a somewhat later date though prior to the time of Earl William Sinclair, the last of the earls under Norse rule.

    (c) Conquest lands. These were likewise private earldom property and liable for skat, but in the early rentals they are frequently stated to have been 'conqueist' by Earl William, and it is probable that all the lands specifically called 'Conquest lands' had been acquired or bought by that earl from previous udal owners.

3. OLD KINGSLAND. It will be recalled that King James III of Scotland, immediately after his acquisition of the King of Norway's *sovereign and other rights* in Orkney (through King Christian's failure to pay all his daughter's dowry on her marriage to James), made an agreement with the Orkney Earl William Sinclair by which he also acquired all *earldom rights and properties* in these isles. Thus by the time of our earliest rentals—some 30 years after that transaction—all the above-mentioned classes of Earldom lands might have been termed Kingslands. But we know that after the Battle of Floruvoe (1194) King Sverrir of Norway confiscated the lands of the Orkney 'Island-Beards' who had risen against him, and he placed sysselmen or stewards thereafter in Orkney to superintend his interests there. It is thus fairly certain that some

at least, if not all, of the lands specifically termed kingslands or *pro rege* lands in the rentals really represent old private property-lands of the kings of Norway.

4. BISHOPRIC LAND. The Bishopric estate was of several kinds. It included the actual ownership of a large part of Orkney from which it drew landmail or rent. These lands represented bequests or endowments from pious earls or udallers, together with some of another type. In a report by Bishop Graham in the 17th century we read: " Understand that the old bishopric of Orkney was a greate thing, and lay sparsim thro'owt the haill parochines of Orkney and Shetland. Besyde his lands he hade the teynds of auchtene kirkis : his lands grew daylie as adulteries and incests increased in the countray." The latter offence no doubt included marriage between persons prohibited by canon law.

In addition to property-lands as referred to above (and of course teinds) the Bishopric had right to the skats payable from many other properties. The grant of such skats must also have represented endowments from some earl or earls, though the rentals hint that since the time of Bishop William Tulloch who "farmed" the isles for some years under James III various skats were claimed by the bishops to which they had no right.

5. KIRKLANDS. These were not bishopric property, but lands which had been bestowed as endowments for one or other of the several local churches or of the various prebends in the diocese. The endowments of some of those prebends or chaplainries were quite considerable.

The foregoing classification of Orkney lands has been on a basis of ownership; a totally different classification falls to be made in regard to their liability for skat. In the rentals we often find a unit entered as ' quoyland but scat,' i.e. free from skat, and 'quoyland' was the technical term for such skat-free land. The term *quoy* will be discussed more fully in Part III; here it is sufficient to state that the generally accepted explanation for such exemption from skat is that quoylands had not been brought into cultivation at the time skats were first imposed. *Quoy* is one of the commonest elements in the formation of Orkney farm-names, and many farms so named, e.g. Grimsquoy in St. Ola, were technical quoylands and unskatted. But on the other hand the element *quoy* in a farm-name is no sure criterion that the farm was 'quoyland,' for several quoy-named farms were fully skatted, e.g. Swartaquoy in Holm, and Sutherquoy in Sandwick. Such farms, though not of the earliest settled, had obviously been in cultivation when skat was laid on.

In regard to quoylands, however, certain difficult questions arise which for lack of information must remain unanswered. In the first place, while quoylands were exempt from the main skats, many of them were valued in pennylands in the rentals, and at the end of each parish account these were included in the total number of pennylands in the parish on which the skat called Wattle was calculated. We have no information at all as to when, or by whom, that pennyland valuation had been extended to quoylands.

Then again the data vouchsafed in the two earliest rentals is sadly deficient. The 1492 R. was drawn up by Henry, Lord Sinclair who had a grant of the earldom from King James IV, and this rental is in a sense twofold, the skats payable from each parish being entered under different heads—Earldom and Bishopric. In the Earldom section are specified not only the skats due, but also the landmail or rents due from such lands as were actual property lands of the Earldom. Quoylands appear here, but only such as belonged to the Earldom, and these mainly because of the rents due from them. In the Bishopric section are entered the lands from which the skats were claimed by the Bishop, and their inclusion would seem to have been mainly due to the fact that Lord Sinclair suspected the Bishopric was not really entitled to these skats. Bishop William Tulloch, who had had a grant of the Earldom for some years in the previous decade, is stated in this Rental to have been responsible for several of those 'transfers' of skat to the Bishopric. But as it was only those suspect skats in which Lord Sinclair was interested, Bishopric quoylands are not specified at all.

In the same lord's Rental of ten years later—the 1500 R.—the Bishopric parts of the county are omitted altogether, but in the Rental of 1595 both Earldom and Bishopric parts are included, and there for the first time we are able to note those quoylands which pertained to the Bishopric.

The question now arises—Were there no udal quoylands? It is hardly credible that *all* these relatively late and thus unskatted lands had been acquired by the Earldom or Bishopric, but the rentals give no indication of any that were udal property. This is the more puzzling, because many quoylands pertaining to the Earldom had been valued in pennylands and were thus liable for wattle. That being so, one would scarcely expect udal quoylands to have been forgotten.

The first comprehensive and really competent survey of these rentals in modern times was made by Capt. F. W. L. Thomas, R.N., in a paper contributed to the Society of Antiquaries of Scotland in 1884 and entitled "What is a Pennyland." That paper was a masterly piece of pioneer work to which later students are deeply indebted, but its conclusions have since had to be modified owing to the fact that Thomas had misinterpreted the monetary standards employed, as well as some other matters. Even yet, however, despite intensive study by various scholars, numerous points in these rentals await elucidation.

THE SKATS CHARGED

Postponing for the moment consideration of their purpose and origin let us look at the amounts of the skats charged, and how they were assessed. At the period of the early rentals with which we are at present concerned,

FARM BACKGROUND 195

there was very little money in circulation, and both skats and rents were usually paid in farm produce of some kind. It is therefore necessary to have first of all some idea of the units of weight in use. Weighing was carried out by means of two types of instrument—the pundlar for large, and the bismar for smaller quantities. These were wooden weighing rods operating somewhat on the steelyard method, and the units of weight reckoned on were as follows:—

$$\begin{aligned}
24 \text{ marks} &= 1 \text{ setting, or settin (O.N. } settungr \text{, a sixth part.} \\
&= 1 \text{ lispund (O.N. } linspund\text{).} \\
\left.\begin{array}{r}6 \text{ settings}\\6 \text{ lispunds}\end{array}\right\} &= 1 \text{ meil (O.N. } mælir\text{).} \\
24 \text{ meils} &= 1 \text{ last (O.N. } lest\text{).}
\end{aligned}$$

Though the setting and the lispund were equivalent weights, setting was the term used in connection with grain products, while lispund was used of butter; mark was used of both. It is impossible to correlate those old units with present-day standards. They survived in use right down into the 19th century, but from time to time there were repeated complaints by Orkney landowners that the standards had been increased. Here it must suffice to say that the mark was supposed to be about ½ pound, but the actual pound referred to is somewhat obscure. In a paper contributed several years ago to the Orkney Antiquarian Society (P.O.A.S. XV) I gave an account of those weights as current in the early 18th century, and from data available from that period I was able to show that then

$$\begin{aligned}
1 \text{ meil (on Malt Pundlar)} &= c. \ 200 \text{ lbs. avoir.} \\
\text{and } 1 \text{ lispund or setting} &= c. \ 33\tfrac{2}{3} \text{ lbs. avoir.}
\end{aligned}$$

But it is probable that at the period c. 1500 A.D. the actual weight of those units would have been a good deal less.

For purposes of reference it may be convenient to enter here also the relative values of the chief media of payment.

Malt ... ... ⎫
Cost (⅔ malt and ⅓ meal) ... ... ⎬ All valued at 6d. per meil
Flesh (nature unspecified) ... ... ⎭ or 1d. per setting or lispund.
Oil (presumably fish- or seal oil)
Butter ... ... 4d. per lispund.
Bere (a kind of barley) ... ... 4d. per meil

Bere was weighed however on a bere-pundlar on which a meil was only = ⅔ meil on the malt-pundlar. Weight for weight, therefore, bere and malt had the same value.

Large quantities of butter were charged by the barrel, which at that period (c. 1500 A.D.) was reckoned as 20 lispunds, and the price 4d. x 20 = 80d. or 6/8d. It will be noticed that butter had four times the value of any of the other produce (weight for weight), and it is interesting to record that this ratio of 4 : 1 held good also in the case of butter and meal in Western Norway in the early 14th century.

Before considering the skats charged we must pause again however to refer to a vital question—the monetary standard of the 1492 and 1500 Rentals. As these isles by then had been part of the kingdom of Scotland for a quarter of a century or thereby, one might naturally expect the currency used to have been Scots money. Capt. Thomas indeed assumed it to have been such—a fact which misled him completely and seriously vitiated his conclusions. Here fortunately we have no need to tarry long over the matter, as the evidence thereon was fully set forth in two articles in the Scottish Historical Review, Vol. VIII (1921)—one by the late Mr. Storer Clouston, the other by the late Mr. A. W. Johnston. As those two scholars did not always see eye to eye on problems of Orcadian history, and as on this matter they were in full accord, viz. that the money standard of those early rentals was approximately, if not exactly, that of English sterling, their conclusion may be accepted as beyond doubt—as indeed the evidence cited proves. In Cochran-Patrick's *Records of the Coinage of Scotland* (1876) the ratio of English money value to that of Scots c. 1500 A.D. was 4 : 1; i.e. £1 sterling = c. £4 Scots.

For the imposition of skat an assessment of arable land in Orkney had been made centuries earlier on an urisland (ounceland) basis, each urisland being reckoned to contain 18 pennylands, etc.* Further consideration of these terms can be deferred until we have dealt with the skats charged. These were as follows :—

1. BUTTER SKAT. This was the most important, and was normally charged at the rate of 6 spans (O.N. *spánn*) per urisland. It is uncertain whether the Orkney span had been a measure of weight or one of capacity, and the actual measure—whatever it may have been—was probably more or less obsolete by the 15th century, as it never appears in any other connection than in this assessment of butter skat. A span of butter about the year 1500 was valued at 20 pence, though for fractional parts its value was reckoned as 21 pence.

As it was evidently impracticable for Orkney farmers to produce the total quantity of butter charged, only a proportion was exacted in kind, the residue being payable in 'pennyworths,' i.e. money or produce of other kinds. The proportion exacted in butter itself was termed the 'stent' butter, and was normally one lispund out of each span. The lispund, as noted above, was a weight, and as that amount of butter was valued at 4d. it follows that the span was accounted to be equal to 5 lispunds. If, for example, the unit of land to be skatted was a 3d. land (= one-sixth urisland), the butter skat payable would have been 1 span, and the amount payable in kind would have been 1 lispund; if it chanced to be a 4d. land, the butter skat would have

---

* *Urisland.* The term is now obsolete in the local dialect. When collecting old Orkney words some 40 years ago, I found only one old man—Alex. McDonald in Harray—able to give me any information on the term, and he pronounced it Ersland (ɛr·sland). That suggests the old normalised (Icelandic) form *eyrisland* (*eyrir*, ounce), but the usual record spelling 'urisland' points rather to the Old Norwegian form *øyrisland*.

been assessed as 1 span and 7d., and the stent butter 1½ lispunds—usually expressed as 1 lispund 8 marks.

It would appear, however, that prior to Henry, Lord Sinclair's time a larger proportion of butter had been 'stented.' In that connection the following entry in the 1492 R. is of special interest, all the more as it has not hitherto appeared in print:

(Westray). "Imprimis to remember that on Sundaye the 28 day of October the zere of God [?1490] Henry Lord Sinclare & Justice of Orkynnaye for the tyme sperit at Will Randell, Brandy Draver, John Draver, Mavius [= Magnus] Maibsoun, Henry Randell, Willzame Young and diverse othirs the best of Pappay & Westra, how the Stent Butter was payit in auld tymes of Pappay & Westra. And thai said faithfully that evir ilk new callit [i.e. callowed, calved] Cow payit half a Lisp. Butter zerly—Nevertheless the said Lord understud thare Poverty and appuntit with thame for his tyme to tak of ilk 3d. terre aye 1 Lisp. and the laif of the Butter scatt of unblawn land to be payit in Pennyworthis sic as growis upon the ground. And give God send Butter or Uly [oil] attour [over and above] the Stent thai sould give it in thair Detts upon the auld Price, viz. other half Lisp. [i.e. 1½ lispunds] for the m[eil] of flesch or victual quhare it wantit. The quhilk thai tuke appoun hand till underly [accept or agree to] the said things. Before Sir Wilzame Duthe, Alexr. Lesk, Robert Haithlly, John Lynclett with othir divers. And of the Cow ut supra the hale Stent was tane in all the North Ilys."

While butter skat was normally charged on the lines above indicated—viz. a span per 3d. land, a fifth of that (1 lispund) being paid in actual butter, there were some inexplicable variations—especially in the South Isles. In South Ronaldsay, for example, a lispund of actual butter was stented from each pennyland, which meant three-fifths instead of the normal one-fifth of the butter skat in actual butter. Then in Walls we note a curious diversity—some places being charged a span from each 2d. land, one farm at 12d. per 1d. land, and others again at the normal rate of a span from each 3d. land (here termed 'richt scat'). Hoy being all bordland paid no skat, but its small neighbour Graemsay was all skatted at the rate of a span per 2d. land, while its stent was at the normal proportion of 1 lispund out of each span. These facts might seem to imply that the South Isles were relatively more fertile than the rest of Orkney, but such a conclusion is certainly not tenable today, and the actual reason for such variations is quite obscure.

2. MALT SKAT. This skat was normally paid in malt at the rate of 4 settings per 1d. land, though the rate was often higher. Sometimes however it was paid in cost (⅔ malt, ⅓ meal) which was priced the same as malt, viz. 1d. per setting or 6d. per meil. In some of the North Isles (e.g. Sanday and Papa Westray) bere was paid instead. Bere, as we have seen, was weighed on a bere pundlar on which a meil was only ⅔ the weight of a malt-pundlar meil. The rate of payment with bere—7 settings

per 1d. land—was slightly higher, as a meil of bere (6 settings on the bere-pundlar) was valued at 4d. : 7 settings were thus worth 4⅔d.

3. FORCOP. This skat was assessed in money, the commonest rate being 30d. per urisland or 1⅜d. per pennyland. But it was the most variable of all, and sometimes omitted altogether. In such cases an explanatory note is entered, indicating that its absence was due to specially high rates of butter or malt skats. But no forcop at all was charged in the South Isles, and for its fluctuations in other areas no general reason can be discovered. As for the curious name, we shall return to that later.

4. WATTLE. This was the only skat which never fluctuated. It is also noteworthy that it was not specified for any individual farm, but when the total skats from a parish were summed up, a note followed indicating the number of urislands therein (with fractional parts), and then was added the appropriate sum for wattle, which was calculated at the rate of 18 pence per urisland or 1d. per pennyland.

While the above were the four main skats, at the end of each parish assessment we find certain extras. Of these the chief were skat 'merts,' i.e. cattle slaughtered and salted down at Martinmas for winter use. These were also assessed on an urisland basis, from the parish as a whole, at the rate of 1 mert per urisland. If there chanced to be 3½ urislands in a parish the rate would be 3 merts one year and 4 the next. Frequently a note follows—" for the price." The meaning is uncertain, but it would seem probable that the price of such merts was to be credited to the parish and placed against other skats. The normal price of a mert c. 1500 seems to have been 40 pence = ¼ mark.

'Halk-hens' (for the King's hawks) were also charged, but the rate cannot be ascertained. It is probable, however, that it was one hen from each inhabited house—what was later known as the 'reek hen.'

In order now to obtain a clearer idea of the data contained in the rentals let us glance at a few specimen entries in the 1500 R. which was a rental of the Earldom only. It was compiled parish by parish, and specified the skats payable, together with the Earldom rents over the ensuing three years.

1. DEERNESS PARISH.

Brabustare ane uris terre ant. in butter scat vi span. Jam tantum. Inde stent vi leisp. Et in malt scat ant. xii m. Jam tantum. Et in forcop ant. xld. Jam tantum. Et in land vii d. terre et uther half farding conqueist per comitem ant. in mail j last iiii ss. Jam to pay bot x m cost tantum flesche et iiii m iiii ss doune.

2. WALLS PARISH.

Kirkbustar ixd. terre ant. in butter scat iii span viz richt scat jam tantum Inde stent ix leisp et heir iiii merk makis the d. terre. Et in malt ant. ix m jam tantum. Et in land iid. terre conqueist per comitem ant. in male XVI m iiii ss. Jam set v part fall, et to pay xiii m Inde viii m iiii ss cost et iiii m ii ss flesche.

3. HARRAY PARISH.

Nerstaith iiiid. terre ½ uthale ant. in butter scat uther half span. Jam tantum. Inde stent uther half leisp. Et in malt skat ant. vi m ii ss ½ and na forcop.

These typical entries may be interpreted as follows :—

1. BRABUSTARE—a whole urisland (18d. land). Ant[iquitus] formerly, the butter skat was 6 spans; now the same. Inde (thence—of that amount) the stent (in actual butter) is to be 6 lispunds. And in malt skat formerly 12 meils; now ditto. And in forcop formerly 40 pence; now ditto. And of this tunship 7d. land and 1½ farthing land was acquired by the earl; of that the rent formerly was 1 last 4 settings; it is now to be only 10 meils of cost and the same quantity of flesh. Thus there is to be a reduction of 4 meils 4 settings.

2. KIRKBUSTAR—9d. land. Former butter skat 3 spans, that is to say— right skat. Now ditto. Out of that amount 9 lispunds are to be paid as stent (i.e. in actual butter). And in this tunship the pennyland has the value of 4 marks. In malt skat formerly 9 meils (i.e. 1 meil or 6 settings per 1d. land, which was 1½ times the normal malt skat); now ditto. And 2d. land here was acquired by the earl; of that the former rent was 16 meils 4 settings : it is now let at a reduction of one-fifth, and the rent will be 13 meils, of which 8 meils 4 settings are to be paid in cost, and 4 meils 2 settings in flesh.

(There is an error here in the 1500 R. which the 1492 R. reveals. The 'conqueist' land amounted to 2¼d. land, not merely 2d. land as stated in the 1500 R. As the 1d. land here was equal to 4 marklands there were thus 10 marklands liable for rent, and the 16 meils 4 settings payable for that amount of land represented the normal rate of 10 settings = 10d. per markland.)

3. NERSTAITH (now Knarston)—a 4¼d. land, all udal owned. Former butter skat 1½ spans; now ditto. Of that amount 1½ lispunds to be paid as stent butter. Malt skat formerly 6 meils 2½ settings. And no forcop.

(Here again a slight error occurs. In the 1492 R. the old malt skat was stated to have been 6 meils 4 settings—" viz. 8½ settings on the pennyland at the full." At that rate the total malt skat should have been not 40 settings but only 38¼, so that the 1500 R. figure of 38¼ is more nearly correct. In Harray 8½ settings per 1d. land was the usual rate of malt skat, more than double the normal rate of 4 settings; hence the 1492 R. note " Et in forcop nil—where hie scatt," i.e. by reason of the high malt skat.)

Before leaving these rental entries we may glance at a couple of items in the summation at the end of the Harray account in the 1500 R.

Summa de stent xxv leisp. xxi mark.

Summa de butter scat preter the stent xxxiiii s. viid. [34/7d.]

From what has been explained above butter stented to the extent of 25⅞ lispunds implies that the total butter skat was 25⅞ spans, which at 20d. per span represents a sum of £2 3s. 1½d. If from that we deduct the value of 25⅞ lispunds (paid in actual butter) at 4d. per lispund—viz. 8/7½d.—we have a balance of £1 14s. 6d. to be paid in 'pennyworths' of other kinds. The rental figure of 34/7d. is only 1d. out, which may well be the result of a misreading of a MS. vi as vii. Here then we see convincing proof of the procedure followed in regard to butter skat, though Capt. Thomas (and no doubt others) misinterpreted it altogether.

As has been already noted, skats were not levied at uniform rates in all cases. Yet certain rates were regarded as normal, or 'richt scat,' and these were as follows :—

Per urisland or 18d. land—

    Butter skat— 6 spans ... ... = 120d.
    Malt skat——12 meils (72 settings) = 72d.
    Forcop ... ... = 30d.
    Wattle ... ... = 18d.

                      Total   = 240d. = £1 stg.

Per pennyland the normal skats thus amounted to 1/1½d.

LANDMAIL OR RENT

(1) MARKLANDS.

Every udal owner of land was assessed for skat on the basis of his property valued as pennylands, etc. But by 1500 A.D. the greater part of the Orkney lands belonged to the earldom or the bishopric or one or other of the various church prebends or 'stukes' as they were called. The occupiers or tenants of such lands had to pay not only the relevant skats thereon, but of course landmail or rent as well.

In that connection we encounter a somewhat confusing situation, for rent was charged—not on the basis of the ounceland-pennyland valuation—but in accordance with an entirely different valuation of the same lands as 'marklands' or fractions thereof. Here acquaintance with the old Scottish marklands tends to make confusion worse confounded, for these were quite different entities. The Scottish markland was a unit of which the *annual value* was reckoned to be a mark (13/4d.), whereas the old Orkney markland represented land which had the *capital value* or purchase price of a mark.

Between a pennyland and a markland there was no fixed ratio. A pennyland in one tunship might be valued as one or two marklands, while in another it might be reckoned as ten or twelve marklands. In Sanday, for example, the general value of a pennyland was only 1½ marks, but over the county as a whole the average pennyland was valued probably at four marks. When the urisland-pennyland valuation was originally made those units presumably were all deemed of approximately equal value, and the great diversity apparent by 1500 A.D. was due no doubt to the fact that in some cases land had been developed or extended to a greater degree than in others. Sanday, having a light, sandy, easily-tilled soil was probably far more extensively cultivated than some other areas when Norsemen first arrived, and thus there would have been less opportunity for development. Hence its very low rate of marklands per pennyland by 1500 A.D.

In view of such increasing variation in the actual value of urislands or pennylands it is obvious that revaluation for purposes of sale or let would have gradually been felt necessary, but we have no information as to when or how it came about. We do know, however, that such revaluation had taken place prior to 1329 A.D., because a document survives from that year dealing with the purchase of certain lands in South Ronaldsay by Katherin, a widowed Countess of Orkney. The lands she proposed to buy are specified by name, together with their markland valuation, and the marks which she has already paid for some (and now offers for other) marklands are specifically described as 'good English money,' i.e. sterling marks. And as late as the 16th century we have records of the sale of marklands " at full land's price "; i.e. they were valued in relation to the English mark, and the price paid in Scots money varied in accordance with the varying exchange rates.

At one time I assumed that this markland valuation had been made 'from above,' so to speak, as the urisland valuation had previously been, but such a view is quite untenable. The real explanation, I believe, may be found in one of the fundamental principles of udal law. Under that law each member of a family (son and daughter) had a claim to a share of the paternal property in land, and while the eldest son succeeded to the headhouse he could not claim the rest of the estate until he had 'outred,' or paid out, the other heirs in proportion to their shares. Often indeed an heir succeeding might have to pay out much more distant relations than sister or brother, and it is obvious that each claimant would demand the full economic value of his share. As lands changed in value fresh valuations would thus have become necessary, and such revaluations would no doubt have been carried out by resort to legal process of some kind, the nature of which is not on record, but would probably have been an *ad hoc* assize of local impartial men.

The above-suggested explanation would imply that the same property might be revalued from time to time, and such a conclusion would seem to be borne out by reference again to the South Ronaldsay sale of 1329. At that date one of the farms—Brough was valued at 3 marks per 1d. land, while in the 1500 R. we find the pennyland there valued apparently at 4 marks, a fact suggesting that improvements or extensions had taken place between the dates in question.

Rents charged on earldom property-lands c. 1500 were paid in kind—

malt, cost or flesh—at a normal rate of 10 settings = 10d. per markland. Frequently, however, the rate was 12 settings = 1/- per markland, and the reason for such difference is uncertain, though it may have been due to the fact that these lands had appreciated in value though no revaluation had taken place. For lands which belonged to the earldom there would *ex hypothesi* have been no need for revaluation unless they lay intermixed with udal lands. The uniformity of rental rates—either at 10d. or at 1/- per markland—is often the only clue we have to the markland value of pennylands, as it is only occasionally that the rentals specify the markland value of individual tunships.

We have now to record a quite extraordinary and apparently inexplicable fact in connection with the rents charged for earldom quoylands. As such lands were free from skat one would naturally expect that they would have been more highly rented than skatted lands, but as far as can be discovered rental rates were alike for both. The plainest evidence of this may be seen in the case of various farms in Paplay (Holm), of which a couple of examples may suffice. In the 1500 R. we find:

1. CORNEQUOY. Here there was $3\frac{1}{2}$ farthing land "quoyland but scat." On that the landmail or rent was 7 meils = 3/6d. But there was also 1d. land skatted, on which the skats amounted to 1/1d., and the rent was 8 meils = 4/-— (i.e. at the same rate as the quoyland).
2. FLAWS. Here 2d. land quoyland was rented at 16 meils = 8/-. But in addition there was 1d. land skatted, on which the rent was 8 meils = 4/-—the same proportionally as for the quoyland—, and in addition skats amounting to 1/1d.; the total for skat and rent being 5/1d.

Different annual charges on quoylands in one farm or tunship from those on skatted lands in another would not appear so anomalous, but in the cases cited above we see the different burdens on lands in one and the same farm. Such a paradoxical situation makes any generalisation in regard to the relation between capital value and rents extremely hazardous, and we can but note the separate ratios for quoylands and skatted lands respectively.

As for skatted lands the commonest capital value of a pennyland c. 1500 A.D. was probably 4 marks = £2 13s. 4d., and the rent, if at 10d. per markland, = 3/4d., if at 1/- per markland = 4/-. The *normal* skats on 1d. land amounted to $1/1\frac{1}{2}$d.; thus skats and rent together would have amounted to either $4/5\frac{1}{2}$d. or $5/1\frac{1}{2}$d. These figures represent $8\frac{1}{3}$% and approximately 10% respectively of the capital value.

It should be remembered, however, that there was a further annual charge on land—church teinds. These are never referred to in the earlier rentals, and first appear in the 1595 R. But there can be no doubt that they were payable at the earlier period also, and from Shetland evidence the normal rate was probably one-fifth of the rent = 2d. per markland. If 8d. per pennyland be added to the above charges for skat and rent, the total annual burdens on 1d. land, if skatted, would have been approximately $5/1\frac{1}{2}$d. or $5/9\frac{1}{2}$d.—representing roughly 10% or 11% respectively of the capital value.

(2) OTHER VALUATION-UNITS.

In addition to the markland valuation of Orkney lands some other denominations of value were in use. The first to be noted was the 'cowsworth,' a term that has frequently been misconstrued in attempts to correlate it *directly* with the pennyland. Such attempts merely confuse the issue, for a cowsworth was a recognised standard of capital value like the markland, not (as the pennyland) an indication of its skattable liability. The commonest ratio between the two was 1 pennyland = 16 cowsworths, but such a ratio obtained only when the pennyland had the value of 4 marks, because the cowsworth in mediæval times in Orkney was reckoned to be = $\frac{1}{4}$ mark.

In the 1739 R. (printed in Peterkin) it is stated in regard to the Innertown of Stromness that it contained 36d. lands, 3 marks per 1d. land, and 3 cowsworths per mark. From that it would appear that at the time that tunship had been valued the cow had risen in value so as to be accounted = $\frac{1}{3}$ mark instead of $\frac{1}{4}$ mark as of old.

In a note prefixed to the 1595 R. it is stated that there are "ten kowsworth of land in ane pennyland," and "There is no kowisworth of land in the Kingis land or Bishopric land, but only in the udall land." As for the first of these assertions it may be stated that while there is no known record of any such ratio between a 1d. land and a cowsworth, it might quite well have been the case sometimes, but only where the pennyland was valued at 2½ marks. And as for the second it was manifestly untrue, as the writer of the note might have discovered had he troubled to read the second entry in that rental.

The cowsworth was a unit of value also in Norway, where it was termed a *kyrverd* or *kyrlag* or *kugildi*, and by the Gulating Law (13th cent.) a 'gild' cow (i.e. one without defect and not over 8 years old) was deemed to have the value of 2½ øre = five-sixteenths Norse brent mark. As that mark was about nine-tenths of the value of an English mark, a cow would thus have been equal in value to c. 0·28 of an English mark as current in Orkney. By the 14th century the cow in Norway had risen slightly in value, and was accounted = $\frac{1}{3}$ of a Norse mark.

Two other land-valuation units—the 'meils cop' and the 'uris cop'—are of much more obscure origin, neither being known in Norway, and in Orkney curiously enough their use seems to be on record from one island only—Westray. The whole southern end of that isle was bordland property, which embraced the Bu of Rapness and two areas known respectively as the Swartmeil Bordland and the Wasbuster Bordland. In the former of these two were nine small farms, each valued in 'uris coppis,' while in the latter there were twenty-two valued in 'meils coppis.' Prefacing the entry of these in the 1500 R. we find the following note:

"Imprimis to Remember that ilk vi m coppis or ilk vi uris coppis in all this said bordland makis the haill d. terre et payis na scat."

'Cop' in these terms certainly represents O.N. *kaup*, purchase or bargain, and 'uris' (as in *urisland*) denotes O.N. *eyris* (or rather the Norw. form *øyris*) genitive of *eyrir* (*øyrir*), an ounce. As the Norse ounce was = $\frac{1}{8}$ mark one might suppose that an 'uris cop' was a term for $\frac{1}{8}$ markland, and that the pennyland here thus had the value of only $\frac{3}{4}$ mark. But a brief glance at the rent payable tells a very different story. As we have seen above the normal rent of a markland was 10d. or at most 1/-, whereas the rent charged for an uris cop was no less than 19 settings of grain = 19d., while a meils cop was rented higher still—at no less than 26 settings = 2/2d. !

From these figures it is obvious that those terms do not refer to the capital value of farms but more probably to their annual rental value. An ounce being = $\frac{1}{8}$ mark (in Orkney in the 14th century, at any rate) = $\frac{1}{8}$ x 160d. = 20d., there would seem to be little doubt that an uris cop, of which the rent was 19d., represented a unit on which originally a rent of an ounce of silver or its equivalent was payable, though apparently a slight reduction of a penny had taken place.

As for the meils cop, however, no satisfactory explanation can be offered. The rent as we have seen was 26 settings = 26 pence but a meil of grain was only = 6 settings = 6d. It may be recalled, however, that butter in Orkney had four times the value of grain, and a meil of butter (= 6 lispunds or settings in weight) would have been worth 24d. at the period of our early rentals, and at an earlier date may well have had the value of of 26 settings of grain (= 26d.). But against such an implied possibility (viz. that the rent may have originally been based on butter and afterwards commuted to payment in grain) it must be stated that, in Orkney, butter never seems to have been reckoned by meils; the only units for butter were mark, lispund, span and barrel. The real origin of the term meils cop must therefore be regarded as still obscure.

Compared with most of Orkney, these bordland farms, if judged by their rents, must have been deemed of exceptionally high capital value. At the common rental rate of 10d. per markland a 1d. land of 6 uris coppis paying 114 settings of rent would have had a capital value of nearly 11½ marks, while a 1d. land of 6 meils coppis paying 126d. would indicate a capital value of 12½ marks.

Though the use of these terms is confined in the rentals to these two bordlands, we know from later land charters that the meils cop at least was also used in another part of Westray—viz. Dykeside.

B. NATURE AND ORIGIN OF ORKNEY SKATS

In the preceding Section an attempt has been made to survey some aspects of the agricultural economy of Orkney in later mediæval times in so far as that is revealed in our early rentals. As might be expected from such a source the information available is chiefly in regard to the ownership of land and the burdens thereon. These, as we have seen, were mainly of two kinds—skats and rents, and the usual rates of both have been examined. Rents are so familiar a feature of life still today that they call for no further notice, but until recently Orkney skats have been a complete mystery, an unpleasant enough burden for landowners it is true, but one for which no satisfactory or convincing explanation has been forthcoming.

To Dr. C. J. S. Marstrander, Professor of Celtic in the University of Oslo, credit is due for being the first to supply a clue to their real nature and origin. It was not however with Orkney that he was dealing, but with another old Norse domain—the Isle of Man. In his study of the Norse settlement of that island—*Det Norske Landnåm på Man*, published in 1932*, he discussed all the surviving Manx place-names of Norse origin (a task for which he—*doctus utriusque linguæ*—had quite unique qualifications), and thereafter went on to discuss those hitherto unexplained Manx terms for certain island-divisions—*sheading* and *treen*. These names he argued were associated with the old Norwegian system of national defence known as *leding* (O.N. *leiðangr*), and though in a later work *Treen og Keeill* (1937)† he stated his reasons for discarding the derivations he had originally suggested for these names, his argument as to their connection with leding has proved to be thoroughly sound.

It was Professor Shetelig of Bergen who first drew my attention to Marstrander's theory. With characteristic insight or intuition Shetelig suspected that the Orkney urislands might be somehow associated with leding also, and he suggested I should look into the problem from that standpoint. The results of my investigation were submitted to a meeting of the Orkney Antiquarian Society in 1935, in a paper entitled " Leidang in the West" which appeared later in Vol. xiii of that Society's Proceedings. As the gist of that paper has also appeared in Vol. xxviii (1949) of the Scottish Historical Review, it is unnecessary to go into the problem again in great detail here; a summary of the main facts may be sufficient.

First, however, some brief account of the old leding system may be desirable. It was a system of national defence, first organised, it is believed, by King Hakon the Good in the 10th century. With that end in view the coastal regions (in parts of Norway at least) were divided into areas known as shipredes, on each of which rested the obligation to provide, equip and

---

* *Norsk Tidsskrift for Sprogvidenskap*, VI.
† id. VIII.

man a war-vessel of prescribed size. In that way Norway in the 13th century was supposed to be able to muster a war fleet of about 300 vessels, though that may have represented the theoretical rather than the actual number.

It was only in the event of imminent or threatened invasion that the full leding fleet could be called out. In the earlier period, however, it came to be a recognised right of a king to call out 'half-leding' each summer, and with that force he could embark for a couple of months or so on private forays or enterprises of his own. As Christianity gradually asserted its influence such Viking raids became less 'respectable,' but kings were reluctant to give up a recognised perquisite, and in time an agreement was reached by which the annual half-leding might be dispensed with, provided the king was paid the equivalent of the annual outfit of food and stores. That payment of half-leding in peace time came to be known as 'bord-leding' (table-leding), and it ceased to be a personal burden and became an annual charge on the farms in each shiprede. In other words it developed into a virtual national land-tax.

In western Norway the shiprede was divided into units called *manngerds*, each of which was liable for the cost of equipping and provisioning one man for the leding ship. Payment was made in kind—usually barley-meal and butter—and as the quotas from each manngerd were alike, the term manngerd gradually came to be a recognised unit of land-valuation. A manngerd usually embraced a group of small farms or holdings jointly liable for the manngerd tax.

In the south-east of Norway, the Borgarting area, the corresponding one-man unit was known as a *lide*. And in the Trondheim region the *fylki* or county seems to have been the unit for the provision of leding ships rather than the shipredes farther south. From the scanty extant references it would appear that in England in Anglo-Saxon times it was also the county that was responsible for the provision of naval vessels, and it may indeed have been from England (where King Hakon had lived as a fosterson of King Athelstan) that the Norse leding system was borrowed. The similarity also between English ship-money and Norse leding skats is significant; but that question does not seem to have been closely studied in either country as yet.

The tax payable from a West-Norwegian manngerd was the clue I sought to follow in my examination of Orkney skats. For data as to the meal and butter paid in Norway, and the current money value thereof, I was mainly indebted to the results of exhaustive research by Dr. Asgaut Steinnes (later Norway's Riksarkivar), whose *Leidang og Landskyld* and *Gamal Skatte-skipnad i Noreg* are quite indispensable sources of information thereon. In Orkney, as we have seen, skats were paid mainly in malt and butter, whereas in Norway the media were barley-meal and butter. I therefore sought to reduce the various media to their equivalents in money, and as a result I found that there was a close correspondence between the value of the skats paid from a Norwegian manngerd and those from a $4\frac{1}{2}$d. land in Orkney.

Now, as it chances, in our early Orkney rentals that unit—a $4\frac{1}{2}$d. land—is sporadically referred to by the term "scatland." Thus e.g. in the 1492 R.

we find an entry "Lyngrow 1 Scattland, viz. 4d. terre et a half . . ."; and again "Camstaith a Scattland," the skats of which show it to have been a 4½d. land; in the 1595 R. "Grobuster 2 scat land"; "Dishes ane Scatland." All Orkney lands liable for payment of skat might be termed in general skatlands, but in the foregoing references we see the term used in a restricted and quite specific sense for a certain definite unit—a 4½d. land = a quarter urisland. Previously the term used in that sense had been a standing puzzle; now, in view of the similarity of its skatting to that of a Norwegian manngerd I believed I had stumbled upon the corresponding Orcadian one-man unit.

On consulting Dr. Steinnes, however, I found that he could not at first agree. Whereas I had based my conclusion on equivalent money values, he felt that any true comparison must be made between the *ipsissimae res* —the actual amount of food paid in kind. But he was good enough to explore the problem further, and in a later letter he informed me that he had found there was such a close parallel between the skats payable from an Orkney 4½d. land and those from a Norwegian *lide* (the one-man unit in S.E. Norway) that he could hardly believe it to be any mere accident: everything appeared to correspond so well. Thus both roads led to the same conclusion—that the Norwegian manngerd and lide and the Orcadian skatland or quarter-urisland all represented one and the same type of unit—that originally having the obligation of providing one man for the leding defence fleet.

There is no memory or record in Orkney of any area corresponding to the Norwegian shiprede, but once the one-man unit had been ascertained it was easy to make an approximate estimate of the total personnel available for manning an Orcadian defence fleet. The total number of urislands in the county cannot now be determined with accuracy, but it is considered there were somewhat fewer than 200 all told. Taking 175 as a fairly safe and conservative estimate, we should have 700 quarter-urislands or 'skatlands,' representing naval personnel of about 700 men. If now we assume (as the meagre Saga evidence suggests) that the local leding vessels were either 20-, or 25-sessers, we should with 700 men have sufficient to man 17½ ships of the smaller type, or 14 of the larger. In a paper on "The Battle of Tankerness" (P.O.A.S. VI) Mr. Storer Clouston had previously discussed the size of the fleet Orkney earls had at their disposal, and from a most acute analysis of the data contained in the *O. Saga* had reached the conclusion that it numbered 16 vessels. Obviously, the almost identical result of two different lines of research, based on entirely different data, was in itself the strongest confirmation of its essential accuracy.

EVIDENCE FROM OTHER NORSE COLONIES.

As the Norse domain 'west over sea' included Shetland and all the north and west coasts of Scotland it would be strange if traces of a taxation system similar to what has been noted in Orkney were not to be found in these other areas as well. Nor has one to search in vain; though much obliterated the surviving evidence is sufficient to show that a similar system obtained from Shetland all round to the Isle of Man.

After the battle of Floruvoe in 1194, Shetland it may be remembered was detached from the Orkney Earldom by King Sverrir and thereafter administered from Norway. Owing probably to that transfer Shetland was skatted on somewhat different lines from Orkney, the markland there becoming the skattable unit rather than the urisland as in Orkney. Nevertheless traces of the old urisland divisions are still extant in Shetland records : Whalsay, e.g., is stated in an old rental to have contained three urislands. And I have suggested elsewhere that the somewhat puzzling Shetland term ' scattald ' may be merely a variant of the Orkney skatland (4½d. land), as the two forms are found actually equated in the famous Complaint of 1576 against Bruce of Cultemalindie.

In Hebridean charters of the 15th and 16th centuries frequent references may be found to a land-unit known as a *terung* or *terunga* or *teirroung*, etc., which was equated with another term *davach*, which in turn was a Celtic or Pictish name for a certain land-valuation unit throughout most of northern Scotland. Thus e.g. in a charter of 1505 (O.P.S. II, 874) James IV of Scotland made a grant of 60 marklands in the North Head of Evist (Uist) including " the davach called in Scotch *le terung* of Yllera, the davach in Scotch *le terung* of Paible . . ." In a Tiree Rental of 1662 (S.H.R. ix, 344) we read : " The extent of Tirie is 20*tirungs* or 120 merkland and 5 shillings more "; and " A *Tirung* is a 6 merkland and is divydit into 48 malies or 20 pennylands." Tirung is easily recognised as Old Celtic *tirunga*, lit. ' land-ounce, and is thus simply a translation of O.N. *eyrisland*. In the Hebrides, however, the tirung or ounceland contained 20 pennylands—not 18 as in Orkney. That curious difference is no doubt to be associated with the difference between the old Norwegian and English mark, but why the Norse mark appears to have been the original standard of reference in Orkney, and the English mark in the Hebrides we do not yet know. That is a problem which demands the attention of coinage specialists. As further proof that the tirung signified ounceland we frequently find it equated with the Latinised term *unciata*.

The Hebrides were transferred to the Scottish Crown in 1266, two centuries before Orkney and Shetland. By the time of most of the early land charters from that region the old term tirung survived merely as a designation for certain areas of territory, any association with the former Norse skats having probably been entirely forgotten. Another unit of value had taken its place—the Scottish markland of Old Extent. Exactly when that valuation was first made is uncertain, and for us relatively unimportant. But, as noted earlier, the Scottish markland was quite different from the Orkney unit of the same name; it was a valuation—not of the capital value as was the Orkney markland—but of the annual value of land. Yet just as the varying value of an Orkney urisland c. 1500 A.D. was shown by its varying content of Orcadian marklands, so the varying annual value of a Hebridean davach or tirung was shown by its varying valuation in Scottish marklands.

From data collected and printed in my Leidang paper (P.O.A.S. xiii) it appeared that the average annual value of a davach or tirung in the Hebridean islands was somewhat over £4 or 6 marks, while on the western seaboard of the Scottish mainland the average value was only a little over £3. An exact average over all cannot easily be ascertained now, but 6 marks or £4 would be a tolerably close approximation. That, it may be recalled, was the value of the tirungs in Tiree.

Now the Scottish Old Extent valuation never applied to Orkney which was still a Norse land when Scottish lands were thus "extended." But it did apply in the neighbouring county of Caithness, and by chance we find in a 1614 Retour a ¾d. land of Ulbster valued at 3/4d. Old Extent. That clue, though tenuous enough, is significant. On that basis a 3d. land would be valued at 13/4d. or 1 mark of Old Extent, and a whole urisland of 18d. land at 6 marks or £4. If then that Caithness ¾d. land had been a 3 markland in *capital value* like the average Orkney ¾d. land, we can regard the normal Orkney urisland to have been of such a value as to have been accounted = £4 of Old Extent had such a valuation been made. That sum, it may be recalled, represents exactly the amount of the normal annual skats (£1), together with the normal rent (£3). Apart then from the names tirung and urisland the lands so designated obviously represented units of approximately equal value.

Along with the tirung or davach another term—'quarter,' or 'quarter-land' often appears in the Hebrides. Thus e.g. in 1309 King Robert Bruce made a grant of six davachs and three *quarters* of land in Uist to a certain Roderic son of Alan (O.P.S. II, 366), and the west end of Coll extended to seven *quarters* land (id. 281). In an Islay Retour of 1615 a whole series of units are entered, each termed *quarteria terrarum*, and in a 16th century *Description of the Isles of Scotland* (*Skene's Celtic Scotland* III, 438) it is recorded that "Ilk town in this Isle (Islay) is twa markland and ane half." According to Skene and Thomas the davach in Islay was valued at 10 marks of Old Extent, and the quarteria terrarum was thus plainly a quarter-davach or quarter-tirung, and thus the counterpart of the Orcadian 'skatland.'

Still farther south, in the Isle of Man, traces of the same old Norse system of taxation still survive and have been exhaustively examined by Prof. Marstrander in the work already referred to. There, the whole island was divided into areas called 'treens,' which were fiscal units, and though the original number is rather uncertain there were approximately 170 all told. Each treen again was composed of four 'quarters' or 'quarterlands,' and the manorial dues of a typical treen about the year 1511 are shown by Marstrander to have amounted to 79/2d.—almost exactly the same as the average Old Extent of a West-Scottish tirung, and as to combined skats and rent of an average Orkney urisland. All those land-units were thus obviously of similar value, and when writing my paper on Leidang I was so struck by the similarity of the two words themselves—*tirung* and *treen*—that I suggested they were really of one and the same origin—both representing O. Celtic *tirunga*. I had not the necessary linguistic knowledge, however, to prove their identity, but not long afterwards I had the satisfaction of seeing how in his later publication *Treen og Keeill* Prof. Marstrander, who had arrived at the same conclusion, was able to demonstrate the phonetic development of *treen* from *tirunga*.

From the preceding brief survey it is now clear that all round the north and west coasts of Britain from Shetland to the Isle of Man a similar basis for skat had been organised—the original main fiscal unit having been the *eyrisland* or ounceland. Now, strangely enough, the application of that term seems to have been confined to the regions around the coasts of Britain which had been occupied or conquered by Norsemen; it is not on record in Norway itself, nor in any of the other Norse lands. Not only so, but the term up to the present time has been consistently misinterpreted in all the standard dictionaries of Old Norse. In the first edition of Fritzner's *Ordbog* (1867) the word *eyrisland* does not appear at all. In Cleasby and Vigfusson's *Icelandic-English Dictionary* (1874) it is defined as 'land giving the rent of an eyrir,' the only literary reference cited being Vol. X of the *Fornmanna Sögur*, which contains the Saga of King Hakon Hakonsson. In the later enlarged Fritzner (1886) it is defined in terms merely translating C. and V.— '*Jord som giver en øre i Landskyld*,' but here a cautious mark of interrogation is added. Again, however, the only literary source cited is that given by C. and V. Lastly, in Hægstad and Torp's *Gamalnorsk Ordbok* (1909) we find the same rendering—'*jord som gjer en øyre i landskyld.*'

Now the occurrence of the term in the Hakon Saga is to be found in the section dealing with that king's stay in Orkney after the Battle of Largs, when, it is recorded, his men were billeted out on an eyrisland basis. And though the word is obviously of Norse origin nothing could more convincingly demonstrate that the unit it denoted was definitely unknown in Norway than the fact that the definition appearing in those dictionaries is quite wrong. In *eyrisland* those distinguished scholars had come face to face with something entirely foreign to them, and with which they were quite unfamiliar. For in point of fact the eyrisland was not a rental unit at all; it was, as we have seen, the basic valuation-unit for imposition of skat.

ORIGINAL IMPOSITION OF SKAT

So applied the term must have originally denoted land in respect of which an eyrir or ounce of money (or its equivalent) had to be paid. There were 8 aurar (ounces) in the Norse mark, and 20 penningar or pence in the ounce. In Orkney, however, we have seen that there were only 18 pennylands in the urisland or ounceland, and consequently the original urisland skat in Orkney must have been 18 pence or the equivalent thereof. But by the time of our early rentals the skats charged per urisland amounted to approximately 240 pence—a vast increase since the date of the original tax. We have seen how by 1500 A.D. there were actually four main skats, two of which—the butter and the malt skats—have been discussed at length. The third was forcop, a tax in regard to which there is still much obscurity. The name fairly certainly represents O.N. *fararkaup* (travel-pay), which in Norway meant payment to representatives having the obligation of attending a 'thing.' In Orkney the incidence of this skat was sporadic; in some areas it was not charged at all, and in most cases its absence would seem to have been attributable to the butter or the malt skat having been charged at a rate higher than the normal. In a few instances in the 1500 R. of Sanday the term forcop is equated with the compound 'girse-male' which is equally obscure. In my Leidang paper I suggested it might correspond to a term *gres-leding* which was similarly used occasionally in the Borgarting area of Norway, but whether that be the case, or whether 'girse-male' was regarded by the compiler of the rental as a skat payable in respect of grassland we cannot now tell. In any case this skat was variable in amount, and offers no basis for comparison or identification with the original tax of an ounce per ounceland.

Only one other regular skat is left to consider—viz. Wattle—, and in this we can in all probability recognise the original urisland skat. This was the one skat which never fluctuated in rate; it was regularly charged at the rate of 1d. per pennyland or 18 pence = 1 eyrir or ounce of silver per urisland. The name derives from O.N. *veizla*, a term used of the guesting or hospitality to which a ruler was entitled when on tour through his domains, an obligation which dates from remote times : and the continuance of this tax at an unchanged and unvarying rate over so many centuries would seem to leave little doubt that it represented the original eyrisland skat.

If that be so, the next obvious question is—By whom was it imposed on Orkney and Shetland and all round the west coast of Britain to the Isle of Man? The following passage from the *Ork. Saga* would seem to remove any doubt as to the answer. I quote from my friend Dr. Taylor's Translation.

> "One summer Harald Fairhair sailed west to punish the vikings as he had grown tired of their depredations; for they harried in Norway during the summer but spent the winter in Shetland or in the Orkneys. He subdued Shetland and the Orkneys and the Hebrides, and sailed west to the Isle of Man and laid waste the Manx homesteads. He fought there many battles, and *annexed the land* farther west than any Norwegian king has done *since*."

That last word 'since' is significant, and can it be mere coincidence that the range of that expedition tallies so noticeably with the regions where eyrisland units can still be traced today? Or is it not necessary rather to conclude that the establishment of those units, and the associated imposition of the original eyrisland skat, date from that time, and are to be ascribed to that great founder of a unified Norwegian kingdom?

Neither from the Hebrides nor the Isle of Man have we any information regarding the individual skats charged in Norse times, but in all probability they would have been similar to those charged in Orkney and Shetland. As we have already seen the vastly increased skats as specified in our rentals can be ascribed to the leding-system of defence first organised by Harald Fairhair's son Hakon the Good. The malt and butter skats in any case are to be referred to that source, and when leding skats were introduced over here the earlier eyrisland tax of 1d. per pennyland may have been taken into account and credited as part of the leding-skat. As for forcop, that is more doubtful; it is just possible that like the Borgarting *gresleding* it may have been an additional skat imposed by King Sverrir (after Floruvoe), but where our data is so scanty it is unsafe to speculate.

Our eyrisland problems, however, are by no means at an end. When leding-skats were imposed, or rather when they became a charge on land, one question of prime importance had to be settled—viz. the unit of landed property which should rank as the equivalent of a Norwegian manngerd in providing the necessary maintenance for one man in the leding fleet. For that purpose no fresh valuation would have been required. From what has been indicated already one can assume that eyrisland units or urislands had been in existence for some time, and all that would have been necessary was to decide what portion of each would be an appropriate one-man unit. And that, as we have seen, was settled by the choice of a quarter-urisland or 4½d. land, the unit which came to be known in Orkney as a 'skatland,' and which had a corresponding unit in the Hebridean and the Manx 'quarter-land.'

Working still backwards we have now reached the stage when we have to ask how the eyrisland unit itself, the urisland of our rentals, the tirung of the Hebrides and the Manx treen, was first determined and established as a basic unit for Norse taxation. It has been suggested above that King Harald Fairhair was responsible for its institution, but, however advanced in ideas that monarch may have been, it is hardly likely that he went about with a District Valuer in his train. The only feasible answer to our question is that which was suggested by Capt. Thomas long ago—that *eyrisland* was merely a Norse name applied to a territorial unit which was already in existence and known by the Celtic or Pictish name *davach*. That name, it is true, is never on record from Orkney, but Thomas quite logically assumed that such units must have existed here as well as on the Scottish mainland. As Orkney also was a Pictish land the absence of davachs here would be more difficult to explain than their presence.

As a territorial unit the davach is not known outside of Scotland, but in the Isle of Man, as also in Ireland, there was a kindred unit the *baile*, by which was denoted the Celtic tunship, an institution of which the antiquity is beyond compute. Skene was disposed to regard the davach as the Pictish equivalent of the baile, but, in Ireland at least, the baile would seem to have been of much greater extent, and McKerral (P.S.A.S. lxxviii, p. 52) more plausibly suggests that the davach may have represented only the arable lands of the tunship. Be that as it may, there can be little doubt that it was the earlier Celtic or Pictish davach unit on which the initial Norse skat of an eyrir or ounce of silver was imposed (presumably by Harald Fairhair), and which thereafter came to be known as a *tirunga* in the west, and an *eyrisland* in Orkney. And it is further noteworthy that the quarter-land was a well-recognised fraction of the Irish baile, just as we have seen it was of the Hebridean davach or tirung; so that when a decision fell to be made as to an appropriate one-man leding unit (that which was to be later known as a skatland in Orkney) the unit was already in existence—a territorial unit, so to speak, ready-made.

URISLAND AND PARISH

In studying a period from which written records are practically non-existent we are at times confronted with facts that at first sight may appear incredible or absurd, and may yet prove to be a window through which new light may enter. Of that we may have an instance in connection with urislands and parishes. The exact date, or even the approximate date at which an ecclesiastical parish organisation was established in Orkney is unknown, but it was pretty certainly not before the 13th century, by which time eyrislands had been in existence for some three centuries or more. Yet we find that parish after parish contained a varying number of whole urislands—without pennyland fractions over. Thomas had noticed this, and pointed out how in a Bishopric Taxt Roll of 1617 Holme and Paplay were stated to include 108d. lands (i.e. 6 urislands), Hoy 54d. lands (3 urislands), Shapansay 108d. lands (6 urislands), half of St. Ola 72d. lands (4 urislands). Such grouping might of course suggest that when parish areas were delimited each was framed so as to embrace a group of whole urislands, but in his *History of Orkney* Mr. Storer Clouston argued that though urislands were admittedly of far greater antiquity than ecclesiastical parishes, yet areas corresponding to, or coincident with, the later parishes must have been in existence and recognised for some purpose even before urislands themselves were established. In support of that thesis he cited four interesting examples of urisland grouping, two of which are so remarkable that his own words must be quoted :

P. 187. " In the early rentals the various townships of each parish are entered in geographical order, making up a succession of eyrislands, or eighteenpenny lands, as one goes along. Thus in Rendall, beginning at the Evie border, we get the successive towns along the coast running, 6d., 3d., 6d., 3d. = 18d. land. This takes us to Quoynameikle, and, beginning with that town, the next three run—3d., 9d., 24d. = 36d. land, or two-eyrislands, ending with Gorsness. Then come Isbister 12d. and Halkland 6d. = 18d. land; and finally we have Ellibister 5d. and Gairsay 13d. = 18d. land. That means we have reached the Firth border, doubled back [inland], and made up the last eyrisland out of two widely-separated places; and from this it seems to follow inevitably that the Firth boundary must have been there when the skat was first laid on the islands. Otherwise, the doubling back and grouping of Ellibister [which is right away inland] with Gairsay [an island] is incomprehensible."

" The fourth case is St. Mary's Parish in Westray, composed both in the 1492 and 1500 rentals of the following townships (given here in their geographical order from south to north) : Brugh 11d. land, Waa 36d. land, Noltland 18d. land, Rackwick 36d. land, Aikerness 25d. land. We have here 7 eyrislands, and to get this exact number, the towns at the two extreme ends, about 3 miles apart, have been added together for skat purposes, and counted a two-eyrisland. Manifestly therefore the whole parish formed one district when skat was laid on."

Additional support for that conclusion is probably to be seen in connection with the Wattle skat. That particular skat, it may be remembered.

was assessed, not on individual tunships, but on the parish as a whole. I have already sought to show that Wattle represented the earliest skat of all, and it continued through the centuries to be levied at a never-fluctuating and probably original rate. In these circumstances it is hardly likely that its basis of assessment would have been changed, and thus by yet another route we are left to envisage "parish *areas*" as recognised territorial units coeval at least with urislands themselves, and thus in existence centuries before parish churches or ecclesiastical parishes were established.

The origin and purpose of such units are alike obscure. They were no doubt of pre-Norse origin, and probably linked up with the primæval tunship economy. It may be that light thereon may come from other parts of the early Celtic world; here it must suffice merely to draw attention to the facts which at first blush seem quite anachronistic.

At a much later date urislands came to be associated with parish churches in a somewhat surprising and fortuitous manner. In his researches on the Isle of Man Prof. Marstrander devoted much attention to the treen chapels or "keéills" as they are called. In former times practically every treen in the island had its own little chapel, and in that respect its counterpart the Orkney urisland afforded a striking parallel. In the first *Statistical Account of Scotland* (c. 1790) the Rev. George Low wrote in regard to his parish of Birsay and Harray: "Remains of Popish chapels are many because every Erysland of 18 penny land had one for matins and vespers, but now all are in ruins." For the areas of which he was writing Low's statement was probably literally true, and for Orkney as a whole it was substantially true. Many of those urisland chapels had also burial grounds attached to them—some of which are still in use today. And there were other curious parallels with the Manx keeills. An old Harray man once told me that each Harray 'ersland,' as he termed it, had its own part of the parish-church graveyard for burial, while in the Isle of Man each treen was responsible for the maintenance of a certain portion of the parish churchyard wall.

There are of course in Orkney various old chapel sites which date from the time of the pre-Norse Celtic mission, but the urisland chapels are to be assigned to a period several centuries later, though still long prior to the establishment of parish churches. There is little doubt that most of them can be assigned to the 11th century, and that they represent the earliest flush of church-building after Orkney became officially Christian at the end of the 10th century. A similar outburst of church-building was taking place in Iceland in the 11th century, in regard to which a statement in the *Eyrrbyggja Saga* is probably not wholly ironic : "This which was promised by the clergy made men very eager in church-building—namely, that a man would have room in the kingdom of Heaven for as many men as could stand in the church that he had built." Such an inducement was well calculated to appeal to a Norseman whose prestige was closely linked with the number of men in his train. And though similar inducement may well have been offered in Orkney also, yet the relatively systematic urisland distribution would rather suggest direction from a central authority—in other words—some earl. And if such were the case, one would naturally think of the great Earl Thorfinn Sigurdson, who, as we know, himself caused Christ Church in Birsay to be built.

Like the early Icelandic churches these urisland chapels were not official church property at all, but belonged in the first instance to the builders, and thereafter to their successors or owners of the ground where these chapels stood. But when parish churches were established, two centuries or so later, it would appear that it was the urisland chapel of the leading family in each parish that was elevated to the dignity of the parish church. In every case that can be checked, at least, that was what happened. We do not indeed know anything definite of local magnates at the time parish churches came into being, but remembering the tenacity of udal possession it is more than probable that the descendants of the leading family in the 12th century would still be the leading men in the 13th. Thus in Birsay and in Orphir we find the parish church adjacent to one of the earl's own residences; in Rousay it was at Skaill, the home of Earl Paul's chief friend Sigurd of Westness; in Deerness at Skaill in Sandwick where Amundi and his more famous son Thorkell Fostri lived; in Holm—at Paplay, the estate of Sigurd who after Earl Erlend's death married his widow Thora—mother of St. Magnus; and in St. Andrews—at Tankerness, the seat of Erling another Saga chieftain. And as further confirmation it may be added that in any urisland where a parish church stands there is rarely, if indeed ever, another chapel site to be found.

C. THE ORKNEY TUNSHIP

When we reflect that agriculture has been carried on in Orkney for over 3000 years, the present type of consolidated, self-contained farm may be regarded as merely—so to speak—a creation of yesterday. Some such farms are little, if at all, over a century old. In districts where landed proprietors or 'lairds' had acquired estates the change had indeed come about somewhat earlier, but where land continued to be owned by udallers the old system of farming was adhered to tenaciously. That system was based on the tunship, a feature of Orkney life which in one form or another ran back to prehistoric times.

The early history of the Orkney tunship, it must be confessed, is shrouded in deep mystery, and one hesitates to refer to it at all. Various problems arising in connection therewith have never, so far as I am aware, been previously discussed in any detail, and that may be sufficient excuse for touching on them here. But in the circumstances any comments or opinions expressed must thus be regarded as tentative only and merely provisional, and they will almost certainly require amendment in the light of further research and criticism.

A tunship unit was of course not peculiar to Orkney, or even Scotland, and though its characteristics have been the subject of study by numerous scholars of the highest distinction, yet its exact individual nature at different periods and in different countries is still far from clear. The term tunship itself is an English compound, and was rarely, if ever, used in Orkney, where the local term was simply 'toun,' or 'toon' (tun). There is obscurity, however, as to how even that term came to be employed for a tunship in Orkney because O.N. *tun* had a different connotation. 'Toun' occurs rarely and sporadically in our early Rentals, though a few tunships or part-tunships appear which are named Overtoun (upper t——) and Nethertoun. In Rousay, however, there is a field on one farm called the West Toon, and on another the South Toon. There the term seems to be used merely in the general sense of a field, while in the generic term *toomal, tumail*, etc. (i.e. O.N. *tún-vǫllr*, 'town-field') the word seems to have had the Norse meaning of a field near the *tún* or farm-houses. O.N. *tún* does not seem to have ever been applied to what we know as a tunship, and the use of the term for that unit in Orkney (if not a relatively late borrowing from Scots) would seem to suggest that a familiar word had been applied in an altered sense to an unfamiliar unit met with in a new land. In actual fact, however, we do not know by what name the Orkney tunship was designated in the early Norse period in Orkney.

The Orkney tunship may be said to have had a parallel of sorts in the Norwegian *bygd*, but its real congener is no doubt to be traced in the Celtic or Pictish unit which under the name *baile* is to be found in all the old Celtic parts of Scotland, in the Hebrides, Man and even in Ireland as well. The Pictish equivalent of *baile* was *pett* or *pit*, and as Orkney was a Pictish area there can be no doubt that the *baile* or *pett* unit existed here as elsewhere in Celtic/Pictish lands. From that relatively certain fact, however, it would be rash indeed to infer that Norse settlers over here adopted the earlier tunship system unchanged, or that an Orkney tunship of—say A.D. 1000 was the exact equivalent of a contemporary *baile* in the Hebrides. The Norse concept of landed property was not that of the Celtic world, but was rooted in the ancestral udal system. Norsemen too were great individualists and in the famous *Landnámabók* we have a clear picture of how they carried out the settlement of Iceland. Each chieftain ' took land ' to *himself* : there is no suggestion of any settlement by groups. And it is most probable that the earlier settlement of Orkney would have been conducted on lines not dissimilar.

Conditions in Orkney however were different. Iceland was an uninhabited, untilled land, whereas in Orkney the newcomers found separate areas of cultivated land—in other words, tunships—already in existence, and we can imagine each chieftain or leader after selecting his more or less extensive ' land-take ' allocating these tunships amongst his followers, the larger units being shared perhaps between two or three. The land thus acquired would become, either immediately or after a certain number of generations, udal property, inalienable except in accordance with the very strict prescriptions of udal law. That law provided *inter alia* for the division of a man's land after death among his family or kinsmen, and in that way original estates became split up among many owners. But in Orkney at least, the tunship unit was not split up into smaller discrete units, but was shared *pro indiviso* by its several owners, each in proportion to the amount of his share. This fact—that the tunship continued to retain its unbroken unity as a *whole*—would seem hard to explain from Norse law or custom, and may probably have been due to the exigencies of the system of agriculture found practised over here. That whole problem is however most obscure; contemporary records do not exist, and, until fresh light can be thrown on the question from some quarter, further speculation is idle.

The form in which our early Rentals are compiled serves but to add to the difficulty of our tunship problems. The various agricultural units entered are doubtless for the most part tunships, but they are not specifically so called. Now tunships of course varied greatly in size and value, but in these rentals along with high-valued units, such as 18d. lands or urislands, we find very much smaller units entered—3d. lands and less, even ¼d. lands, such small parcels of property in fact that it is quite incredible that they could have constituted separate tunships of the normal kind at all. And at this late date it is quite impossible to tell to what extent these smaller units may have been farmed independently or as parts of larger tunship areas.

There are indications, also, that by the date of these rentals the older tunship system over much of Orkney may have been radically modified. Large areas of the county had become the actual property lands of either earldom or bishopric. How far that may have gone to modify the old tunship system we cannot tell, but the existence of the many 'bus' (see further Part III) would seem to suggest that a number of estates were being farmed on an individual rather than a communal basis. On one bishopric estate also—Sorwick in Rousay—the large number of uniformly skatted units gives the strongest impression of a systematic splitting up of a larger tunship (see also Part III).

Perhaps the most puzzling evidence of all appears in connection with the tunship of Swanbister in Orphir. That was a ½ urisland tunship, and in the two earlier rentals the skats due from the whole tunship are first entered, and thereafter follows a list of nine constituent farms or 'placeis' (as they are termed) therein—Lerquoy, Akirris, etc. When that occurs one normally expects to find only the earldom landmail or rent entered against each, but here strangely enough we find the skats attaching to each entered as well. In the 1492 R. it is stated that 4½d. lands in this tunship were 'conqueist' (bought or acquired by the earldom), and these lands were scattered among six of the nine units referred to, the other three being entirely udal property. Now the apportionment of skats among the constituent farms, though most unusual in the rentals, might be understood, but it is in the highest degree surprising to note that the markland value or content of the pennylands pertaining to the various farms or 'placeis' varied greatly, *though all were in the same tunship.*

What conclusion are we to draw from such facts? Here plainly we have no homogeneous tunship, since some pennylands are valued much higher than others, and it would thus seem that those 'placeis' were already, to some degree at least, discrete and self-contained farms within the confines of a single tunship. True, this tunship seems to be unique in that respect so far as we can judge from the rentals, but it puts us on our guard at least from assuming that the tunship organisation in Orkney was always of a uniform pattern.

From such considerations one cannot but conclude that in the course of centuries great changes must have taken place in the agricultural economy of Orkney since the time Norsemen first settled on the Pictish tunships. And apart from anything else the area under cultivation must have been greatly extended. The numerous rental entries of unskatted quoylands indicate very clearly extensions that had taken place since skat was imposed, presumably c. A.D. 900; but by then Norsemen had been resident here for something like 1½ centuries, and during that long period many other areas must have been brought into cultivation. The skatted quoylands are almost certainly to be assigned to that period, and perhaps a good deal more. But to what extent such new outbreaks meant enlargement of previous tunships, or the creation of quite new ones, we simply cannot tell.

TUNSHIP-URISLAND RELATIONSHIP.

We now pass on to consider another aspect of the tunship question. There can be little doubt that when skat was first imposed on Orkney the main unit of taxation aimed at must have been the urisland or ounceland—that is the area of arable ground in respect of which an ounce of silver

FARM BACKGROUND 219

had to be paid. And it would appear that, so far as at all feasible, Orkney lands were graded as urislands, or perhaps half-urislands : Rousay e.g. was skatted evidently as 6½ urislands. Now in a previous section it has been pointed out how the Celtic davach corresponded to the Orkney urisland, but, it may now be asked—in what relation did the urisland stand to the unit we are at present considering, the Orkney tunship? In that connection we are faced with a somewhat puzzling state of things.

Today, over much of Orkney the old tunship areas are practically forgotten. It was on the Mainland that the old udal system survived longest, and probably on that account it is there that the former tunship areas are best remembered. In the West Mainland, in particular, even today when all farms are separate individually cultivated units, and the former tunship boundaries rarely visible or exactly known, even yet the old tunship names are familiar, and the areas so denoted, e.g. Mirbister, Gorseness, etc., are still recognised districts, though such a name may be borne by no present-day farm.

Now in the heart of that West Mainland region, on the western slope of the Harray hills, lay a group of five more or less adjacent tunships, all well remembered still. Stretching in sequence for about a mile from north to south they were skatted as follows : Knarston (the most northerly) 4½d. land; Mirbister, 3d. land; Garth with Midgarth, 4½d. land; Corston, 4d. land; and lastly Corrigill, 2d. land. Together, it will be noted, these made up 18d. land or an exact urisland, and at first blush one might be tempted to conclude that here we have five sub-divisions of an earlier undivided urisland tunship.

A little reflection will show such a conclusion to be quite untenable. Assuming that all five had constituted an original tunship which had been split up into separate parts among different udal heirs or owners, why then, one may ask, had not each in turn been similarly divided up? Yet we know that in fact such division had not taken place, but that on the contrary each of the several owners of land in—say—Corrigill was obliged to share it merely on a *pro indiviso* basis with his fellow-owners. A more correct inference would therefore seem to be—not that an urisland tunship had been split up into five smaller tunships,—but that these five adjacent tunships had been grouped together as an urisland, and their individual skats adjusted accordingly.

If that inference be sound, as it indeed seems to be, it would follow that the tunship and not the urisland was the prior unit; and that while the former was the basic settlement, or agricultural, unit the latter is to be regarded rather as a superimposed fiscal unit for purposes of taxation. And if such be the case, it might well be that the forerunner of the urisland—the Celtic/Pictish davach—was a unit of similar nature, instituted originally for the collection of tribute by Celtic/Pictish overlords.

SOME FEATURES OF THE ORKNEY TUNSHIPS

From the two or three centuries prior to the final disappearance of the tunship system a good deal of information in regard thereto is to be found in legal processes, the majority of which are preserved in the Sheriff Court Record Room in Kirkwall. The most informative deal with disputes between portioners as to their respective shares of a tunship. In such cases the settlement usually took the form of a so-called 're-planking'—a procedure undertaken by a panel of competent and impartial men appointed by the Court. These made a perambulation of the tunship and measured off the 'town-lands' into planks or squares of 40 fathoms a side, and apportioned to each of the owners such number of planks as might correspond with the amount of his share. In such apportionment due consideration was also given to the relative quality or fertility of the various planks.

The late Mr. Storer Clouston made a very thorough study of such records, and the results of his research were published in a series of invaluable articles contributed to the Scottish Historical Review, and were afterwards summarised in his *History of Orkney*. It is hardly an exaggeration to say that our present knowledge of the old tunship economy in Orkney is almost entirely due to Mr. Clouston's outstanding work in that field, and it is thus unnecessary to go over it again in detail here. We shall restrict ourselves to some of the more general features, touching incidentally on a few aspects which he did not discuss.

Originally, the choice of location for a tunship had no doubt been largely determined by the nature of the terrain. Patches of territory which appeared most suitable for cultivation would have been selected, and though, in later centuries, at anyrate, the lands of one tunship were sometimes contiguous with those of another, yet in general each was separated to some extent from its neighbour by a tract of uncultivated ground. In Orkney, for the most part, each tunship abutted on a beach—either of the sea or of a loch—and its lands ran up to the 'hill' behind. The arable lands were unenclosed, and in summer, when crops were growing, all animals other than such as might be tethered, were driven on to the 'hill' or other common grazing lands outside the tunship. In order to keep them from straying back it is obvious that some barrier was required, and in consequence a tunship dike or 'hill-dike' was built separating the tunship lands from the hill or common. In some cases a tunship by reason of its situation required a dike all round it, but frequently, as just indicated, sea or loch served instead for part of the circuit.

Where the lands of two tunships met it is uncertain whether there was a dividing dike between, but various facts suggest there was not. On Mackenzie's Charts (1750) many, if not all, of the hill-dikes are shown, and we can observe that several contiguous tunships were separated from the hill by one continuous hill-dike, but mutual partition or dividing-dikes are not apparent. The same inference would seem to follow from entries in a MS. Bailie Court Book of St. Andrews parish (17th cent.), where we find that each spring a definite day was appointed by which date all dikes in the parish had to be repaired and animals sent outside to the hill. And it may be noted more specifically that between the tunships of Inner and Outer Stromness there was such uncertainty as to the dividing line that one had to be laid down afresh in an official planking in 1765 (p. 164).

From data collected by Mr. Clouston in connection with many 17th and 18th century plankings, etc., it is evident that the property rights of the several owners of a tunship could be grouped under four main heads, as follows :—

1. HOUSE AND HOUSE-FREEDOMS. In addition to his farmhouse itself, each portioner had exclusive right to what were termed his 'house-freedoms.' These presumably included his outbuildings—byre, stable, barn, etc., together with the stackyard and all the immediate area in which these were contained.

2. TUMAIL LAND. Similarly he had sole and exclusive right to a field more or less adjacent to his farm-buildings which was called his 'tumail,' a term found variously spelt—towmail, tumal, etc.—but pronounced toomal (tu·məl), and representing O.N. *tún-vǫllr*, town-field. The name is still common today as a proper noun, applied to a field on many farms, and in general it is the most fertile field on a farm. That is due no doubt to the fact that tumails were not liable to be shared with other portioners, and would thus have received the best manuring all down the centuries.

3. TOWNSLAND. Under this name was included the great bulk of the arable land in a tunship. It lay in the form of fields, or 'sheeds' as they were called, distributed all over the tunship and of no regular size or shape. Each portioner had a right to a share of each sheed in proportion to his share of the tunship as a whole. In some parts of Britain where fields were held in common they were divided into rigs or ridges which were shared in annual rotation by the respective owners, but in Orkney it would appear from Mr. Clouston's research that once a portioner had his share of a field allotted to him he held that continuously until such time as a replanking took place.

4. MEADOWLAND. By this term was understood all uncultivated portions of a tunship (especially wet or boggy land) where the rough native grass grew tall enough to be mown for cattle food in winter. Meadowlands were divided into portions also, but—unlike the townlands—these meadowland divisions were shared annually in rotation by the various owners in accordance with what was called meadow-shift or "kappa-skift." The origin of the latter term is uncertain, but it may be cognate with Norw. *kappa*, to chop off (in the sense of to mow grass).

At this late date it is impossible to discover or understand in detail all the rationale of that obsolete system of agriculture, and many of the old technical terms that survive in records, or even in human memory, are largely unintelligible. For example, many years ago an old Rousay man, whose recollections ran back to the middle of last century, told me that the 'auld folk' had many queer names for different kinds of fields such as 'flaas,' 'ties' and 'skuttos,' but even he was quite unable to define these more precisely. Considering the many centuries through which the tunship organisation was the basis of the farming economy in Orkney, it is indeed surprising how few traces or memories of it survive today.

THE PASSING OF THE TUNSHIP SYSTEM

Most of the data collected by Mr. Clouston was based on evidence relating to tunships on the Mainland of Orkney where it was that the udal system survived longest. Comparatively little is on record from the North Isles. That does not imply that North Isles folk were less litigious by nature; it is more probably due to the fact that there the udal system disintegrated sooner, much of the land being gradually bought up by a new class of landed proprietors. Some of these had acquired feus of bishopric land, while others seem to have bought out udallers and established large farms or estates for which they in time obtained feu-charters from the Stewart Earls. Thus, for example, Gilbert Balfour obtained a feu of extensive lands in Westray from Bishop Bothwell (his brother in law) in 1560, and the first James Fea of Clestran obtained a feu of that Stronsay estate from Earl Robert Stewart in 1590. Such feudal transactions, one would suppose, were quite illegal in Orkney at that period, as the old Orkney laws were not abolished until 1611.

Under such landowners many large consolidated farms were established, some at least of which, such as Stove and Elsness in Sanday, must have represented whole earlier tunships. On such farms one finds in the 17th and 18th centuries a totally different order of things from what obtained in the udal tunships of the Mainland. The udal owners of those much-divided udal tunships were still in a way hereditary gentlemen, and independent in the sense that they owned their lands, though they were relatively small fry as compared with the new class of 'lairds,' or even with the tenants of the large new farms. But they were in a different class altogether from the men who provided labour for the large consolidated farms. Yet in the course of the two centuries mentioned the old udal class everywhere was gradually being bought out or crushed out by economic pressure, and in consequence helped to swell the ranks of the landless, or at best the small tenant class.

On the larger farms of which we are speaking labour was mainly provided by a class of men who might be termed cottars, but who were known in the isles as "on-ca's," a Scots term = English 'on-calls.' In a valuable paper entitled "Remarks on the Agricultural Classes in the North Isles of Orkney,"* the late Walter Traill Dennison gave the following account of that class.

> "'ONCAS.' The word cottar is of more modern use in Orkney, but may be regarded as synonymous. The onca held from the large farmer a house, a piece of cultivated land called a haerst-fee, one, two or three 'coogils' of grass land—a coogil was a cow's grazing—for which he paid in 'crish butter,' that is coarse butter, in spring, and in fowls. He had right to keep a certain number of sheep on the 'hagi' or out-

---

* Printed in the "Report of H.M. Commissioners of Inquiry into the Conditions of the Crofters and Cottars in the Highlands and Islands of Scotland," 1884.

pasture common to the district, for which he paid a tenth of the wool and a tenth of the produce of his sheep. The onca wrought to the tenant [of the large farm] in harvest, helped to thatch the steading, to thresh Yule straw, to take up ware (seaweed) in Vore (spring), mell clods in bere seed, to weed thistles and cut peats in summer. In short he was called on whenever his master required him. And for his occasional labour he received an allowance of bere."

The term 'onca' itself is of interest. It is a Scots translation of two O.N. words—*á*, on, and *kvæða*, to call—which in Norse are found combined in the term *ákvæðisverk*, n., which denoted a certain kind of labour contract, full details of which may be found in Østberg's *Norsk Bonderet*, II, p. 13 ff. The origin of such a class as oncas in Orkney is rather obscure, but it is possible that their status may have corresponded in some measure to that of the occupants of the 'umsetts' attached to the old Orkney 'bus'; see *sub* Bu in Part III.

It was on those large consolidated farms that improved methods of agriculture gradually crept in, but even on those the old practices lingered long. Potatoes and turnips appear not to have been cultivated at all before the middle of the 18th century, and their general adoption was very slow. And as for crop rotation the following extract from " Answers from Orkney to Queries proposed by the Board of Agriculture " in 1794 speaks for itself.

" The Grains cultivated on these Islands are almost universally Gray Oats and Big or Bear; and these in alternate Crops without intermission, Bear has succeeded to Oats and Oats to Bear invariably on the same land for centuries. No fallow or other Crop has intervened, unless we except a few acres annually croped with Potatoes or Flax. Fallowing is of late a little practised by the larger farmers, but not at all by the smaller ones, tho' those who have tried it have found its great advantages."

Such was the state of things at the close of the 18th century, and as I have previously discussed* various other aspects of Orcadian agriculture as it was in the earlier 18th century it would be superfluous to dwell on it here. In the course of the hundred years from 1800 to 1900 greater advances were made than in the course of the whole previous millenium, while during the half-century just ended Orkney farmers have earned for themselves a reputation for scientific methods second to none in Britain.

---

* *Merchant Lairds of Long Ago*. Part II. Kirkwall 1939.

Part III.---Chronology

Part I of this book includes a representative and fairly comprehensive list of the main farm-names in Orkney, and in Part II an attempt has been made to outline the background against which these names have to be viewed. From a strictly linguistic standpoint that background might be regarded as quite irrelevant, but in reality it is by no means so, for in any attempt such as this on which we now embark—the placing of those names in their historical setting and chronological sequence—an intimate acquaintance with their general background is essential. Even when one is so equipped the task is beset with extreme difficulty; paths are obscure and signposts hard to recognise. The present attempt must thus be regarded merely as a rough blazing of the trail, and the suggestions made as being provisional in the meantime, and subject to amendment or correction in the light of later research.

In Orkney today there are some farms, relatively few in number, bearing names of Scots origin, but these can be ignored in the following survey. Similarly, there are a very few—regrettably few—with names which are of pre-Norse origin—Airy, Arian, Treb, Trave and one or two others—, but the farms so named cannot be regarded as having been in existence and so named in pre-Norse times; their names merely denote words which had been borrowed from the earlier race, and applied to parts of the countryside which later became farms. The vast majority of Orkney farm-names, probably about 99 per cent., are of Norse origin, and it is with these we have now to deal.

Many of these are nature-names so to speak, Old Norse terms such as *nes, vik, haugr, fjall,* etc., appearing either alone or in compound form— e.g. Ness, Fea, Howe, Elwick, Leafea, etc. Some names of this type may well indicate original settlements (as will be mentioned later), but as a class they do not provide any data for chronological placing. Having such an end in view we are obliged to rely on names containing as substantive element one of the several generic terms denoting a farm, habitation-site or human settlement of some kind. Of these there are in Orkney some half score or so which have to be considered in detail.

1. O.N. *kví.* This term might be roughly defined as a cattle-fold, but it was applied in a somewhat wider sense also of a place where animals were wont to assemble, or be gathered together for milking, etc. Though now spelt *quoy* in Orkney, and pronounced kwai, the old pronunciation in farm-names was usually 'kwee' (kwi). Of all farm-name elements in Orkney this had the longest life, and even today the word can hardly be deemed obsolete: a field or park might still perhaps be referred to as a quoy by elderly people, though such an occurrence would be rare indeed now.

In 17th and 18th century records we find sporadic references to a 'bowed quoy' or 'ringit quoy' within a tunship, and though no exact explanation is now accessible these terms must have denoted enclosures of a sort: an old tunship dike was frequently referred to as the "auld bow." And though quoys might sometimes be found *within* the confines of a tunship, normally a quoy-farm would appear to have been an 'intake' from the hill or common outside. When animals were banished outside the tunship during the summer months they would naturally tend to be attracted by the sight or smell of the more succulent crops growing inside, and at night would thus assemble at the more sheltered spots immediately outside the dike. In consequence of the constant manuring such places would receive they would gradually become more and more fertile, and if a newly-wed couple required a little farm and there was none available within the tunship (*innangarðs*) the natural course would have been to select one of those still untilled but fertile 'quoys' outside (*utangarðs*). There a new house could be built and a new small holding or quoy-farm come into existence.

Such a procedure is of course hypothetical, but it would seem to be the most feasible explanation of the fact that in general quoy-farms have to be regarded as more or less peripheral, that is—situated on the fringes or more outlying parts of a tunship. A striking illustration of that is to be found on the north side of the valley in Marwick tunship in Birsay. On the lower slopes near the bottom of the valley lie two closely adjacent farms —Langskaill and Netherskaill, names clearly indicating an earlier undivided Skaill, the head-house of the tunship. To the west of these and nearer the beach is another farm Flaws, a name which again indicates old arable land. Above these the ground slopes steeply up to Marwick Head, and more or less on a contour some 50 feet or so above the afore-mentioned old farms we find a series of small quoy-farms on the fringe of the moor— Stedaquoy, Cumblaquoy, Quackquoy and, farther round to the east, Leaquoy. The situation of those small quoys leaves no doubt as to their relative chronological setting in the tunship.

Further evidence of *relatively* late date is to be found in the early rentals. There quoyland is used as a generic term for such lands as were exempt from skat, the obvious explanation being that such lands were still uncultivated when skat was imposed. But while most farms bearing quoy-names were thus skat-free, we nevertheless find that a considerable number—some of them quite large units such as Tuquoy in Westray and Southerquoy in Sandwick—were fully skatted, and as such must represent very early extensions of older settlements. And as a farm-name element *quoy* continued in use for many hundreds of years, even after the old Norn language had been largely superseded by Scots. Of that the best evidence is to be found in Westray where there was an astonishing plethora of quoys which are referred to at some length *sub* Quoy (Westray) in Part I. Some of these were compounded with Scots names such as Chalmers, while one —Palliequoy or Palacequoy—must have dated from sometime after the erection of Noltland Castle in the latter half of the 16th century.

Quoy may thus be regarded as the longest-lived and latest-used of Norse farm-name elements in Orkney. The manner in which the word appears varies greatly. It frequently appears uncompounded as Quoy or with English plural—Quoys. Sometimes it is the suffix as in Cornquoy, but probably just as often it is the prefix, as in Quoyostray, Quoygrinnie, Quoyberstane, etc. Instead of Quoys we sometimes find the O. Norse plural *kviar*, now spelt Queear (kwi·ər), and oblique forms of the O.N. words are also numerous, usually with the definite article attached—Queean, Queena, etc.

2. O.N. *setr*. In his *Indledning* Rygh states that in farm-names this term in most cases signifies merely a dwelling-place or homestead (*Opholdsted, Bosted*), and he notes that while it is a very common element in Norwegian farm-names it is curiously absent in Icelandic names. It may also be easily confused with a different O.N. word *sœtr*, Mod. Norw. *sœter*, which is used of mountain pastures to which cattle are driven in summer, and which corresponds roughly with the Celtic *airigh*. In Orkney, however, there is no evidence of sæters, and accordingly in the present work no derivation from *sœtr* is suggested.

As a farm-name *setr* occurs sometimes alone and is then generally spelt Seater, the old pronunciation being set·ər. More often it forms the second element of a compound, in which case it appears as -setter, -ster, or -set. Though there are various other examples of the name we shall here confine ourselves to those appearing on record in the early rentals—some 25 examples in all. These are as follows: Blanksetter, Colsetter, Cursetter, Fulsetter, Garsetter, Grymesetter, Grottsetter, Ingsetter, Melsetter, Morsetter (twice), Mossater, Mussater, North Setter, South Setter, Okilsetter, Quoyneipsetter, Setter (four times), Setiskarth, Stangasetter, Winksetter and Yernasetter. Fifteen of these are on the Mainland of Orkney; the remainder are distributed among the other isles, Westray and Rousay being the only two of the larger isles without any example.

I have little to add to what I wrote about these names in a paper on Orkney Farm-Names* several years ago, and which I may quote here. " In three cases—Colsetter, Grymesetter, and Winksetter—the prefix may with confidence be assumed to be a personal name—Kolr, Grímr and Víkingr respectively. In one or two other cases the prefix *may* be a personal name, but we cannot be sure. Garsetter indicates probably a setter adjacent to a garth or hill-dike. Melsetter may be either 'sand-setter' or 'mid-setter' (O.N. *meðal-*). Morsetter and the two other variants indicate pretty certainly a setter on the moor. Quoyneipsetter indicates indirectly a setter adjacent to a headland. Setter uncompounded needs no comment, and Setiskarth represents no doubt a setter in a 'skarth,' in this case the well-known gap on the hill-ridge between Rendall and Harray over which the Lyde road passes.

"As an indication of size we may note that one—Stangasetter—was a 11d. land, three were 6d. lands, five were 3d. lands, and the rest were all smaller. The 11d. land is in Sanday where the pennyland valuation was two or three times as high as in most other places, and there is no example of a setter figuring as one of the common type of urisland, two-urisland or half-urisland tunships. The general impression one gathers is that these setter farms are relatively smallish, secondary settlements."

* P.O.A.S. **IX.**

"That impression is strengthened by a [further] review of their skatting. Of the 25 examples we are considering 12½ or exactly half were quoylands, i.e. unskatted in the strict sense and thus on an equality with the previous 'quoys.' The fractional case is interesting and suggestive. Garsetter in Birsay was a 6d. land of which 3d. land was skatland and 3d. land quoyland. This seems to indicate pretty clearly that it was situated near the 'garth' or hill-dike, and that the quoyland part represented land reclaimed from outside after the imposition of skat.

"The secondary or derivative character of these farms is confirmed further by a scrutiny of their location. It is not always possible now when tunship-dikes have been so often levelled out of existence, and inter-tunship areas brought under the plough, to determine exactly the relation of those setters to the parent tunships. But in most cases it is clear enough. The Morsetters speak for themselves; Seatter in Firth is over the hill from the parent tunship, and (with Wasdale) forms a little island among the heather; the position of Cursetter suggests strongly an outgrowth from the adjacent Grimbister; Setiskarth lies away up on the slope of a hill and Winksetter is tucked away on the opposite slope—most obviously late settlements. Other examples might be adduced, but here we may content ourselves by referring finally to the three Warsets or Warsetters, i.e. setters at a wart or beacon hill in Sanday, Shapansay and Egilsay—the most unlikely spots for early settlements: (these three examples however are outside our rental list).

"After what has been recorded above there can, I think, be little doubt as to the conclusion to be drawn. The setter farms are very definitely settlements of *relatively* late date. That does not imply that they necessarily date from the last half-dozen centuries; but it does imply that they are not primary farms. They are plainly farms that were established after the main tunship farms were taken up. . . . The essential point to be recognised is that all the facts regarding these farms indicate that when they were first settled the [bulk of the] old arable land in the primary tunships had all been appropriated.

"Before leaving these setters, however, I wish to draw your attention to one example which may be illuminating. Just beyond the Wideford Burn [from Kirkwall], and on the outskirts of the parish of St. Ola, lies the ¾d. skatland of Grymesetter. Next adjacent lies the 'quoyland' of Grymesquoy. There can hardly be any doubt surely that both names point to an original farmer—*Grimr*. He apparently settled there on a 'setter' just before skat was imposed on the Orkney lands, for the 'quoy' which we assume to have been an extension he made to his settlement was late enough in origin to escape. If it were not that I realise the possibility of Grymesquoy being a truncated form of a 'Grim-setter-quoy,' and thus of perhaps later date, I should emphasise much more strongly this striking and suggestive juxtaposition of names."

The only point of any consequence that I should now like to amend in the foregoing is in regard to the Birsay Garthsetter. That farm was in the so-called Barony of Birsay, a creation apparently of one of the two Stewart Earls. Under Birsay Parish Name in Part I, I have referred to a noteworthy feature revealed by the 1595 R.—viz. that in this Barony area a parcelling out of quoyland among the several farms seems to have taken place. The skatted lands are entered as 'townlands,' and most of the farms have a certain amout of quoyland assigned to them as well as of townland. (A similar state of things, it may be noted, was apparent in the Paplay district of Holm, where a 'grange' had been established by Earl Wm. Sinclair.) Hence we cannot offhand assume that the 3d. land quoyland of Garsetter represented a normal extension; it may not indeed have been adjacent at all. But the name itself indicates that this setter was adjacent to a 'garth' which was almost certainly the tunship dike, and it thus forms no exception to the general setter-class as being relatively outlying, secondary or peripheral settlements.

LAND, GARTH, BISTER.

We pass on to consider next a group of three elements which seem to belong to one and the same phase of settlement, and to have approximately the same status in the scale of what may be termed farm-dignity.

3. O.N. *land*. This being a Scots as well as a Norse term, and one which has been constantly used in naming a stretch of ground all down the centuries, we have to be on guard against assuming all -*land* farm-names in Orkney to be of early Norse origin. But by restricting ourselves again to such as appear in the early rentals, and having regard to the skatting of the farms in question, we are unlikely to go wrong. A few even of the rental names have to be ignored but for a different reason; in North Sandwick the early rentals specify five substantial units having names compounded with -*land*—Mobisland, Brekisland, Eriksland, Sowieland and Boysland—but as these appear also as Mobisyord, Eriksyord, etc., we shall leave them out of account here.

Apart from these there are some 35 -*land* names, and these are fairly evenly distributed over the county. There are no fewer than 9 Hollands; 2 each of Weyland, Mousland, Midland, Redland and Noutland; and one of each of the remainder—Raynland, Ireland, Rusland, Twaitland, Butterland, Houndland, Halkland, Trumland, Browland, Bigland, Sutherland, Leyland, Burrowland, Tarland, Sotland and Quoymirland. The majority were tunship names, though the tunships varied greatly in size. One, Noutland Bewest, in Westray was a 2-urisland; five—Ireland, the Evie Redland, Leyland, the Sanday Noutland and the Stronsay Holland—were whole urislands; seven or eight were 6d. lands and there was a similar number of 3d. lands, the rest being all smaller.

These names having all been discussed in Part I, it is unnecessary to refer to their origin here. But with the exception of Sotland (a $\frac{3}{4}$d. land), the Stromness Mousland (1d. land) and part of Twaitland in Birsay, all these -land farms were skatted and thus, as a class, of definitely early settlement. Even Quoymirland, the quoy of a -land farm, was itself skatted. And, so far as one can calculate, only about 1 per cent. of the lands designated by -land names in Orkney was unskatted.

The contrast with the quoys and setters is certainly striking, and (to quote again) "no less significant is the fact that so many of those -land names are tunship names. Though a few are situated on the outskirts of a parish they are not as a whole characteristic ' outliers ' as most of the quoys and setters are. In general, they are substantial fertile farms (or tunships) lying in the body of the parish, so to speak. They are without any doubt very early and venerable settlements. And yet—not I think the earliest!"

4. O.N. *garðr*, a farm, etc. Here again, as I have nothing fresh to add or any reason to amend, I quote from my former paper : " In most respects the *garth* farms are very similar [to the -*lands*]. Confining ourselves again to the early rentals we have 45 names with which to deal. One of these— Housegarth in Sanday was a 27d. land; seven—Gardemelis in Sanday (twice), Garth (Sanday), Garth (Evie), Stansgar (Birsay), Negarth and Housgarth (Sandwick) were 18d. lands or whole urislands; Suthirgarth (Graemsay) was 12½d. land; four—Leivsgarth, Cuthilgarth and Brusgarth (all in Sanday) and Thurkinsgarth (Sandwick) were 9d. lands; Harmannisgarth (Sanday) was apparently a 7d. land; Nistagar in Redland (Evie) was a 6d. land; Garth and Midgarth (Harray) and probably Calgair (Deerness) were 4½d. lands; Garth (Stronsay and South Ronaldsay) each 4d. lands; Midgarth (Stronsay), Garth (Westray), Negarth (Evie), Growagarth (Evie), Hamagarth (Evie), Midgarth (Rendall), Hawgengarth (Sandwick), Kirkgair (St. Andrews) and Suthirgarth (S. Ronaldsay) were 3d. lands; and 17 were of smaller dimensions . . . .

" Of all these, only Midgarth (Stronsay), Quoygarth (Stronsay) and Garth (St. Ola) seem to have been unskatted quoyland, approximately 1½ per cent. of the total valuation. And the comments already made on the general character and situation of the -land names hold good equally of these garths. With a few exceptions they are undoubtedly early settlements, though, again, not I think primary. Perhaps the most notable feature of this group is the number of personal-name prefixes in the Sanday and Sandwick examples. That is unquestionably significant if we could but read the riddle aright.

" In one respect, however, they differ considerably from the -land names —viz. their very uneven distribution. No fewer than a fifth of the whole number are in Sanday, and the skattable value of those nine farms was more than two-fifths of the whole garth total. Even allowing for the high valuation of Sanday as a whole, this is a remarkable proportion. Further, we may note that practically all the larger garth farms (or tunships) were situated in Sanday, Evie and Sandwick. Over two-thirds of the total garth skats came out of these three areas. The South Isles are relatively barren of garths."

5. O.N. *bólstaðr*, a farm settlement. This name, the third of the abovementioned trio, appears in Orkney usually in the form -bister, -buster or -bist. In one or two cases, e.g. the North Ronaldsay Busta or Buistie-tun (bøst·i) the origin may have been the synonymous O.N. *bústaðr*, but that hardly affects the general position.

Confining ourselves as before to the rental examples we have some 50 farm-names to consider, as follows :—

Of Kirbister 8 examples, of Braebister 4, of Wasbister 3; 2 each of Coubister, Easterbister, Evribister, Grimbister, Hobbister, Skelbister, and Warbister; and one example of each of the following : Aikerbister, Bimbister, Buista, Ellibister, Foubister, Grobister, Hensbister, Inyebister, Mirbister, Musbister, Nybister, Osbister, Rennibister, Sellibister, Simbister, Sketebister, Stembister, Swanbister, Tuskerbister, Westerbister and Ystabister.

Of these, eleven were whole urislands, one was 12d. land, ten were half-urislands, one was 8d. land, four were 6d. lands, one was 5d. land, three were 4½d. lands, eight were 3d. lands, and the remainder smaller or indeterminate.

In regard to their distribution, while practically every parish was represented, two in particular—Holm and Orphir—had a noticeably high proportion. In Orphir the various bisters totalled up to over 40 per cent. of the whole skattable valuation, and in Holm proper (excluding Paplay) the bisters amounted to practically 85 per cent. of the whole. On the other hand, the only -land name in these areas was Midland, an 8d. land in Orphir (the name itself implying a mid-division of a larger unity); and there was no garth in either parish. In Sandwick, where garths are common, the only bister lies at one extreme end of the parish. Such mutual exclusiveness affords strong confirmation of the validity of our previous impression that these three types of name are, in the main, more or less contemporary. Two of the bisters must of course be somewhat later in date because they were unskatted quoyland, but one of these—Kirbister (Birsay)—lies right away inland among the hills, and the other—Skelbister (Orphir)—was a small 1d. land on the outer fringe of Groundwater tunship —both obviously also from their location relatively late settled. The general location of -bisters, however, is very much the same as that of the -lands and -garths. They occupy relatively central positions in their various parishes; at all events they embraced good old fertile ground, and the large size of so many of them shows clearly that they were relatively early settlements, not fragmentary or residual areas which could be utilised at a later period. Individual examples of each type may indeed represent *original* Norse settlements, but, as we shall see presently, other facts tend to show that on the whole these three types, old and venerable as they are, must still be regarded as, to some extent at least, secondary.

Before leaving the bisters, however, we must refer to one very peculiar fact which has probably been noticed—the relatively large number having a Kirk- prefix. In addition to the rental eight mentioned above there are at least two which do not appear, one in Egilsay, the other in N. Ronaldsay, making a total of 10 all told. Orkney did not become officially Christian until Earl Sigurd the Stout was caught unawares by King Olaf Tryggvason and compulsorily 'converted' in or about the year 995 A.D. As all but one of the ten Kir(k)bisters were however fully skatted, they must date from before the period of King Harald Fairhair, a century or so earlier. How then is the existence of so many 'Kirk-bisters' to be explained in a land still heathen?

For an answer we may look farther north. In the *Landnámabók* we read that the Icelandic *Kirkjubólstaðr* was settled long before the official conversion of that land by a Christian settler Ásolfr who had emigrated thither from Ireland. And in the same earlier period another Christian emigrant Ketill inn Fiflske, who went out from the Hebrides, settled at a place which he named Kirkjubœr.

There is no reasonable doubt that the Orkney Kirbisters had similar origin, and the larger number in Orkney was probably due to its being so much nearer to the centres from which Christian influences were disseminated. The founders of these farms may indeed have been converted by Celtic clergy who were already in Orkney when the Norse first arrived, and who (if we may judge from the history and traditions associated with the Celtic church-sites) were certainly not exterminated. Or it may be that like the Icelandic Ásolfr and Ketill they had adopted the new faith in other Celtic lands. The exact place of their conversion is relatively unimportant; the point to be noted is that those Kirbister farms are not to be interpreted as late settlements, but rather as evidence that there were Christian Norsemen living in Orkney long before the official conversion of these isles. A possible further inference will be referred to later.

STATH, SKAILL AND BU

We have next to consider a second trio of names, each of which presents problems peculiar to itself, and is to be regarded as standing somewhat apart from what we may regard as the normal chronological sequence of farm settlement.

6. O.N. *staðir*, farm. This word, curiously enough a plural form, was the commonest of all elements in the formation of Norse farm-names at the period when Iceland was settled. In Norway itself there are reckoned to be some 2,500 examples, in Iceland 1165, and in each of these countries staðir-names are characteristic in two ways: (a) in the vast majority of cases the term is compounded with a personal name; (b) the farms so named are in general secondary or peripheral in the various districts where they are situated, i.e. for the most part they do not have the stamp of primary settlements.* With these facts in mind we may now examine the Orkney examples.

Today, with two exceptions, the term is represented by the spelling *-ston*, the exceptions (Costa and Yinstay) having *-sta* or *-stay*. In the early rentals, however, the spelling was *-stath* or *-staith*, the 'ai' being a Scots spelling for a long 'a' sound. In the 1595 R. the spelling was usually *-stane*, which indicates that the term was then being regarded as = English stone. The local form (pronounced -stən) is most probably to be explained as derived from the O.N. dative plural—*stǫðum*.

* *Nordisk Kultur: Stednavne.* 1939.

CHRONOLOGY 235

There are 23 examples in all, three of them obsolete, and they are distributed as follows:—

North Isles:—Knarston (9d. land) in Rousay.
Mainland:—Costa (18d. land) in Evie parish; Hourston (6d. land), Tronston (6d. land), Tenston (18d. land) in Sandwick; Knarston (4½d. land), Corston (4d. land), Grimeston (18d. land) in Harray; Germiston (6d. land), Colston (9d. land), Tormiston (9d. land), Clouston (6d. land), Unston (1½d. land) all in Stenness; Cairston (?d. land) in Stromness; Hatston (6d. land), Knarston (4½d. land), Berston (3d. land, *Jaddvararstaðir*—in St. Ola; Yinstay (8d. land), Campston (9d. land) in St. Andrews; Flenstaith alias Sands (9d. land), *Skeggbiarnarstaðir* in Deerness.
South Isles:—Herston (3d. land) in South Ronaldsay.

The first point to be noted about these names is the fact that in each case the first element is either certainly, or most probably, a personal name. Secondly, the great majority of the farms in question are either inland, or otherwise so situated that they give the impression of not being 'first settlements' in Orkney. Thus in both respects, nomenclature and situation, they are very much on a par with their namesakes in Norway and Iceland; they appear in fact to be as a class secondary.

A word of caution is however necessary as to the use of the word secondary here. It will be suggested later that the primary Norse settlements in many, perhaps most, cases embraced comparatively extensive blocks of territory which were later broken up into smaller units. Several of the farms bearing stath-names may well have been included in such original land-takings, but from their situation it is improbable that any one of them indicates the original 'focus' or parent house of such a settlement. And even if the lands of some of them did form parts of original settlements, their names at least would be of later date when such parts became separate farms.

In so far, however, as certainty is possible in such matters it may be deemed certain that the Orkney stath-names do not date back to the time of the original settlement. A glance at their distribution is sufficient to make that clear. No one would care to dispute that the North Isles were among the first to be settled from Norway; not only are these the nearest to Norway, but the unusually high skatting of Sanday is further evidence thereof. But neither in Sanday, or Stronsay, or Westray or Shapansay—all very fertile islands—is any stath-name to be found, nor in any of the North Isles at all except Rousay where there is one solitary example—Knarston. Similarly in all the South Isles only one example is known—Herston in South Ronaldsay.

With these two exceptions stath-names are confined to the Mainland of Orkney, and even there their distribution is surprising. Though the lands of a few do indeed abut on the sea-coast, they are singularly absent from the immediate vicinity of bays or natural harbours where one might expect original settlements to have been made. Costa, indeed, forms the extreme outer end of Evie parish, having a steep slope down northwards to a beach which is one of the roughest and most exposed in Orkney. On the other hand, there is an astonishing concentration in the middle of the West Mainland around the inland Lochs of Harray and Stenness. Here are no fewer than eleven stath-names, more than half of the present extant total, three in Sandwick, three in Harray and actually five in Stenness parish. Such concentration can hardly have been fortuitous, and of itself provides one of the strongest arguments against their dating from the earliest period: had they done so, it would be incredible that such large areas of Orkney should be without a single stath-name.

Nevertheless, they were all skatted units, and as such their *lands* must have been settled before Harald Fairhair's time. For a clue to their history we may turn to the three examples noted above as obsolete. In the 1500 R. we find in Deerness parish a tunship entered as Flenstaith; in the 1595 R. it appears as Flenstaith alias Sands. Thereafter Flenstaith vanishes, but Sands is still represented by a group of farms—Wester-, Easter-, and North-Sands. The name Sands—O.N. *Sandar*—is a characteristic early name-type, and it may be suggested that it was the original name of a settlement which was later renamed with a *staðir-name* which developed into Flenstaith, a name that survived for a time but never wholly displaced the earlier name, which has also long outlived it.

In the *O. Saga* we read of another farm—*Skeggbiarnarstaupum*—apparently in Deerness also, but the name is nowhere else on record, and it is quite unknown locally. No such farm can now be traced, either in Deerness or elsewhere, and we may thus suspect that what happened in the case of Flenstaith happened again here, and that the farm in question —wherever it was—now again bears, and for many centuries has borne, its original pre-stath name.

The data regarding the third obsolete name is not a little confusing, but the inference to be drawn therefrom is again the same. In Chapter 56 of the Saga (Nordal's edit.), which tells of various people of earls' kin who lived in Orkney in the time of Earl Paul Hakonsson (c. A.D. 1130) we read that Jaddvǫr, an illegitimate daughter of Earl Erlend, and half-sister of St. Magnus, lived together with her son Borgarr at Knarston (St. Ola). In cap. 66, however, we find that c. 1135 a man Arnkell and his two sons lived there. Then in cap. 76 we are told that Borgarr, who lived *at Geitabergi*, saw Sweyn Asleifson's ship (presumably passing through Scapa Flow) on its return south with the kidnapped Earl Paul. And finally, in cap. 94, which deals with events only a few years later, we find a man Botolfr as the 'buandi' at Knarston.

So many changes of occupants at Knarston within so short a time arouses the suspicion that confusion has crept into some Saga MSS., and that is confirmed when we find that in the old Danish Translation of the Saga Jaddvǫr (in the first reference above) is said to have lived—not at Knarston, but at *Jadvarstodum*. That name is otherwise unknown, but as Borgarr is said to have lived soon afterwards at *Geitabergi* we may suspect that these two names indicate one and the same farm. The latter name is now also unknown, but there is no doubt as to its being the farm now called Gaitnip, a synonymous name (O.N. *gnípa* and *berg* were both used of bold headlands). And Gaitnip, situated also in St. Ola, but on the other side of Scapa Bay from Knarston, and on a high promontory commanding an excellent view of nearly all Scapa Flow, would have been by far the most likely place in the whole district from which Sweyn's ship might have been noticed; whereas Knarston is on quite low ground near the shore with a comparatively restricted view.

We should thus seem fully justified in regarding Gaitnip and *Jadvarstodum* as alternative names for the same farm, and in that case we would have a third instance of a stath-name temporarily superseding an older name, which, however, had vitality enough to survive and outlive its competitor. And it may be incidentally noted that, as this stath-name presumably referred to Jaddvǫr (Borgar's mother), the formation of such a name as late as the 12th century was extremely unusual.

In the foregoing paragraphs an attempt has been made to set forth all the accessible pertinent facts about the Orkney *staðir* names; what are the inferences to be drawn as to their date and origin? It has been suggested that they indicate a second wave of emigration from Norway some considerable time after the first, but their restricted and peculiar distribution tells heavily against such a theory. In his *History of Orkney* Mr Storer Clouston expressed a different and more feasible theory. After referring to the two chief seats of the early Norse earls—at Birsay and Orphir—he wrote:

"Between these two 'bus' lay the *staðir* country, the one district of Orkney where personal names are found in plenty, the tract of land obviously —from the place-names—resettled at the period of the creation of the earldom. The presence of the two chief earls' seats, one at either end, provides at once the key to this late settlement here, and invests it with greatly heightened interest. Here evidently the earls' followers settled down and his hirdmen were given land, either as their odal own or in *veizla* (fief) as vassals, for their lifetime. In Grim, Geirmund, Thormod, Kol and the other names which are commemorated in the West Mainland townships, we see the long-forgotten swordsmen of Sigurd the Mighty and Torf-Einar."

That theory, so picturesquely expressed, is plausible indeed, and has much to recommend it, but it may be doubted whether it would cover all the facts. It might explain a renaming of farms, but why did not Sigurd or Einar place some of their trusted hirdmen in the more outlying and vulnerable islands of their domain? As yet, it must be confessed, no fully satisfactory explanation can be offered for those stath-names. The farms they denoted were manifestly settled at a relatively early date because they were skatted; and some at least were apparently re-named. But when or why that renaming took place must be deemed as yet quite uncertain.

7. O.N. *skáli*, a hall (see also below). The problems confronting us here are quite different, but hardly less obscure. First, however, let us look at the various occurrences of the term, most of which are still extant today, though a few are known only from records.

NORTH ISLES.

Sanday:—Langskaill (in Noutland? and in Gardemells tunships), Northskaill (in Burness), Backaskaill (in Southerbie), Skaill (in Evirbist), Skelbister (in Leyland).
Westray:—Skaill (in Aikerness), Langskaill (in Noutland Bewest and in Swartmeil bordland), Breckaskaill (in Rackwick).
Papa Westray:—Backaskaill and Breckaskaill (adjacent).
Eday:—Skaill (in Backaland).
Egilsay:—Midskaill and Netherskaill (adjacent).
Rousay:—Skaill (in Westness), Saviskaill (in Wasbister), Langskaill (in Libuster?).
Gairsay:—Langskaill and Skelbist.

MAINLAND.

Rendall:—Skaill (in Isbister).
Birsay:—Langskaill and Netherskaill (adjacent in Marwick), Langskaill (in Birsay Besouth).
Harray:—Langskaill (in Grimeston).
Sandwick:—Skaill (in an old tunship of Sandwick?), Langskaill (in Skorwell).
Stromness:—Langskaill (in Kirbister).
Orphir:—Skaill (in Hobbister and Tuskerbister), Skelbister (in Groundwater and Orphir tunship).
Holm: Skaill (in Paplay and Aikerbister), Skailltoft (in Hensbister).
St. Andrews:—Langskaill (in Toab).
Deerness:—Skaill (in Sandwick).

SOUTH ISLES.

S. Ronaldsay:—Rockerskaill (in Aikers?).

In three cases it can be assumed with complete confidence from the close juxtaposition of two skaill-names that they indicate a division of a single original skaill-farm: these are the two Papa Westray skaills, the two in Egilsay, and the two in Marwick, Birsay. Allowing for these we are left with a list of 35 names, fairly evenly distributed among the North Isles and the Mainland; the S. Ronaldsay example stands in sharp isolation as the solitary example in the South Isles.

It may appear strange that only about a quarter of those names appear in the early rentals, but the explanation is that they formed parts of tunships which are entered as whole units without mention of the component farms. It is indeed significant that the term Skaill seems never to have been applied as a tunship name. In point of fact, however, many or perhaps most of the skaills represented the chief farms in their several tunships, and a skaill farm is still today associated with the idea of good, old, fertile land. There is little doubt that most of them indicate some of the earliest lands in

Orkney to be cultivated by Norsemen, and the very name itself has still a certain air of dignity about it.

The next aspect of the Orkney skaills to be mentioned may perhaps cause surprise in some quarters, but the facts must speak for themselves. In his invaluable *Indledning* Rygh states that in O. Norse *skáli* was used sometimes in the sense of a hall or large room, sometimes of a detached building (alongside the regular dwelling-house) which might be used for diverse purposes, and sometimes of a house for temporary occupation some distance from a farm. And in Norway today the modern term *skaale* is used of a shed or suchlike building. What Rygh says of Norway applies more or less to Iceland as well, and when the word occurs as an English place-name Ekwall suggests no other meaning than hut, shed, or temporary building.

In Orkney the contrast is indeed striking. Obviously, when applied as a farm-name, the term must in the first instance have referred to the *farm-house*, to a building rather than to the land. But farms of the type just mentioned would have been disgraced by the application of a name signifying merely a hut or shed. Something far different is implied by the use of the name Langskaill, which occurs no fewer than twelve times. And the matter is placed beyond doubt by consideration of the saga-time occupants. Langskaill in Gairsay was the seat of Sweyn Asleifson who is said to have kept a bodyguard of 80 men about him! Sigurd of Westness, Earl Paul Hakonson's closest friend and counsellor, had his seat at Skaill in Westness, where the earl himself was on a visit at the time of his kidnapping. Amundi and his more famous son Thorkell Fostri lived in Sandwick in Deerness where the old head-house, adjacent to the parish church, is still called Skaill. Several of the head-houses in other tunships also had skaill-names, such as Skaill in Aikerness in Westray, Skaill on Skaill Bay in Sandwick, Skaill in Paplay—the estate of Sigurd who married Thora, widow of Earl Erlend and mother of St. Magnus. Men of the class just mentioned were not likely to have resided in huts, sheds or temporary buildings. In other words, an Orkney skaill indicated a dwelling-place of real dignity.

Mr. Storer Clouston in his *History* discussed the problem of the skaills also, and argued for a pre-earldom date on two grounds in particular—their practical absence from *staðir* tunships (where only two can be traced), and also from earldom bordlands (where only one is known). A *skáli*, he held, was evidently some new-fashioned kind of building, and he suggested that it represented the type of timber-building which superseded the older *dyngja* in Norway, and that the Orkney skaills were to be assigned to an early period when such buildings were something new and remarkable. By the date of the settlement of Iceland, he suggested, the *skáli* had ceased to be a novelty, and so hardly ever appears as a place-name there.

My own interpretation of the facts would be somewhat similar. Whether Mr. Clouston's idea that the *skáli* represented the innovation of a timber-hall is well-founded or not Norwegian archæologists are best qualified to determine, but that it denoted a new building of some distinction can scarcely be doubted. And I would suggest that most of the Orkney skaills indicate places where such buildings were erected in the earliest settlements immediately after the first phase of ' land-taking ' was over, or when the first sub-division of such settlements was taking place. The first settlements themselves would have been named *before* fresh buildings were erected thereon at all; hence the absence of skaill tunship-names. But when a new homestead was established it may be suggested that the new house not infrequently was of the type to be called a *skáli*.

Finally, reference must be made to the pronunciation of the word in Orkney. O.N. *skáli* should normally become ' skoli ' in the local dialect, and Scollay is in fact an old Orkney surname. But the local pronunciation of -skaill in farm-names is ' skil ' or ' skjil,' forms which are difficult to reconcile with an original *skáli*. The medial vowel must have passed through an intermediate stage of ' e ' (skel), and the subsequent change of e>i (or ji) presents little difficulty: the English word kale (cabbage) becomes in Orkney keel (kil, or more often kjil), pail becomes peel, hale—heel, tale—teel, etc. The prior change of O.N. á to e is more obscure, though it may be noted at *skáli* appears in English place-*n*ames as Scale (skel). But there is another possibility. Modern Norw. *skjæle*, shed, hut, etc., presupposes an O.N. *skæli, a form which may have been current in Orkney, and which would readily pass into skel (skjel), and thence into skil (skjil). But it has to be confessed that the whole history of the phonetic change is obscure.

---

8. O.N. *bú*, farmstead, estate, etc. In old records and rentals this word was often anglicised to ' bull ' on the false assumption that it was a Scots form like ' foo ' for ' full,' ' poo ' for ' pull,' etc.; locally, however, the pronunciation was, and is, simply Bu. This word, which is common as a farm-name in Orkney, differs from other farm-names in being practically a generic term. It is always accompanied by the def. article—*The* Bu, and to complete the name the words " of Rapness," " of Hoy," etc., are added, though these are normally omitted when one is referring to the farm in its own immediate vicinity where there is no other Bu.

A hint as to what the term signified is to be found in the 1614 Rental printed by Peterkin. Under the parish of St. Ola we read : " Nota— Newbigging, Foirland, Corss and Archdeanis Quoy labowrt in ane *bow*, and never payit gersum (grassum)." These were adjacent farms which had been united by Earl Robert Stewart to form The Bu of Corse which is mentioned in that earl's Testament (1592) as one of the two Bus actually in his own hands. And an éven more explicit statement appears under South Sandwick : " Bull of Skaill, 8d. terre pro Epo. . . . This was sett be the ald Earle in 19 yeirs takis to umql. Mans Sinclair, and yreftir laborit in a *bull* be the Earle, swa never payit gersum." In other words, at that period a bu seems to have been a farm of some considerable size, farmed as a whole unit and not in runrig with others, and (as shown by immunity from grassum) not even let to a tenant.

The two bus just mentioned were born, however, out of due time, so to speak, and were in a sense artificial. For the genuine old bus we must turn to the earlier rentals. There, among the bordlands or private earldom estates, we find no fewer than eleven : The Bu of Orphir, of Cairston, of Hoy, of Burray and of Rapness, and actually six in the one island of Sanday—Halkisness, Brough, Tresness, Lopness, Waa and Tofts. In the Bishopric part of Orkney were the Bu of Sound in Shapansay, of Noltland (Westray), of Roithisholm (Stronsay), of Flotta, and of Egilsay; while of udal-owned bus may be noted those of Skaill in Paplay, of Ireland, of Rendall and of Wyre. Besides the above there are a few others, known today simply as The Bu—without territorial addition—though each no doubt must formerly have been The Bu of some particular area.

Reference should here be made to a paper by Mr. Storer Clouston on *The Orkney Bus*,* in which the subject is discussed with the author's accus-

* P.O.A.S., V.

tomed acuteness and insight. From a careful analysis of the data he distinguished four features which he regarded as characteristic of the various Bus :—

1. Relatively large size. Each of those in Sanday was a whole urisland, while most of the others were exact half-urislands. That means that in former days each was a very substantial farm.
2. Wherever the full circumstances are known each had a few smaller satellite farms attached to it. In the 1492 R. those attached to the Bu of Hoy were termed 'unsetts,' and those of the Bu of Rapness 'Quyunsetts'; in the 1500 R. those were all termed 'umbesettis,' and Mr. Clouston believed the real name to be 'umbesetters,' though there I think he was mistaken. These satellites were probably designed to furnish additional farm labour at special seasons. (Cf. the 'oncas referred to above, p. 222.)
3. and 4. His third and fourth criteria were less important—the presence of a kirk or chapel in each, and proximity to the sea.

Perhaps the most remarkable fact about the name is that it seems to be practically confined to Orkney. From Iceland Finnur Jónsson could cite only three *bú*-names, and in these the term was a suffix; Matras cites none from Færoe; in Shetland Bu appears only as a prefix, and would seem to imply farm-stock rather than a farm; even in Norway itself Bú is by no means abundant. And in none of those areas does one seem to find the peculiar Orkney formula—The Bu of X.

Of the bishopric bus in Saga times we know nothing, but each of the udal bus was the seat of a *gœðing* or man of chieftain class, and the various Earls' bus were no doubt granted in *veizla* to trusted followers or hirdmen. In that connection Mr. Clouston's remarks in a Postscript to his paper must be quoted verbatim.

" When this paper was read before the Society the question was raised by Dr. Marwick : How did the word 'bu' come to be employed with this peculiar significance in Orkney? After thinking over this point, and discussing it several times with Dr. Marwick, the following seems to be the likeliest explanation :—

"There were two kindred words, *bú* and *bœ*, used interchangeably in such texts as Heimskringla and Hakon's Saga: e.g. one finds the *bú* of Husabœ, and the *bœ* of Vettaland—both baronial seats. But in Orkney there seems to have been a distinction. In the Orkneyinga Saga we have *bú* 17 times, always applied to an earl's or chieftain's manor; while *bœ* is found 6 times, in every case applied to a place not belonging to an earl or chief. This can scarcely be a coincidence: the instances are too numerous. It seems to point very decidedly to the conclusion that in Orkney *bú* was the aristocratic, fashionable word, and was used only of big places. Thus as early as the Saga period it had come to have a significance very much equivalent to 'manor' in England.

"But there is nothing corresponding to the prefix "Bu of —" forming part of a place-name. This must have been a Scotch designation invented by the Scottish chamberlains or factors of the earldom estate (probably soon after 1379 when the Sinclairs became earls) in order to distinguish these large manorial farms. They found bu in use in Orkney as the regular term for such places, and they used it just as they would have said "Mains of," "Place of," etc., in Scotland.

"One final addendum must be made to this paper. Only after it was written did I come across the following passage in Prof. Taranger's *Udsigt over Den Norske Rets Historie*, II, p. 313, in the section dealing with the King's and Crown estates in Norway:—

"'The Crown's permanent estate was partly king's-farms (*konungs-bu, hǫfuðból*), partly leasehold farms, partly town-property. . . To the king's-farms (head-farms), which were originally managed by bailiffs, there were attached evidently from ancient times a greater or less number of leased farms which were let out by the manager of the king's-farms, and the profits of which were paid to him.'

"Here we have precisely the same arrangement found in Orkney in the form of the great Bu with its umbesetters. It is satisfactory to have dug this Orkney arrangement out of our local records independently, and to find it thus confirmed."

While Mr. Clouston's diagnosis is sound in the main, his deductions as stated above from the use of *bú* and *bœr* in the O. Saga are rather too sweeping. Actually, each term occurs about 30 times in the Saga, though he was perhaps referring to the instances when the words were used of places in Orkney only. But with regard even to these a careful scrutiny reveals that *both terms* are actually used of the earl's residence in Orphir, and of Sweyn's home in Gairsay. The several instances are worth noting.

1. Orphir. "i Aurfioru at bui Harallz jarls." (O.S. p. 126); "at bui sinu þvi er heitir i Aurfuru" (p. 166); "ok ætlaði i Aurfuru til buss sins" (p. 285); "i Aurfuru til buss jarls" (p. 287).

    But on the famous occasion when Sweyn slew his namesake there it is mentioned that "þar var mikill *husabœr*" (p. 168); and Sweyn escaped "a bak bœnum" (p. 171).

2. Gairsay. "þa for hann heim i Gareksey til buss sins" (p. 202); "ok for heim i Gareksey til buss sins" (p. 229); "til Garekseyjar til bus

Sveins " (p. 294); " Sveinn for heim i Gareksey til bws sins " (p. 318).
But when Sweyn was seeking Earl Harald who had seized his *bu* in Gairsay he was so angry that he wanted to set fire to the hall " ok brendi *bœinn* ok jarl inni " (p. 295); he ultimately became aware however " at jarl var eigi á *bœnum* " (p. 295).

From these instances of the use of the two words may be drawn a distinction which holds good in all the Saga references. *Bú* is used in a general sense of 'residence' or 'estate' or even 'home,' while *bœr* is applied more concretely to the actual buildings at the place. And just as we should not normally speak of a residence without the associated idea of its owner or occupant, so when a *bú* is mentioned it is always *somebody's bú*. A *bœr* can be mentioned, however, apart from the thought of the occupant, just as we might speak of seeing or coming to a 'farm.' But it should be added that, in the Saga, *bœr* is used also in a wider sense sometimes; it is used of the houses that collectively go to form a town, e.g. Bergen or Oslo.

In conclusion it should be noted that this further scrutiny of the Saga use of the words in no way conflicts with, but only confirms, Mr. Clouston's contention as indicated above—that a *bú* was a residence or estate of somebody, and *that* a somebody of importance. Originally it was, strictly speaking, not a farm-name at all, but rather a descriptive term. Mr. Clouston's comparison with " The Mains of " or " The Manor of " is exactly to the point, though the latter is an over-pretentious term to use in regard to Orkney farms.

In our discussion thus far of Orkney farm-name classes we have seen how at least two—the quoys and the setters—may be definitely regarded as relatively late settlements; and as for the remainder various facts have been adduced which seem to indicate that while certain individual farms so named may represent original Norse settlements, yet all the name-types referred to denote farms which are to a greater or less degree of secondary or derivative origin, usually representing sub-divisions of earlier undivided units. We have now reached a stage where we have to cast around for settlements which have no such secondary or derivative stamp, in other words—territorial units which may with some measure of confidence be regarded as original Norse settlements. Of such there are two main types now to be considered.

9. O.N. *bœr*. This, if not the oldest of all, was a very old type of farm-name in Norway, and it has been already referred to in connection with *bú*. The old Danish form of the word was *býr*, a term whish appears so often in the parts of England overrun by the Danes, and which is now usually represented by the ending -by, e.g. Whitby, Grimsby, Kirkby, etc. Though such names may now denote a town or village, the original sense was that of a farm-settlement.

In Orkney such names are by no means rare, but it is significant that many do not appear in the early rentals at all. Some still denote large important farms, but others now survive merely as names applied to fields, an old house-site, a boat-noust or even a small district—the farm association being entirely forgotten. The explanation of course is that the original units have been broken up, fresh names have been applied to the component parts, and the original name has survived only by chance, if at all, to denote some feature of the former unity.

The spelling of the name now varies greatly—Bae, Bea, -by, -bay, -bie, etc.; here, however, when referring to the name in a generic sense we shall keep to the form 'bae.' In Stronsay there are three certain, and possibly four, examples of the name, and the areas covered include the bulk of the old arable land in what we may term the main block of the island. In its south-east quarter is Houseby, which formed a whole urisland by itself, and is still one of the largest and best farms in Orkney. In the south-west quarter lies the present large and excellent farm of Holland, which was another urisland. On this farm a field situated near an old boat-noust goes by the name of Erraby, which would seem to represent an O.N. *eynar-bœr*, 'beach-farm,' and may possibly indicate an original name of the settlement here. The north-east quarter of this island block forms a district locally known as Everby, i.e. O.N. *yfri-bœr*, 'upper b——,' and though that name does not figure in the rentals the constituent farms—Odness, Airy, Hescum, Dirtness, Musseter and Kirbister—totalled up to exactly 18d. land or an urisland. There was also an old house in this area called Everby which is found in records, and was important enough to be entered on Blæuw's Map of Orkney c. 1650. And in the remaining fourth or north-west quarter is a farm still called 'The Bay.' That name is similarly absent from the rentals, the farm having been included in the 24d. land tunship of Aith. But adjacent to this farm was the old parish church of St. Mary's, a clear enough indication of its status in earlier Norse times.

Sanday can also boast four examples of the name. Bae itself no longer survives as a farm-name, but Bae Ness, Bae Sand, Bae Loch and the Mill of Bae—all in the immediate vicinity of the present pier of Kettletoft, the chief landing-place in the island—unite in testifying to a former Bae settlement, the main part of which figured in the 1500 R. as the 1½ urisland of Housgarth. To the west or south-west of this was the adjoining urisland tunship of Southerbie, a name now entirely forgotten locally, the greater part of it composing the present farm of Backaskaill. As pointed out in Part I there is a strong presumption that these adjacent units Bae and Southerbie originally composed one large settlement unit, which would have embraced all the lands around the bay now called the Bay of Backaskaill.

On the east coast of the island the long sandy isthmus which connects the peninsular farm of Tresness with the mainland of Sanday is known as the Sand of Erraby—the same name, it may be remembered, as is applied to a field of Holland in Stronsay. Here there has been a vast amount of sand-blowing, not to speak of land-submergence, and it is impossible to tell how much arable land has thus disappeared. But the name Erraby is suggestive, and may have denoted an early settlement which embraced the present farm of Tresness also. If that be so, the fourth Sanday bae might form a parallel, as a park on the present large farm of Lopness—a former urisland unit near the Start Point—is still known as Okkamby, the first element of which is of uncertain origin though it might probably represent the personal name Hákon.

In Westray the extensive peninsula running north-east from Pierowall included two old tunships, Rackwick a 2-urisland unit, and Aikerness a 25d. land. Where these two meet is a large, sheltered, sandy harbour, or 'oyce' (O.N. *óss*) as it is called locally, where vessels could lie in safety, and immediately above is the farm of Trenaby which formed the north-eastern end of Rackwick. Here again we have a strong presumption of an early bae settlement, which might have included the two tunships just mentioned. And on the opposite side of the island on the southern slopes of Fitty Hill lies the old tunship of Midbea (O.N. *mið-bœr*), a name which plainly indicates the middle portion of a larger unit. Adjacent on the west was the tunship of Noutland-Bewest, and that on the east can only have been the tunship appearing in the rental as Tuquoy and Are. I have discussed these units at some length *sub* Westray in Part I, and further comment is thus unnecessary here.

Passing over to Rousay we come upon what is probably the best attested example of a vanished bae. In the Sourin district, which appears to have included 1½ urislands, the 1595 R. specifies an unusual number of neighbouring farms each skatted as a 3d. land unit—no fewer than seven in all. Nothing could more plainly suggest sub-divisions of a larger unit, and when we find that certain fields on the present farm of Hurtiso in the heart of this area still go under the name of Husabae there is little room for doubt as to the old name of that larger unit. Furthermore, in one of those fields there used to be in my boyhood days a small mound left uncultivated and called the Taft o' Husabae (*taft* = *toft* or old house-site). That mound has since been ploughed over, and I was informed by the farmer's son that once when thinning turnips at this place his hoe accidentally slipped down through a hole in the ground, and he was able to push it down the full length of the hoe-handle. Here with little doubt lie the ruins of the old head-house of Husabae, a territorial unit which was divided up so long ago that the name even is not mentioned in 16th century records.

In Shapansay a farm Housebay is situated amid rich old fertile lands on the south side of the island, but today it is only a small farm, and it was not mentioned in the rentals either. It represents in all probability only a fractional part of what once bore the name. And in this island there is a third example of the name Erraby which was noted in Stronsay and Sanday. In this case it is not applied to a farm any more than the others, but here, significantly enough, to an old boat-noust near the northern tip of the island—The Noust of Erraby.

The examples so far noted all occur in the North Isles, but on the Mainland there are many others. In the heart of the valley known as 'Abune the Hill' in Birsay (which together with a neighbouring valley of Swannay composed an urisland) is an old 3d. land farm called Bea. From its situation there is little doubt that it represents the old head-house of the valley, an inference strongly reinforced by the presence of an old chapel-site. In

Birsay also, on the southern slope of Greeny Hill, is the old tunship of Beaquoy. This is an interesting example of how the name of an obviously secondary and outlying quoy-farm has supplanted the parent name of the tunship, for the earlier bae of which this was a quoy-extension may still be found in Houseby, a farm more centrally situated on the shore of the Loch of Sabiston with the old urisland chapel adjacent.

In Harray parish we have another farm Bea to record, situated centrally in the populous old tunship of Nether Brough, while over the border in Sandwick at least four baes can be still identified. In Skaebrae tunship there is one named simply Bea, and elsewhere in Sandwick are two farms, each called Bain (pronounced as a disyllable—bɛ·ən), a name which points to an O.N. *bœ[r]-inn*, 'the b——.' And on the exposed western seaboard of Sandwick is the fourth bae—Yesnaby (formerly Yescannabie), of which the first element is obscure but perhaps a personal name.

In the midst of the large 4-urisland area of Inner and Outer Stromness there was a 3d. land farm Bea, and another of the same name is situated in the heart of the urisland tunship of Ireland in Stenness. In Firth is the 4d. land of Binscarth (Benyeskarth 1500) situated in a very prominent 'skarth' or gap in a hill-ridge, the name indicating almost certainly 'the bae in the skarth.'

I am unable to adduce any example of a bae from the parishes of Evie, Rendall, St. Ola, Orphir, or Holm, but in St. Andrews we still have the famous old half-urisland farm of Sebay (sea-bae). In Deerness in the tunship of Skae is a Netherby and the site of a former Cataby—parts no doubt of a larger unit, while on the farm of Stove in Sandwick tunship a field is still known as The Bay.

From the South Isles only two cases can be cited—Chuccaby in Walls, and Sebay in Herston, South Ronaldsay. The former is situated near the border between two old tunships Wards and Kirbister, and the latter would seem to have been overshadowed by a later stath-name—Herston.

There are several other farm-names which might more doubtfully contain the element *bœr*, e.g. Bimbister, Beboran, Beafield, etc., but such names we may for the time ignore. Restricting ourselves to those mentioned above we have a couple of dozen or more which may with confidence be referred to a *bœr* origin, distributed fairly evenly through the North Isles and Mainland, and with two representatives at least in the South Isles.

Between those on the Mainland and those in the North Isles a certain distinction may be noted. Most of the Mainland baes are now comparatively small farms; but in general each is so situated that it gives the impression of having been the original core, so to speak, of the tunship or district where it lies. And it is probably significant that such tunships or districts do not in general bear names of the settlement-type—bister, land, garth, etc.; for the most part the names are topographical—Abune the Hill, Stromness, Skaebrae, Nether Brough. Ireland is a notable exception, but even there the fact that the farm in question is named simply Bea, without any qualifying adjunct, would seem to imply that when the name was given there was no other farm there from which there was any need to distinguish it.

In the North Isles on the other hand the bae names generally speaking are either still applied to large units such as Houseby and Everby in Stronsay and Midbea in Westray, or we may reasonably infer that they once did so, as in the case of Bae in Sanday and Husabae in Rousay. And of bae-names as a whole it may be said that wherever such a name occurs we can in nearly every case recognise the existence of a comparatively extensive early settlement which either has had a bae-name itself, or in which one of the earliest homesteads bore such a name.

As to their *relative* age, it is instructive to note how other early types of names appear so often in an obviously secondary relationship to the baes. In Everby, Stronsay, one of the constituent farms is a Kirbister, situated on the outer fringe of the tunship. In the same isle the tunship of Aith, in which we noted The Bay as the principal farm, is bounded on the south by a small bister tunship—Grobister, which formed the southern boundary of St. Mary's Parish. In Sanday, immediately adjacent to and inland from the old Bae settlement is a farm Skelbister, and though that farm was part of a -land tunship—Leyland, it is not improbable that Leyland itself may have been included in the original Bae. In Westray, another Kirbister lies in an isolated position at the extreme outer end of a cultivated coastal strip, and just beyond the confines of the very large settlement of which Midbea was the middle portion. In Rousay two -land farms—Bigland and Broland—form parts of what we concluded was the old Husabae. In Sandwick one of the two Bains was surrounded on one side by four -lands—Brekisland, Eriksland, etc.—, and on the other by two garths—Housegarth and Negarth; while the second was partly ringed by several garths. The only instance I have noted where a bae stands in a secondary relationship to another type of name is Bea in Ireland (Stenness), referred to already, and in that case there is the possibility that Ireland has been applied as a topographical rather than a specifically farm-settlement name. In any case, that one possible exception serves but to throw into sharper relief the indubitable fact that the Orkney baes must be regarded as the most characteristic of the primary settlement-names.

Before leaving this class reference must be made to one circumstance which may perhaps be significant. In discussion of the bister group attention was drawn to the surprising number having the prefix Kirk-, and an explanation was suggested for the fact that such names must date from a period prior not only to the official conversion of Orkney to Christianity, but even to the imposition of skat a century earlier. We saw too that, in Iceland, besides a *Kirkju-bólstaðr* there was a *Kirkju-bœr*, but among the numerous Orkney baes there is no 'Kirkby,' and one may wonder why. No certain explanation is possible, for in Orkney farm-names a Kirk- prefix is, with one possible exception, confined to bisters—a very remarkable fact in itself. But it may perhaps be suggested that the baes, dating as it would appear from the very earliest phase of settlement, were established before any Norseman had been long enough in touch with Christianity to adopt

the new faith; whereas the bisters may in general date from a period
slightly later, but before the imposition of skat, in other words—roughly
the 9th century, during which time Christianity slowly began to take root
among the Norse settlers.

It should not be assumed, however, that all primary settlements in
Orkney are to be associated with the term bae. It has above been pointed
out repeatedly that individual cases of the bister, land and garth types may
date back to the earliest stage of settlement also. But apart from such
possible cases we have to take note of a much more certain group. Many
of the old seaboard tunships bearing names of a geographical nature have
undoubtedly to be classed among the very earliest settlements. As examples
of such we may mention Elsness (Sanday), a peninsula by itself, Rothies-
holm (Stronsay), a similar peninsula unit, Huip (Stronsay), Waa (Westray),
Westness (Rousay), Sound (Shapansay), Sandwick (Deerness and South
Ronaldsay), Ronsvoe (South Ronaldsay), the old tunships of Orphir,
Stromness, Firth, etc. In two of these the 'focal' farm was a Skaill, and
in another there was a Bea. All these, and no doubt others of the same
character, must be regarded as having an ancestry every whit as ancient
and venerable as the baes.

In the preceding pages an attempt has been made to examine all the
available data in regard to the main types of farm-settlements in Orkney,
and to ascertain so far as possible their relative, if not absolute, chronology.
In such a quest it is obviously futile to expect to attain complete accuracy
or precision. Apart from the lack of adequate data, in the application of
names the human element enters, and there was no doubt not a little over-
lapping. Any single name-type would have had its period of greatest
efflorescence, though individual examples might well date both from before
and after that phase. But the facts, such as they are, must speak for themselves, and the sequence suggested cannot be very far wrong.

Working backwards in time we have seen that of all the types considered
the quoys must be regarded as the latest, though the use of the term
extended over 500 years or more. The setter group, judged by the fact that
only about half of them were skatted, would appear to be next youngest—
dating both from before and after the beginning of the 10th century. One
interesting group, the staths, appear to indicate a renaming of farms
previously settled, and if that be the case they probably date from the 10th
or 11th century, though one such renaming would appear to have taken place
as late as the 12th century. The 'bus' also form a group by themselves
and can hardly be reckoned in any normal sequence: we have really no
data on which to determine exactly when the peculiar formula "Bu of x"
first came into use, though the farms so named had no doubt been settled
from a much earlier period.

The remaining types—lands, bisters, garths, skaills, baes—must all with
rare exceptions be assigned to a period prior to the 10th century, and it is
impracticable to generalise as to their relative ages, as individuals of each
class may date back to the time of the first Norse settlement or immediately
afterwards. It has been suggested that the skaills may represent the earliest

homesteads in such original settlements, but it might be unwise to stress that view unduly. As for the bae group and sundry more or less self-contained seaboard units, we have seen that there is every reason to regard these as primary Norse settlements, or at least the core of such settlements, and as such they must be ranked highest of all in the scale of ancestral dignity.

Having reached this stage in our survey we may now ask ourselves: Are those relatively large settlements just mentioned, the baes and their seaboard contemporaries, to be regarded as actual settlements of individual Norse invaders similar to the 'land-takings' of individual *landnámsmenn* in Iceland of which we have an authentic record in the invaluable *Landnámabók*? There is good reason for believing that some of them—perhaps a good many—were. But Iceland was a much bigger apple to bite into than Orkney, and the individual 'land-taking' of an Icelandic settler was huge as compared with an average Orkney tunship having a coastline of only a mile or two in length. Conditions however were very different. Iceland was a relatively barren and uncultivated land, whereas Orkney was already (to a considerable extent at any rate) a fertile arable land in which a much smaller area would have been sufficient to maintain men and beasts. That fact of itself would go far to explain smaller settlements in Orkney.

Yet even here in Orkney, underlying individual tunships we can occasionally catch glimpses of larger units. Reference may be made to entries in Part I under Bae and Stove (Sanday), Noutland Bewest (Westray), Westness (Rousay), Evie, Rendall, Stromness and Orphir Parish Names, etc. In each of these cases we have hints of a unit which embraced much more than a single rental tunship, and there may well have been others which are no longer discernible. Such larger units have not been noticed, or at least not referred to, by any previous writer, and in a field where inference plays so large a part it is unsafe to dogmatise. But keeping in mind the practical certainty of the existence of prior Pictish tunships in Orkney it would be foolish to imagine that every Norse invader would have been content to restrict his 'land-take' to one tunship. Those Norsemen would not have been all of the same status; some would have had many more followers in their train than others, and their several land-takings or settlements would have varied in size accordingly.

Nor can the native Picts be left out of account. Some scholars have indeed held that these isles were practically uninhabited when the Norse arrived, but such a conclusion is quite unwarranted and based largely on a *de silentio* argument as to any opposition encountered by the invaders. But as I have written elsewhere the Ninianic missionaries were not sent to Orkney to preach to seals or seagulls. We must assume that there *was* a Pictish population here, and by reason of that fact the Norse settlement of Orkney would have differed considerably from that of Iceland. But the latter settlement was headed by men of chieftain or semi-chieftain class, and leaders of the same type would have been all the more necessary to obtain mastery in Orkney.

Apart however from fighting altogether, circumstances of themselves must have prescribed such folk-leaders. In my *Place-Names of Rousay* I have already referred to that aspect of the settlement. "How indeed," I wrote, "could it have benn otherwise? Few, if any, small farmers could have had the means to transport their goods and families overseas by themselves. To have a vessel capable of such an undertaking would of itself imply a man of some wealth and substance." It was men of that class who were the Icelandic *landnámsmenn*, and it is only reasonable to suppose that the settlement of Orkney would have been undertaken by men of similar type. Such men would have been accompanied by lesser kinsmen and dependants, who, once a settlement had been staked out, would no doubt have had land therein allotted to them on which to build houses and grow crops for their families, but the original 'land-taking' would have been the responsibility of the folk-leaders or chieftains themselves. And it is probable that those larger units, of which glimpses only can be caught in the pages of our rentals, may indicate the land-takings or settlements of some of the more outstanding Norse *landnámsmenn* in Orkney.

Before concluding this survey some reference must be made to two Old Norse terms *vin* and *heimr* which occur frequently in Norwegian farm-names dating from a very early period. Jakobsen considered that both terms were represented in the place-names of Shetland, and argued in consequence that those isles had been settled as early as 700 A.D. Commenting thereon in the *Stedsnavn* volume of *Nordisk Kultur* (1939) Professor Magnus Olsen states: "It also deserves attention that *vin* appears in a series of Shetland place-names, most frequently as first element . . . once uncompounded, and twice as second element in field-names but not in farm names. . . .' And he adds: "What interests us most of all is *-vin* as the second element in farm-names. Later investigations have confirmed Oluf Rygh's opinion that farm-names in *-vin* [i.e. as second element] belong to the prehistoric period." Rygh indeed believed that, so used, *vin* had become obsolete by the beginning of the Viking Age.

In Orkney I have been unable to discover any instance at all of *heimr* (home) in an old farm-name, but the position is different with regard to *vin*. That term signified good pasture or grassland, and as a first element it is by no means rare in Orkney farm-names as a glance at the index may show. Sometimes it may appear as Win[d]-. It is doubtful, however, whether any such farm-name dates back to the earliest period, and it may be suggested that as a generic term *vin* continued in use long after it had ceased to be applied in the old manner as the substantive second element in farm-names. But in Orkney we have at least one old farm-name—Lyking (Sandwick)—where it can hardly be doubted that *vin* has been applied in the early manner. The name must certainly represent O.N. *leik-vin*, 'sports or games *vin* or field,'—a name that occurs over forty times as a farm-name in Norway. Another example which may possibly contain *vin* as second element is Greeny (Grenying, Grenyng, Grynning (1492); Grenie (1500)), an old tunship in Birsay; but that is more doubtful, as it may probably represent an O.N. **grœningr* (green place).

There are actually two other Lykings in Orkney, one in Holm, the other in Rendall parish, but neither can claim to be classed with the really early farm-name group. They are both smallish farms situated on the outskirts of a tunship and almost certainly of relatively late date. These I should be disposed to place on the same footing as Lakequoy, Leaquoy, etc. (lek·wi), a name to be found in several parishes and islands. In such names *kvi* would seem to have taken the place of the earlier *vin*, and they probably denote places where sports of the kind favoured by Norsemen in olden time —ball-games, horse-racing, etc.—were carried on, long after the days of the early settlement.

INDEX I

FARM-NAMES

(Island and Parish Names entered in thick type)

A

Affall, 187
Aglath, 111
Aikerbister, 92
Aikerness, 31, 124
Aikers, 87, 104, 169
Aikerskaill, 76
Air, 182
Airafea, 23
Airan 17
Airy, 6, 23, 38
Aisedale 96
Aith, 23, 148, 182
Aithsdale, 182
Ancum, 1
Anderswick, 111
Angusquoy, 90
Annynsdale, 90
Antabreck, 5
Appiehouse, 17, 112, **117**, **143**, 157
Appietoon, 121, 144, 157
Archdeansquoy, 54
Are, 31
Arian, 165
Arneip, 169
Arp, 187
Arsdale, 124
Arstais, 6
Arwick, 127
Atla, 169
Auskerry, 23
Avaldsay, 61
Avidale, 121
Ayre, 76, 94

B

Backakelday, 94
Backaland, 48
Backarass, 38
Backaskaill, 17, 45
Backbighouse, 96
Bain, 148
Banks, 29, 61, 90, 104, **120**, **131**, 187
Barnettsdeall, 92
Barnhouse, 112
Barswick, 170
Bay, 29
Bea, 112, 131, 144, 148, 162
Beafield, 18
Beaquoy, 131
Beboran, 144
Beneath the Hill, 179
Benzieclett, 148
Benziecot, 18, 38
Benzieroth, 114
Berriedale, 38
Berstane, 96

Biest, 144
Bigbreck, 139
Bigging, 38, 54, **139**, **154**
Biggings, 87, 175
Bigland, 61
Bigswell, 110
Bimbister, 141
Binnaquoy, 114
Binscarth, 114
Birsay, 130
Bisgarth, 127
Blackywall, 187
Blanster, 170
Blomuir, 92, 187
Blubbersdale, 121
Boardhouse, 132, 144
Boloquoy, 18
Boondatoon, 29, 87
Booth, 182
Boray, 75
Borwick, 148
Bossack, 87
Botulfsyord, 148
Bow, 132
Bowbreck, 162
Bowbustirland, 162
Braebuster, 6, 31, 76, 179
Braeswick, 18
Brain, 175
Brance, 175
Brandyquoy, 80
Brawel, 54
Breck, 5, 6, 45, 61, 104, **122**, **132**
Breckan, 18, 66, **117**, **127**, **144**
Breckaskaill, 38, 45
Breckawall, 38
Breckness, 162
Breckquoy, 90
Brecks, 24, 38, 59, 76, 175
Bredakirk, 48
Breek, 66
Brekisyord 149
Brendale, 61
Bressigar, 6
Brettabreck, 122, 132, 162
Brettoval, 144
Breval, 66
Brims, 182
Brinnigar, 162
Brockan, 132, 149, 165
Brogar, 112
Broland, 61
Broomquoy, 132
Brough, 7, 31, 62, 170
Bruar, 127
Brunthouse, 187
Bu, 31, 73, 90, 104, 162, **167**, **175**, **179**, 181, **184**
Bu-house, 59

Buckquoy, 132, **144**
Bught, 121
Burgar, 127
Burness, 7, 39, 115
Burray, 167
Burrogarth, 31
Burrowland, 170
Burrowston, 54
Burrowell, 170
Burwick, 170
Busta, 1
Butterland, 132
Button, 94, 112

C

Cairston, 162
Caldale, 96
Calgarth, 76
Calset, 87
Cameraljoy, 187
Campston, 83
Canker, 18
Cannigill, 96
Cantick, 184
Caperhouse, **144**
Cara, 170
Carabreck, 83
Carlquoy, 90
Carness, 101
Carpaquoy, 48
Carrick, 48
Caskald, 54
Castlehoan, 73
Castlewell, 187
Cataquoy, 29
Cattaby, 81
Cava, 109
Cavan, 139
Cavit, 73
Chenziebreck, 90
Chuccaby, **184**
Citadel, 162
Claisbreck, 179
Clamer, 7
Classiquoy, 66
Cleat, 7, 24, 32, **96**, **170**
Cleatfurrows, 133
Clerksquoy, 54
Clestran, 24, 104
Cletts, 171
Clifton, 39
Cloke, 133
Clouk, 124, 162
Clouster, 39, 165
Clouston, 110
Clova, 97
Cloviger, 163
Clowally, 107
Clumlie, 149
Clyver, 66
Cockburnsquoy, 87
Cogar, 62
Coldamo, 112
Colligar, 7
Colster, 77
Colston, 110
Comely, 87
Congesquoy, **166**

Conglibist, 1
Consgar, 150
Conyer, 141
Conziebreck, 150
Cooan, 39
Copenago, 87
Copinsay, see Addenda
Cornquoy, 90
Corrigill, 141, 181
Corse, 62, 97
Corston, 142
Costa, 124
Cott, 32, 52, 66, 71, **92**, **133**, **168**
Cottaskarth, 120
Coubister, 104, 115
Coullis, 171
Crantit, 97
Creabreck, 94
Crearhowe, 95
Creya, 81, 104, 127, **163**
Crismo, 127
Crofty, 88
Crotrive, 18
Croval, 150, 166
Crowald, 39
Crowall, 104
Crowrar, 122
Crowsnest, 187
Cruanna, 29
Cruan, 117
Cruannie, 66
Cruar, 66
Cruddy, 18
Cubbygeo, 39
Cufter, 128
Culdigo, 71
Cultisgew, 54
Cumlaquoy, 133
Cumley, 150
Cummaness, 112
Cupper, 128
Cuppin, 45, 112, 144
Curcabreck, 122, **144**, **150**
Curcasetter, 18
Curcum, 133
Curquoy, 39, 66, 125
Curries, 187
Cursetter, 115
Cusvie, 49
Cutclaws, 67

D

Dale, 67, 128, 163, 179
Dam, 187
Damsay, 39, 117
Deerness, 76
Deasman, 163
Deesbreck, 139
Deldale, 77
Denwick, 81
Digro, 67
Disher, 2
Dishes, 24
Doehouse, 150
Doggerboat, 52
Don, 163
Dounby, 157
Dowscarth, 110

INDEX 255

Dritness, 24
Ducrow, 90
Durkadale, 133
Durrisdale, 128
Dyke, 128
Dykeside, 32, 182

E

Eastabin, 112
Eastabist, 139
Eastabister, 92
Eastafea, 62
Eastaquoy, 144
Eastbister, 184
Eday, 48
Egilsay, 71
Ellibister, 120
Elsness, 7
Elwick, 54
England, 175
Eriksyord, 150
Erraby, 19, 29, 59
Errigarth, 29
Ervadale, 62
Essaquoy, 62
Essonquoy, 83
Ettit, 122
Everbist, 8, 133
Everby, 29

F

Fald, 133
Falldown, 67
Faraclett, 62
Faraval, 39
Fara(y), 52, 186
Farewell, 175
Fea, 8, 83, 90, 97, 104, 133, 150, 163, 168, 182
Fea[l]quoy, 67, 91, 125, 144, 163
Feaval, 133, 150
Feswell, 163
Feelicha', 71
Fenzieland, 93
Fersness, 49
Fiddlerhouse, 150
Fidgarth, 139
Fidge, 157
Finyarhouse, 5
Finyo, 67
Fiold, 39
Firth, 114, 115
Fisligar, 5
Flaws, 91, 128, 133, 144, 150, 176
Flawsquoy, 182
Flenstaith, 77
Flotta, 187
Folsetter, 134
Foreland, 97
Forse, 163
Forswell, 151
Foubister, 83
Fribo, 32
Frotoft, 63
Frow, 8
Frowattin, 8
Frowantun, 8

Frustigar, 55
Furrowend, 49
Furse, 63
Fursebreck, 144
Fursin, 128
Furso, 144
Fuxtoun, 55

G

Gairbolls, 45
Gairsay, 75
Gairsty, 55, 134, 157
Gairy, 39
Gaitnip, 97
Gallowha', 128
Gara, 104
Garay, 171
Garbo, 19
Gardemeles, 8
Garricot, 158
Garsetter, 134
Garson, 67, 120, 134, 151, 163, 179, 181, 184, 187
Gart, 24, 84
Garth, 8, 32, 49, 55, 81, 97, 142, 163, 171
Gears, 88
Gebro, 59
Georth, 125, 128
Geoth, 144
Geramont, 19
Gerbo, 2
Germiston, 111
Geroin, 144
Gerraquoy, 176
Gerwin, 107
Gilbroch, 167
Gill, 40
Gimps, 176
Girnigo, 55
Girsay Schottis, 167
Glaitness, 97
Gleat, 19
Glenna, 32
Gloup, 134
Gloupquoy, 77
Goarhouse, 67
Goir, 8
Goldigar, 151
Gorn, 40, 59, 95, 122, 145, 151, 181
Gorsness, 120
Gossaquoy, 104
Gossigar, 171
Gowrie, 46
Graemsay, 181
Grain, 98
Grandon, 117
Grassquoy, 32, 55
Graves, 93
Gravity, 2
Green, 135, 182
Greentoft, 49
Greenwall, 55, 91
Greeny, 134
Grew, 134
Gridgar, 135
Grimbist, 32
Grimbister, 115

Grimeston, 142
Grimness, 171
Grimsetter, 98
Grimsquoy, 98
Grind, 81, 88, 122, 151
Grindally, 9, 40, 68, 105, 135, 151, 176
Grindigar, 81
Grinlaysbreck, 68
Gritley, 81
Grobister, 24
Grother, 32
Grotsetter, 24
Groundwater, 105
Grudgar, 125
Grudwick, 63
Grutha, 171
Grutquoy, 84
Gruttill, 9
Guithe, 49
Gutterpitten, 122
Gyre, 105, 182
Gyren, 158

H

Habreck, 40, 73
Hacco, 122
Halkland, 120, 151
Hacksness, 9
Halcro, 176
Hallbreck, 151, 176
Halley, 77, 179
Halliwell, 187
Hammer, 40, 52, 63, 139, 151
Hammerbrake, 19
Hammercleat, 151
Hammiger, 125, 163
Hamrifield, 68
Handest, 145
Hannatoft, 59
Hanover, 68
Haquoy, 55
Harga, 2
Harpisquoy, 141
Harray, 141
Harroldsgarth, 55
Hass, 139
Hatston, 98
Hawell, 84
Haybrake, 184
Hayon, 122, 139
Hayin, 135
Headgeo, 55
Heatherquoy, 98
Heddle, 115
Hellihowe, 10
Hellicliff, 128
Helyie, 73
Hensbister, 93
Hermisgarth, 10
Herston, 171
Hescombe, 24
Hestikelday, 95
Hestimuir, 95
Hestival, 60, 68, 128
Hestwall, 95, 151
Heyland, 151
Hildival, 40
Hindatun, 145

Hinderayre, 122
Hobbister, 10, 105, 112
Hogarth, 122
Holin, 29
Holland, 2, 10, 24, 46, 49, 52, 55, 78, 88, 98, 115, 171
Hollandswick, 55
Holm, 89
Holm, 2, 56
Holmes, 91
Hookin, 5, 46
Hool, 11
Hools, 176
Hoosay, 11
Hooveth, 158
Horraldsay, 116
Horrie, 84
Horriequoy, 93, 145
Horries, 81
Horsick, 84
Hottit, 171
Hourston, 151
Houserow, 11
Housebreck, 167
Houseby, 25, 60, 135
Housegarth, 11, 152
Housequoy, 112
Housenia, 152
Houstais, 12
Housteith, 68
How, 12, 32, 56, 71, 93, 120, 128, 135, 142, 152, 163, 176, 187
Howa, 125
Howaback, 158
Howally, 135
Howan, 40, 139
Howaquoy, 120, 135
Howar, 2, 19
Howatoft, 2, 68, 176
Howbell, 19
Howbusterland, 163
Howland, 19
Howth, 105
Hoxa, 172
Hoy, 178
Hozen, 140
Huan, 152
Huip, 25
Hullion, 68
Hunchaquoy, 135
Hunclet, 64, 93
Hunday, 29
Hundland, 135
Hunscarth, 142
Hunto, 136
Hunton, 25
Hurkisgarth, 152
Hurliness, 184
Hurtiso, 64, 93
Husabae, 68
Hussaquoy, 136
Hybreck, 145
Hyvel, 136, 152

I

Ingamyre, 105
Ingsay, 136
Ingshowe, 117

INDEX 257

Inkster, 105, 123
Innister, 68
Instabillie, 98, 152
Inyequoy, 172
Iphs, 40
Iquiver, 59
Ireland, 111
Isbister, 120, 136, 172
Isgarth, 19

J

Jaddvorstodum, 99
Jubidee, 128

K

Kebro, 108
Keigar, 78
Kelday, 93
Kethisgeo, 112
Kettletoft, 20
Kewing, 123
Kierfiold, 152
Kingshouse, 142, 166
Kirbest, 2, 32, 72
Kirbister, 25, 56, 78, 105, 136, 163, 182
Kirk, 172
Kirkgair, 84
Kirkhouse, 32
Kirkness, 152
Knarston, 64, 99, 142
Knockhall, 176
Knugdale, 40

L

Laith, 158
Lairo, 56, 64
Lakequoy, 40
Lamaness, 12
Langadie, 136
Langomay, 20
Langskaill, 20, 32, 64, 75, 88, 136, 145, 164
Langtas, 12
Laverock, 56
Leafea, 166
Leager, 164
Leaquoy, 29, 52, 56, 136, 172
Leavsgarth, 13
Lee, 158, 164
Leean, 40
Leith, 167
Lerely, 158
Lerquoy, 105, 164
Lettaly, 117
Lettan, 20
Leyland, 13
Lidda, 84
Liddell, 172
Lie, 136
Linahow, 153
Linda, 153
Linga, 25
Lingo, 188
Lingro, 56, 99
Linklet, 2
Linkletter, 153, 172

Linksness, 30, 54
Linnabreck, 30, 140
Linney, 3
Linnieth, 145
Linton, 56
Livaness, 60
Lobady, 140
London, 49
Looath, 153
Lopness, 13
Lurdy, 167
Lyde, 117
Lyking, 95, 123, 153
Lyness, 182
Lythe, 172, 180
Lythend, 180
Lythes, 172, 184

M

Maeback, 46
Maesquoy, 145
Mains, 188
Maizer, 20
Manclett, 183
Marwick, 136
Masseter, 173
Meal, 78, 93
Meaness, 56
Meiklequoy, 137
Melsetter, 183
Messigate, 84, 137
Midbea, 40
Midbigging, 118
Midgarth, 25, 121, 143
Midhouse, 57, 78, 118, 125, 128, 137
Midland, 105, 121
Midtown, 176
Miffia, 166
Milgarth, 33
Mirbister, 143
Mirkady, 81
Mistra, 128
Mithist, 145
Mithvie, 64
Mo, 33
Moa, 112, 123
Moan, 118, 145, 153
Mobisyord, 153
Moclett, 46
Morsetter, 56
Mosound, 183
Mossetter, 121
Mousland, 33, 164
Muddisdale, 99
Mugly, 68
Murra, 180
Musbister, 25
Mussaquoy, 105
Mussater, 25, 50
Myre, 105
Myres, 20

N

Natural, 99
Naversdale, 106
Neagar, 56
Nearhouse, 88

INDEX

Nears, 57, 69
Neebister, 13
Neigarth, 20, 34
Neo, 64
Nesodden, 46
Ness, 3, 52, 57, 85, 145, 153
Netherbigging, 112, 118
Netherbrough, 143
Netherby, 82
Netherhouse, 145
Netherskaill, 137
Netherton, 93
Netlater, 143
Nettilhill, 99
Newark, 21, 40, 78
Newbigging, 99, 153
Newgarth, 153
Newhouse, 78
Niggly, 88, 128
Nistaben, 112, 145
Nisthouse, 57, 112, 118, 140, 145
Nistigar, 41
Nistoo, 123
Noltland, 13, 34, 82
Northbigging, 145
Northdyke, 153
Northgarth, 21
Northquoy, 78
Norton, 137
Noup, 41, 50
Noust, 35
Nouster, 21, 46

O

Oback, 78, 145, 176
Occlester, 85
Odinstone, 60
Odness, 26
Okamby, 21
Olad, 176
Onziebust, 72, 73
Orakirk, 106
Ore, 184
Orgill, 180
Ork, 57
Orklandquoy, 91
Orphir, 103
Orquil, 99, 106, 123, 129
Orraquoy, 184
Ortie, 21
Osmundwall, 183
Osquoy, 173
Ostoft, 57
Ottergill, 111
Ouseness, 41
Out apoune the Ile, 181
Outerdikes, 69
Overbigging, 113
Overbrough, 143
Overhouse, 153
Over the Water, 21
Owendsætir, 57
Oyce, 21

P

Papa Stronsay, 28
Papa Westray, 45

Papdale, 99
Paplay, 89, 173
Paplayhouse, 50
Perth, 35
Pole, 41, 167
Pool, 78, 173
Pow, 41, 65, 129, 146, 153, 164
Puldrit, 123
Pulkitto, 129
Pultisquoy, 164
Purtabreck, 88

Q

Quackquoy, 157
Quandal, 65
Quanterness, 101
Quatquoy, 116
Quean, 158
Quear, 118
Queena, 113, 118, 123, 158
Queenafinieth, 146
Queenalanga, 158
Queenamarion, 113
Queenamoan, 113
Queenamuckle, 121
Quindry, 177
Quivals, 21
Quoy, 41, 52
Quoyangry, 30, 173
Quoybanks, 99, 173
Quoybellock, 82
Quoybernardis, 91
Quoyberstane, 101
Quoybewmont, 85
Quoybora, 26
Quoybrown, 173
Quoyburing, 78
Quoyburray, 88
Quoycanker, 82
Quoyclerks, 106
Quoycrusda, 78
Quoycuise, 173
Quoydandy, 102
Quoydoun, 173
Quoyfaulds, 50
Quoyfea, 78
Quoyfree, 123
Quoyleith, 173
Quoygarth, 26
Quoygelding, 137
Quoygrana, 26
Quoyhokka, 137
Quoyingabister, 91
Quoykea, 85, 146
Quoykindness, 26
Quoylanga, 140
Quoylanks, 82
Quoyloo, 154, 177
Quoymorhouse, 60
Quoymos, 78
Quoymyirland, 78
Quoynaknap, 181
Quoyneipsetter, 183
Quoyness, 188
Quoyolie, 26, 46
Quoyostray, 65
Quoyrush, 78
Quoyscottie, 140

Quoysharps, 173
Quoyshorsetter, 173
Quoyskega, 35
Quoysmiddie, 173
Quoythom, 91
Quoytob, 85
Quoyturben, 78
Quoyer, 113
Quoys, 69, 106, 126, 143, 153, 164, 167, 173, 179, 181, 183

R

Rackwick, 36, 179
Ramray, 181
Ramsay, 36
Ramsquoy, 113, 173
Rango, 154
Ranisgarth, 36
Ransgarth, 143
Rapness, 36
Redland, 116, 126, 164
Rendall, 119, 121
Rennibister, 116
Rerwick, 88
Rickla, 146
Riff, 123
Rinansey, 1
Ring, 42
Rinnabout, 57
Ritquoy, 140
Rockerskaill, 177
Roeberry, 177
Ronaldsay (North), 1
Ronaldsay (South), 169
Ronsvoe, 173
Rossmire, 118
Rothisholm, 26
Roundadee, 158
Rousay, 61
Roveland, 42
Roy, 91
Runa, 146
Rusland, 42, 143
Rusness, 21, 74, 75
Russamoa, 95
Rysa, 183

S

Sabay, 85, 177
Sabiston, 137
Saintear, 42
Sand, 3, 13
Sanday, 6, 79, 86
Sandgarth, 57
Sandisend, 181
Sands, 58, 79
Sandside, 79
Sandwick, 79, 147, 174
Sanger, 30, 36
Savaquoy, 58, 138
Savedale, 113
Saverock, 99
Saverton, 42
Savil, 21, 116
Saviskaill, 65
Scapa, 100
Scarpigar, 88

Scarrataing, 158
Scarrigar, 21
Scarton, 183
Schusan, 177
Scockness, 65
Scofferland, 21
Scottigar, 3
Scoulters, 30
Scows, 108, 177, 185
Scuan, 113
Seal Skerry, 50
Seatter, 22, 100, 116, 138, 156, 164, 183
Sellibister, 13
Selliland, 168
Selwick, 179
Sennes, 3
Seraquoy, 188
Serrigar, 174
Settiscarth, 116
Shaltaquoy, 58
Shapansay, 53
Sholtisquoy, 3
Shut Behind, 188
Simbister, 16
Skae, 80
Skaebrae, 69, 155
Skagarth, 36
Skaill, 13, 22, 43, 50, 65, 72, 82, 91, 108, 123, 154
Skailtoft, 93
Skanaquoy, 138
Skeggbiarnerstauþum, 80
Skeithva, 158
Skelbist, 75
Skelbister, 13, 106
Skelbrae, 13
Skelday, 138
Skennist, 46
Skenstoft, 58
Skerp, 185
Skerpie, 168
Skethaquoy, 113
Skethouse, 146
Skethquoy, 158
Sketquoy, 69
Skiddy, 123
Skidge, 138
Skidgabist, 22, 106
Skitho, 22
Skooan, 69, 113, 146
Skorn, 138
Skorwell, 155
Skrutabreck, 140
Slack, 180
Smerquoy, 102, 158
Smerskeal, 43
Smittaldy, 43
Smogarth, 118
Smoogro, 106
Snelsetter, 183
Snippigar, 82
Snukesbrae, 43
Sorpool, 107
Sorquoy, 174
Sotland, 91
Soulisquoy, 100
Sound, 58, 72
Sourin, 69
Sourpow, 166

INDEX

Southwall, 13
Sowlie, 106
Sowlisyord, 156
Spannisquoy, 36
Spithasquoy, 126
Sponess, 43
Spurdagrove, 140
Spurquoy, 27
St. Andrew's, 83
St. Ola, 96
Stagaquoy, 138
Standing Stone, 188
Stangasetter, 14
Stanger, 138
Stank, 166
Stara, 140
Staves, 30
Stembister, 86
Stenaquoy, 51
Stenigar, 164
Stenness, 110
Stennisgorn, 70
Stensigar, 174
Stenso, 126
Stensy, 28
Stews, 174
Stiglister, 22
Stockan, 158
Stoddisyord, 156
Stokaquoy, 30
Stout Farding, 92
Stove, 14, 80, 138, 156, 166
Straither, 165
Strenzie, 27
Stretigar, 22
Stromness, 160
Stronsay, 23
Stye, 58, 82
Stymbro, 129
Suckquoy, 174
Sugarhouse, 3
Sulland, 37
Surrigarth, 37
Sutherbie, 13
Suthergarth, 181
Sutherquoy, 92, 156
Swanbister, 107
Swandale, 65
Swanney, 138
Swartabreck, 107
Swartaquoy, 58, 93
Swartifield, 70
Swartland, 159, 185
Swartmeil, 37
Sweenalay, 123

T

Taftnica, 167
Tafts, 16, 37, 70, 72, 100, 168
Tankerness, 86
Tenston, 156
Testaquoy, 74
Thickbigging, 118
Thrave, 22
Thurrigar, 174
Thurvo, 183
Tiffyha', 82
Tifter, 43, 47

Tingwall, 121
Tirlot, 43
Toab, 86
Todds Zir, 37
Too, 70
Tormiston, 111
Toung, 175
Tratland, 70
Trattletoun, 58
Treb, 3, 32
Trebland, 43
Trenaby, 43
Tresness, 16
Trinnigar, 156
Trofer, 168
Tronston, 156
Trumland, 44, 65
Tufta, 146, 159
Tufter, 140
Tuquoy, 37
Turriedale, 127
Tuskerbister, 107
Twatt, 113, 138
Twattland, 138
Twinness, 37, 87

U

Unigar, 156
Unston, 111
Upperbigging, 118, 146
Upsale, 156
Urback, 188
Urigar, 129
Uttesgarth, 92

V

Vaday, 72
Valay, 94
Valdigar, 88, 94
Veantrow, 60
Vedder, 87
Vedesquoy, 59
Vell, 44
Veltigar, 88
Velzian, 123, 140, 146
Velzie, 159
Vere, 44
Vetquoy, 159
Veval, 182
Via, 47, 159
Vigga, 92, 139
Vinbreck, 140
Vindon, 118
Vinikelday, 88
Vinquin, 129
Vola, 146, 159
Volunes, 16
Voy, 87, 157

W

Wa, 37
Walbroch, 16
Wald, 116
Walkerhouse, 129, 140
Wallis Farding, 92
Walls, 182, 17

INDEX 261

Waltness, 59
Warbuster, 100, 179
Wards, 184
Warebanks, 168
Warset, 72
Warsetter, 17
Warth, 157
Wasbist(er), 38, 66, 157
Wasdale, 116
Waswick, 121
Watriehall, 188
Watten, 72, 80
Wattle, 139
Wattyn, 38
Weardith, 166
Weatherness, 38
Weddell, 168
Weems, 177
Weethick, 88
Weland, 59, 72, 100
Westerbister, 94
Westermele, 100, 168
Westness, 60
Westove, 22
Westray, 31
Whanclet, 188
Whistlebare, 22, 51, 72, 177
Whiteclett, 22, 38, 72, 87
Whitehall, 30

Whome, 60, 165, 188
Whomslie, 166
Whunderless, 5
Wideford, 100
Widewall, 175
Windbreck, 107, 140, 157, 177, 181, 188
Windwick, 175
Windywa'(s), 51, 52, 146, 181
Wing, 185
Winksetter, 143
Withaquoy, 94
Woo, 38, 60
Woodwick, 127
Work, 101
Wyre, 73

Y

Yairsay, 101
Yarpha, 80, 108
Yeldabreck, 139, 168
Yeldavill, 146
Yernasetter, 28
Yesnaby, 157
Yinstay, 87
Yorbrands, 175
Yuildadee, 159

INDEX II

PERSONAL FARM-NAMES

In the absence of recorded forms of farm-names from the pre-Scottish period there must obviously be some uncertainty as to the exact personal names represented in the majority of cases where they do appear; e.g. whether Brandi, or Brandr or Guðbrandr, etc. On that account a list of the actual personal names occurring is not attempted here. In several cases also, there is dubiety as to the presence of a personal name where it has been suggested. Accordingly two lists are here appended, the first of which includes names in which the presence of personal names may be deemed practically certain, and a second in which the presence of such names is more doubtful.

(i) Names in which the presence of a Norse personal name is considered certain.

Anderswick, page 111
Avaldsay, 61
Botulfyord, 148
Brandiquoy, 80
Brekisyord, 149
Bressigar, 6
Cairston, 162
Campston, 83
Carlquoy, 90
Clouston, 110
Colligar, 7
Colster, 77
Colston, 110
Copinsay, Appendix
Crantit, 97
Elwick, 54
Eriksyord, 150
Frustigar, 55
Flenstaith, 77
Frotoft, 63
Gairsay, 75
Geramont, 19
Germiston, 111
Gossigar, 171
Graemsay, 181
Grimbister, 115
Grimeston, 142
Grimness, 171
Grimsetter, 98
Grimsquoy, 98
Hannatoft, 59
Haroldsgarth, 55
Hatston, 98
Hensbister, 93

Hermisgarth, 10
Herston, 171
Horraldsay, 116
Hourston, 151
Hurkisgarth, 152
Hurtiso, 64, 93
Jadvarstodum, 99
Kettletoft, 20
Knarston, 64, 99, 142
Leavsgarth, 13
Mobisyord, 33, 153, 164
Osmundwall, 183
Owendsatir, 57
Ramsay, 36
Ramsquoy, 113, 173
Rinansey, 1
Rognvaldsey, 169
Ronsvoe, 173
Rousay, 61
Skeggbiarnerstauþum, 80
Soulisquoy, 100
Sowlisyord, 156
Stoddisyord, 156
Tenston, 156
Thurrigar, 174
Thurvo, 183
Todds Zir, 37
Tormiston, 111
Tronston, 156
Unigar, 156
Unston, 111
Winksetter, 143
Yinstay, 87
Yorbrandis, 175

CHIEF SOURCES CONSULTED

R.E.O.—*Records of the Earldom of Orkney*, by J. S. Clouston (Scot. Hist. Soc.). 1914.

Sas.—Sasines: (i) *Orkney Sasines*, printed by Viking Soc.
(ii) David Heart's MS. vol. in Sheriff Court Record Room, Kirkwall.
(iii) Other MS. sasines in same repository.

Watson.—*The Celtic Place-Names of Scotland*, by W. J. Watson. 1926.

OTHER CONTRACTIONS USED

O.N. = Old Norse.

No. = Modern Norse (Norwegian).

Icel. = Icelandic.

Fær. = Færoese.

Gael. = Gaelic.

Val. = Valuation of Orkney (1653).

INDEX

(ii) Names probably or possibly incorporating personal names.

Arstais, page 6
Berstane, 96
Blanster, 170
Boloquoy, 18
Cataquoy, 29
Cattaby, 81
Classiquoy, 66
Corston, 142
Coubister, 104, 115
Cut Claws, 67
Dowscarth, 110
Egilsay, 71
Ellibister, 120
Fribo, 32
Goldigar, 151
Grobister, 24
Grudgar, 125
Hundland, 135
Hunscarth, **142**
Ingsay, 117
Innister, 68

Mirbister, 143
Muddisdale, 99
Naversdale, 106
Onziebust, 72, 73
Ottergill, 111
Ranisgarth, 36
Ransgarth, 143
Shaltaquoy, 58
Sholtisquoy, 3
Snelsetter, 183
Spurquoy, 27
Swanbister, 107
Tankerness, 86
Testaquoy, 74
Trenaby, 43
Trinnigar, 156
Tuskerbister, 107
Vedesquoy, 59
Wing, 185
Yernasetter, 28
Yesnaby, 157

INDEX III

(to Parts II and III)

Bae-names (O.N. *bœr*), 243
Baile, The Celtic, 216 f
Bister-names (O.N. *bólstaðr*), 232
Bu-names (O.N. *bú*), 240
Chapels (Urisland), 214 f
Clouston, J. Storer, 196, 207, 213, 220, 239, 241
Davach, 208 *et passim*
Dennison, W. T., 222
Ekwall, E., 239
Extent, Old, 208 f
Eyrisland, 210
Eyrrbyggja Saga, 214
Farm Background, 190 ff.
Floruvoe, Battle of, 192, 208
Garth-names (O.N. *garðr*), 232
Halk-hens, 198
Harald, King (Fairhair), 211
Hebrides, 208
Heimr, as farm-name element, 250
Johnston, A. W., 196
Jónsson, Finnur, 241
Katherin, Countess of Orkney, 201
Landmail—see Rent
Land-names (O.N. *land*), 231
Landnámabók, 217, 249
Lands—Denominations of
  Bishopric, 193
  Bordland, 192
  Conquest, 192
  Cowsworth, 203
  Earldom, 192
  Kingsland, 192
  Kirkland, 193
  Markland, 200 f.
  Meadowland, 221
  Meils Cop, 203 f
  Ounceland, 196 *et passim*
  Quoyland, 193 f, 202
  "Scatland," 206 f
  Townsland, 221
  Tumail, 221
  Udal, 192
  Uris Cop, 203 f
  Urisland (ounceland), 196 *et passim*
Leding System, 205 f

Low, George, 214
Mackenzie, Jas., 191
Marstrander, Prof. C. J. S., 205, 209, 214
Matras, Prof. Chr., 241
McKerral, A., 212
Monetary Standards, 195 f
Olsen, Prof. Magnus, 250
"Oncas," 222 f
Parishes, 213 f
Parish Churches, 215
Payment, Media of, 195
Peterkin, A., 191
Pett, 216 f
Quoy-names (O.N. *kví*), 227 f
Rent, 200 f
Rentals, 191 f
Scattald (Shetland), 208
Setter-names (O.N. *setr*), 229 f
Sheading (Manx), 205
Shetelig, Prof. Haakon, 205
Shetland, 207 f
Skaill-Names (O.N. *skáli*), 237
Skats, 191 ff
  Butter, 196
  Forcop, 198, 210
  Halkhens, 198
  Malt, 197
  Merts, 198
  Wattle, 194, 198, 211
  Normal, 200
Skats, Origin of, 205 ff
Skene, W. F., 212
Staðir-names, 234
Steinnes, Dr. A., 206 f
Stent Butter, 196 f
Sverrir, King, 192
Swanbister, 218
Teinds, 202
Thomas, Capt., 194, 196, 200, 212
Tirungs, 205
Treens (Manx), 205, 209
Tulloch, Bishop, 193
Tunship, The Orkney, 216 ff, 222
Vin-names, 250
Weights and Measures, 195

CHIEF SOURCES CONSULTED

Aasen—*Norsk Ordbog*. Christiania, 1873.

Bœjanöfn á Islandi, eftir Finn. Jónsson. 1911.

C. & V.—*An Icelandic-English Dictionary*, by Cleasby and Vigfússon. Oxford, 1874.

Clouston.—*A History of Orkney* by J. Storer Clouston. Kirkwall, 1932.

Dipl. Norv.—*Diplomatarium Norvegicum* II.

Ekwall, E.—*The Concise Oxford Dictionary of English Place-Names*. 1940.
    The Place-Names of Lancashire. 1922.

English Place-Name Society's vols. 1924 et seq.

Færøsk-Dansk Ordbog, av Jacobsen & Matras. Torshavn, 1927-28.

Fritzner.—*Ordbog over det gamle Norske Sprog*. 1886-96.

Gamalnorsk Ordbok, av Hægstad og Torp. 1909

Gamle Personnavne i Norske Stedsnavne, af O. Rygh. 1901.

Indl.—*Forord og Indledning til Norske Gaardnavne*, af O. Rygh. 1898.

Jakobsen.—*The Place-Names of Shetland*. (Eng. Trans.) 1936.

Lind.—*Norsk-Islendska Dopnamn*, &c. 1905-15.
    Supplementband. 1931.

Matras.—*Stednavne paa de Færøske Norðuroyar*. 1933.

N.G.—*Norske Gaardnavne*. 18 vols. with Register. 1897-1936.

Nordisk Kultur—Stedsnavn. Utgitt av Magnus Olsen. 1939.

Norske Elvenavne, af O. Rygh. 1904.

N.N.O.—*Ny-Norsk Etymologisk Ordbok*, av A. Torp. 1919.

The Orkney Norn, by H. Marwick. Oxford Univ. Press. 1929.

O. Saga.—*Orkneyinga Saga*, edit. S. Nordal. Kjøbenhavn. 1913-16.

O. & S.R.—*Orkney and Shetland Records* (Viking Soc.). 1914.

Peterkin.—*Rentals of the Ancient Earldom and Bishoprick of Orkney*. (The rentals therein included are cited by their respective dates—1500, 1595, etc.)

Place-Names of Rousay, by H. Marwick. Kirkwall, 1947.

P.O.A.S.—*Proceedings of the Orkney Antiquarian Society*. Vol 1923-1939.

P.S.A.S.—*Proceedings of the Society of Antiquaries of Scotland*.

ADDENDA ET ERRATA

p. 8, line 12. For Part II read Part III.

pp. 28, 30. Whitehall. Later information would seem to indicate that the name Whitehall here dates from before P. Fea's time.

p. 77. Add—**Copinsay**: Colbanisay (Fordun, c. 1375); Cobinshaw 1595; Kolbeinsøi (P. Claussøn, c. 1600); Copinschaw 1627. A small island (with single farm) lying a short distance off the Deerness coast. Skattable value uncertain, but probably a 3d. land. O.N. *Kolbeinsey*, isle of a man Kolbeinn.

p. 80. Stove (line 4). There was actually a sixth Stove—in Birsay-Besouth.

p. 81. Mirkady (line 4)—for 'settlement' read 'element.'

p. 114 (line 14 from foot of page): delete whole sentence beginning "But in the same tunship . . . ."

p. 115—line 6: for 'which' read 'while.'

p. 169 (line 25)—*Kui Kobba*. Though the site of that farm is now unknown the name is obviously one of the 'quoy' class (O.N. *kvi*), indicating the quoy of a man Kobba, a colloquial form of Kolbeinn (cf. *Cubbie* Roo's Castle in Wyre erected by Kolbeinn Hruga). The post-fixing of the qualifying element in so many of the Orkney quoy-names may perhaps be an indication of Celtic influence.

Printed at
"The Orcadian" Office
Victoria Street
Kirkwall